OPTICAL MATERIALS

Optical Materials

Joseph H. Simmons

*University of Florida,
Gainesville, Florida*

and

Kelly S. Potter

*Sandia National Laboratories, Albuquerque,
New Mexico*

ACADEMIC PRESS

A Harcourt Science and Technology Company

San Diego   San Francisco   New York   Boston
London   Sydney   Tokyo

ACADEMIC PRESS
A Harcourt Science and Technology Company
525 B Street, Suite 1900, San Diego, CA 92101-4495, USA
http://www.apnet.com

Academic Press
24–28 Oval Road, London NW1 7DX, UK
http://www.hbuk.co.uk/ap/

Library of Congress Catalog Card Number: 99-65137
International Standard Book Number: 0-12-644140-5

Printed in the United States of America
99 00 01 02 03 MB 9 8 7 6 5 4 3 2 1

Dedication

With deep appreciation to our spouses, Cate and B. G., for their enduring support and indispensable assistance with this project.

Contents

Preface

Today, the field of optics is expanding at an explosive rate. The advent of optical communications, personal computers, video-on-demand television, and network interconnections across the globe has placed a heavy burden on materials and devices for signal transmission and processing. Clearly, current state-of-the art technology is being driven, in large part, by advances in both the design and the implementation of complex optical systems. Applications, ranging from optical telecommunications (in which gigabits of encoded optical data are transmitted down hair's-width glass fibers) to orbiting satellite systems, rely heavily on optical materials, optical systems, and lasers. Undoubtedly, the worldwide political and economic changes of the last decade are a harbinger of the increased need for advanced, compact, multifunctional technologies capable of receiving, processing, storing, and transmitting massive amounts of information faster, over longer distances, and more efficiently than ever before as we move toward a truly global economy. The major role of optics in these advances puts the subject of optical materials in the forefront for the beginning of the twenty-first century.

Those of us who have been in science over the last several decades have seen the tremendous impact that the development of a suitable material and process can have on a wide range of industries. The demonstration of the ruby laser in 1964, for example, opened the door for progress that has revolutionized the way we live. Everything from bar-code scanners to compact disk technology to telecommunications to new medical procedures rely on the use of a broad variety of laser sources, each with specific wavelength and bandwidth requirements. Optical materials have played an important role in these advances and promise even greater impact in the future.

In signal processing and transmission, the benefits of optical over electronic techniques have already changed our lives in a major way by giving us access to the information superhighway. Optical-fiber communications are already just outside our houses and will cross the threshold within the next decade. Again, this advance results from the development of novel optical materials that not only can handle larger signal bandwidth but also can transmit a great number of communication channels over global distances. The linearity of the transmission media allows the superposition of optical signals without mixing, thus making possible the processing of massive amounts of data simultaneously and in parallel. The promise of materials that exhibit a strongly nonlinear response to optical radiation makes possible the development of large optical memories, optical switches, and computational logic operations. Complex logic processes, like image analysis, can be done more quickly and accurately optically than electronically, as can signal-processing operations such as amplification, wavelength-division multiplexing, and switching. An intense research effort is currently under way to integrate multiple optical technologies on-chip. One can easily envision a system in which optical signals are generated by microlasers, modulated by light-signal modulators, transmitted and shaped by thin-film digital lenses, coupled into optical waveguide channels by nonlinear optical switches, analyzed by optical logic gates that work as part of complex neural network logic systems, and, finally, stored as information in three-dimensional holographic data-storage media, all within the space of a computer chip.

This book addresses the underlying mechanisms that make optical materials what they are and that determine how they behave. The book strives to group the characteristics of optical materials into classes with similar behavior. We believe that by presenting a broad range of optical materials behavior, we can show the reader what properties are held in common and what properties differ between various classes of materials. In treating each type of material, we pay particular attention to atomic composition and chemical makeup, to electronic states and band structure, and to physical microstructure. We then strive to relate optical behavior and its underlying processes to the chemical, physical, and microstructural properties of the material so that the reader will gain insight into the kinds of materials engineering and processing conditions that are required to produce a material exhibiting a desired optical property.

The book is aimed at the intermediate or advanced reader (or student) and, in order to achieve accurate and quantitative descriptions, presents the principal equations underlying the processes of interest. If only a qualitative understanding is sought, however, then it is fully possible to read the entire book and completely ignore the mathematical treatment. To this end, we have filled the text with explanations and discussions of the physical principles associated with the optical behaviors described. Many of the insights presented here have come from the broad range of specialized literature cited and from the authors' own extensive research efforts with optical materials and the authors' interactions with students in the laboratory and in the classroom.

Each chapter is essentially self-contained, so the book may be used as a text for any level of course on optical materials. In addition, throughout the book we have tried to tie the material to the research arena by including discussions of equipment and experimental techniques relevant to the topics of each chapter. As such, the book may be used as a reference source for the experimentalist or as a guide for the student. Since we have sought to present a complete picture of the behavior of optical materials, an introductory course may wish to leave out some of the later chapters and the chapter appendices.

Chapter 1

Wave Propagation

1.1 Introduction

The optical properties of materials arise from the characteristics of their interactions with electromagnetic waves. In particular, the ability of a material to exhibit an induced polarization or magnetization at a selected wavelength, or over a selected wavelength band, provides the potential for it to change the character of light propagating through it. Such changes can take the form of loss or gain of intensity, shifts in wavelength, and narrowing, broadening or filtering of bandwidth, for example.

Different classes of materials will, in general, differ in their response to optical radiation. Insulators and conductors, for example, each exhibit a unique response to electric polarization. While insulators exhibit local induced polarization of *bound* charges, dipoles, etc., conductors exhibit induced currents from the movement of *free* charges. In some cases, materials will act as insulators at frequencies or temperatures where only local polarization is possible, and they will act as conductors when conditions are suitable for charge transport. Thus, glasses and semiconductors act as insulators at low temperatures and at frequencies below those needed to excite free carriers, and they act as conductors at temperatures and frequencies high enough to excite free carriers and to allow induced electric currents to form. In between these conditions, materials exhibit a variety of changes in local polarization mechanisms, including (1) molecular or dipolar, (2) ionic or atomic, and (3) electronic polarization, and these mark large changes in their optical behavior. This topic is discussed in more detail in Chapter 3.

This book will cover the behavior of materials as grouped by the classifications of conductor, insulator, semiconductor, nonlinear materials, and so on, in order to identify characteristics common to all members of the same group. This will allow us to identify the unifying themes that run through the behavior of different groups of materials and to relate them to the underlying physics.

This first chapter is a terse review of the principles of optics. We emphasize here the principles that will be used throughout the book. The reader who is not familiar with elementary optics is encouraged to support reading of this chapter with books that cover the subject in more detail. Several such books (Born and Wolf, 1980; Hecht, 1990) that use nomenclature similar to that used in this book are listed at the end of the chapter.

1.2   Waves

Light propagates by electromagnetic waves. Therefore, there are certain characteristics of waves, and, in particular, electromagnetic waves, that must be reviewed in order to understand the behavior of light and its interaction with matter. In this chapter, we will review the mathematical structure of wave propagation that will be used throughout the book.

We are surrounded by many kinds of waves. For example, acoustic waves propagate the sounds we make and hear. Light that we see comes from electromagnetic waves at very high frequencies. Radio and television signals are carried by electromagnetic waves at much lower frequencies, and the microwaves that heat our food are electromagnetic waves at intermediate frequencies. The cell phone of Fig. 1.1 receives and emits both electromagnetic and acoustic waves. The water waves that entertain us at the beach are mechanical-displacement waves. Because of their constant pounding of the shore, causing erosion and tidal flows, the action of these waves has been recognized as an essential ingredient in the development of advanced life forms on Earth.

Waves are different in the way they disturb certain properties of the propagating media. For example, acoustic waves correspond to variations in the local pressure or density of the medium and electromagnetic (EM) waves correspond to variations in the electric and magnetic fields. Since EM fields can exist without the presence of matter, light waves propagate

Electromagnetic Wave

Acoustic Wave

Figure 1.1: Acoustic and electromagnetic waves from a cell phone.

through empty space, whereas acoustic waves require a propagating medium. While all these waves are perceived to behave totally differently, they actually follow the same mathematical formalism.

A traveling wave is a disturbance that propagates both in space and in time (Fig. 1.2). Usually, a wave will decay as it propagates, as shown in the figure. By contrast, a standing wave has spatial extent, with an amplitude that oscillates in time but is stationary. As such, standing waves are characterized by fixed nodes that are constant in time. Resonant systems are represented by standing waves (e.g., a vibrating guitar string). If there is loss in the system, the standing wave's amplitude will, in fact, decrease with time, but its spatial form will remain unchanged (Fig. 1.3).

Traveling waves can be either longitudinal or transverse. Longitudinal waves oscillate along the direction of propagation (Fig. 1.4); transverse waves (Fig. 1.2) oscillate perpendicular to the propagation direction. Longitudinal and transverse waves can be formed in solids and in viscoelastic media, but liquids and gases can support only longitudinal waves in the bulk. Liquids can support transverse waves only at their surface; it is the manifestation of surface transverse waves in water that

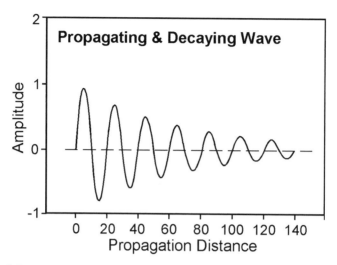

Figure 1.2: Diagram of the change in wave amplitude with distance for an attenuated, propagating transverse wave.

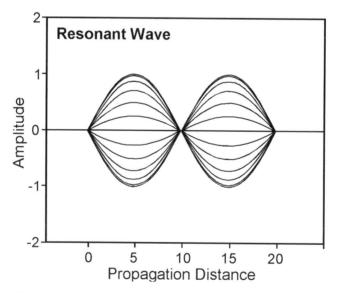

Figure 1.3: Diagram of the change in wave amplitude with time for a damped resonant wave.

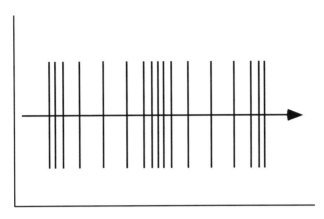

Propagation Direction

Figure 1.4: Diagram of a longitudinal wave oscillating along the direction of wave propagation. Lines close together show an increase in pressure; those far apart indicate regions of lower pressure, as in an acoustic wave.

we see at the beach. If an earthquake were to take place deep under the ocean, a large pulse made of longitudinal waves would propagate in the water. Once it reached the surface, a transverse surface pulse (tsunami, or tidal wave) would form and propagate outward along the surface (Fig. 1.5). *Electromagnetic waves are only transverse.*

1.3   The Electromagnetic Spectrum

The electromagnetic spectrum encompasses all waves that involve electromagnetic fields. The full range is shown in Fig. 1.6. Different regions of the electromagnetic spectrum have been identified at different times in history. Consequently, the broad range of wavelengths, frequencies, and energies that make up the various spectral ranges have different names and, historically, have been treated differently. For example, γ-rays and x-rays are usually reported in electron volts, visible light in angstroms or nanometers, and infrared light in microns. Microwaves, radar, and radio waves are generally reported as frequencies. Following is a listing of relations of interest between the

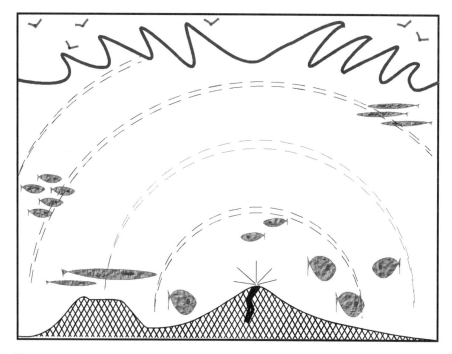

Figure 1.5: Schematic showing the formation of tidal (transverse) surface waves from a deep disturbance propagated to the surface by longitudinal waves.

various units used in these different regions of the electromagnetic spectrum.

ν = frequency (Hz, kHz, MHz, GHz, THz)

λ = wavelength ($\overset{\circ}{A}$, nm, μm, m) $\lambda = c/\nu; \lambda(\text{nm}) \cong 3 \times 10^{17}/\nu(\text{Hz})$

$\bar{\nu}$ = wave number(cm^{-1}) $\bar{\nu}(\text{cm}^{-1}) = 1/100\lambda\,(\text{m}) = 10^{7}/\lambda\,(\text{nm})$

$h\nu$ = energy (J, eV) $h\nu\,(\text{eV}) = 1,242/\lambda\,(\text{nm}) = 4.14 \times 10^{-15}\nu\,(\text{Hz})$

Visible light is matched to the human eye sensitivity spectrum, which in turn is matched to the black body radiation spectrum of the sun. The solar radiation spectrum corresponds to the Sun's surface temperature, which is between 5700° and 5900 °C (i.e., the emitted solar spectrum as seen in space is equivalent to the spectrum of EM radiation emitted by a black-body radiator held at that temperature). As viewed on Earth it is

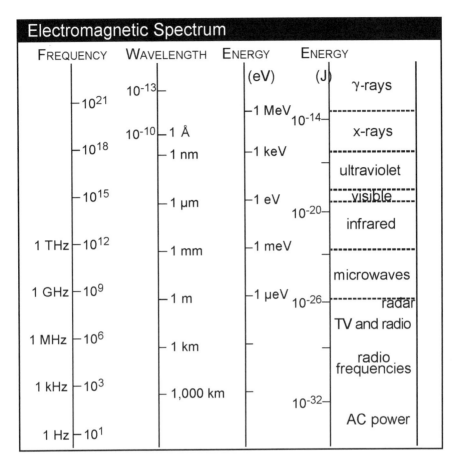

Figure 1.6: The electromagnetic spectrum.

shifted slightly from the black-body radiation spectrum, due to the scattering of blue light and the absorption of red light by the atmosphere. This shifts the "standard daylight" illumination as used in photography to a black body emission temperature of 5800–6200 °C. Figure 1.7 shows the extraterrestrial solar radiation spectrum as well as the solar spectrum observed through the atmosphere. One sees that the sea level spectrum has lost ultraviolet (UV) components, due to ozone absorption in the upper atmosphere. Blue light is also lost, due to ozone and Rayleigh scattering, but this loss is partially compensated by the blue light isotropic scattering of the nitrogen-rich atmosphere. Also visible are regions of severe

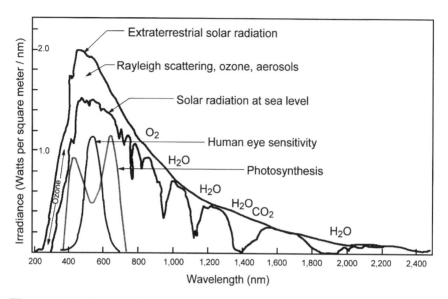

Figure 1.7: Solar radiation spectrum in space and as viewed at sea level through the atmosphere. Also shown are the spectral sensitivity of the human eye and the region of photosynthesis.

decrease in transmission due to oxygen, water, and carbon dioxide absorption. Figure 1.7 also plots the eye detection sensitivity spectrum. Note that the maximum in eye sensitivity matches the peak in the transmitted solar spectrum. Also shown is the photosynthesis region, which also fits well under the solar spectrum. Note the minimum at green wavelengths (530–550 nm); that color is reflected by plants, and it matches the region of maximum eye sensitivity. It is also instructive to discuss the difference in radiation between a clear and an overcast sky (not shown). Under an overcast sky, solar radiation reaching sea level is increased in the UV region due to increased scattering in the atmosphere, while water absorption reduces the transmitted infrared (IR). This explains why sunburns are more severe under an overcast sky, since one feels cooler due to the decrease in IR light and generally will stay out longer, increasing the exposure time while receiving a greater dose of UV light.

The name *black-body emitter* is used to indicate that all light incident on the source is absorbed and that the only light emitted by the black-body source is the light generated by the oscillators of the material. All these oscillators are presumed to be at the same temperature and consequently emit the radiation characteristic of a single temperature.

Planck's equation for spectral radiant excitance as a function of wavelength, $R(\lambda)$ (in J/cm$^3 \cdot$ sec) for the black-body radiation spectrum, is given next, where T defines temperature in degrees Kelvin (K), h is the Planck constant $(6.63 \times 10^{-34}\,\mathrm{J \cdot s})$, and k_B is the Boltzmann constant $(1.38 \times 10^{-23}\,\mathrm{J/atom \cdot K})$:

$$R(\lambda) = \frac{2\pi c^2 h}{\lambda^5}\left(\frac{1}{e^{hc/\lambda k_B T} - 1}\right) \tag{1.1}$$

The black-body radiation spectrum defined in Eq. (1.1) yields the maximum radiant energy emitted per unit volume from a black body at a given temperature. The wavelength corresponding to the peak emission intensity for each temperature can be found from the experimentally determined Wien displacement law:

$$\lambda_m T = 0.28978 \times 10^{-2} m^o \mathrm{K} \tag{1.2}$$

Table 1.1 shows a list of maximum-emission wavelengths corresponding to interesting temperatures. Figure 1.8 shows a series of black-body radiation curves for various interesting temperatures.

Since the radiant energy emitted is a function of temperature only, the

Temperature (°C)	Maximum-emission wavelength, λ_m	Perceived color
20,000–40,000	70–143 nm	Blue-white
6,000	462 nm	White (daylight)
5,700	485 nm	White
2,500	1.045 μm	Yellow (incandescent lamps)
1,600	1.547 μm	Orange
400	4.305 μm	Red
100	7.768 μm	IR (invisible)
37	9.348 μm	IR (invisible)
20	9.890 μm	IR (invisible)

Table 1.1: Equivalent temperatures for black-body radiation.

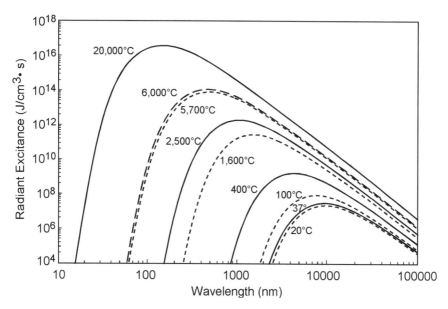

Figure 1.8: Black-body radiation curves for selected temperatures between 20 and 20,000 °C.

expression of Eq. (1.1) can be used to calculate accurately the temperature of a body whose emission can be measured. This method is good only at high temperatures, where the black-body curve sharpens sufficiently for the peak temperature to be easily deduced. Table 1.1 and Fig. 1.8 give a rough idea of the relationship between the emission colors and temperatures. Body temperature (37 °C) has a broad maximum emission at 9.348 μm. Consequently, photodetection requires a semiconductor with a lower bandgap energy, such as mercury–cadmium–telluride. That same material is used in night-vision glasses to detect human emission of black-body radiation. Since the surroundings are at a temperature of 20 °C, their emission peaks at 9.89- μm. Consequently, a detector that can differentiate between human black-body emission and that of the surroundings must be sensitive to 9.348- μm emission and not to 9.89- μm emission. Semiconductors (as we shall see in Chapter 5) are low-pass (frequency) detectors, so if a composition can be tuned to have a bandgap between the two wavelengths, it will respond to 9.348- μm radiation and not to 9.89- μm radiation. Often in scenes viewed through night-vision lenses, one sees bright buildings. This is due to the daytime heating of the concrete that has not yet dissipated during the night.

Below about 400 °C, an object will not emit visible light. Our eyes, therefore, are not good detectors of objects too hot to handle. The visible spectrum extends from about 390 to 780 nm, and it breaks down into rough color ranges, listed in Table 1.2 with other useful wavelengths. The human eye has three types of color detectors (cones) that cover the visible range. The black-and-white detectors (rods) have a slightly wider wavelength detection range and are more sensitive, so all things look gray in very low light.

The eye response, divided between rods and cones, is sketched in Fig. 1.9. Note that the eye sees only 3 colors. Thus, if red and green are mixed together, the eye will not see yellow, but the brain will incorrectly deduce yellow. By contrast, a spectrometer will correctly detect the mixture of red and green. If, on the other hand, a pure oscillator has a wavelength of 585 nm, then the eye will again see a mixture of red and green, and the brain this time will correctly deduce yellow but may presume the light to be made up of red and green. The spectrometer, however, will correctly identify the difference between a single color and a mixture of colors (Fig.

Color	Wavelength (nm)	Frequency (THz)	Energy (eV)
Ultraviolet*	10–400	750–30,000	3.1–12
UV-C*	10–280	1,000–30,000	4.4–12
UV-B*	280–320	940–1,000	3.9–4.4
UV-A*	320–400	750–940	3.1–3.9
Violet	390–455	660–770	2.7–3.2
Blue	455–492	610–660	2.5–2.7
Green	492–577	520–610	2.15–2.5
Yellow	577–597	500–520	2.08–2.15
Orange	597–622	480–500	2.00–2.08
Red	622–780	384–480	1.6–2.00
Infrared*	780–10,000	30–384	0.1–1.6

* Outside the visible range.

Table 1.2: Approximate range of values for various colors.

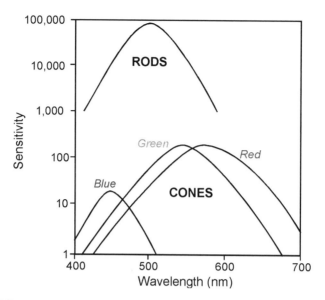

Figure 1.9: Human eye detectors showing the sensitivity of the rods and the blue, green, and red cones (from Nassau 1983).

1.10). In general, most colors correspond to the transmission or reflection of a broad range of wavelengths (e.g., the peaks are broad), and here the eye and the spectrometer will agree. Subtractive colors, such as those caused by the absorption of white light, generally consist of a single peak. The brain and the spectrometer will agree, although the former does not have as high a resolution as the latter.

1.4   Mathematical Waves

If a local disturbance in a field (electric, magnetic, density, pressure, etc.) is expressed by the symbol $\Psi(\mathbf{r},t)$, then the wave equation for that field is written as

$$\nabla^2 \Psi(\mathbf{r},t) = \frac{1}{c^2}\frac{\partial^2 \Psi(\mathbf{r},t)}{\partial t^2} \tag{1.3}$$

where the first term in $\mathbf{r}$ corresponds to the spatial disturbance, the

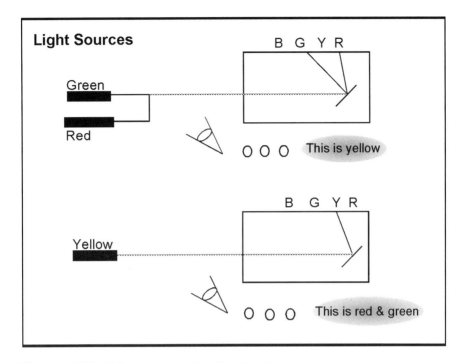

Figure 1.10: Color mixing and ocular detection vs spectroscopy.

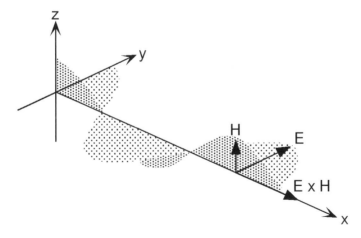

Figure 1.11: Diagram of the amplitudes of the electric and magnetic fields in a propagating wave.

second term in t is the temporal disturbance, and c is the propagation velocity in vacuum (to be replaced by v for the velocity in a medium).

The solution to the wave equation in one dimension, with no boundary conditions, is easily obtained and represents a plane wave:

$$\Psi(x,t) = A_0 \sin(kx - \omega t) \quad \text{or} \quad \Psi(x,t) = A_0 e^{i(kx-\omega t)} \qquad (1.4)$$

Here x is the propagation direction, k is the wave vector in units of inverse distance (m^{-1} or cm^{-1}), and ω is the wave angular frequency in radians. The relationship between the wave vector and the wavelength of the propagating plane wave is given by

$$\lambda = 2\pi/k \qquad (1.5)$$

Similarly, the angular frequency of the wave is related to its cyclic frequency by

$$\omega = 2\pi\nu \qquad (1.6)$$

In addition, the wave vector can be related to the angular frequency by

$$k = \omega/c \qquad (1.7)$$

If the period, or time difference between repeating cycles of the EM wave is given by τ, then the angular frequency may also be written as

$$\omega = 2\pi/\tau \qquad (1.8)$$

Finally, the product of the wavelength and frequency gives the velocity of propagation of the wave, v. In free space this velocity for electromagnetic waves is expressed as c, the speed of light:

$$\lambda\nu = c \qquad (1.9)$$

For electromagnetic waves, the electric and magnetic field disturbances are always normal to the propagation direction, x, as pictured in Fig. 1.11, since they are transverse waves.

If a plane wave propagates along some arbitrary direction $\mathbf{k}$ (vectors are expressed in **bold** letters), the solution may be written in three dimensions as follows:

$$\Psi(\mathbf{r},t) = A_0 e^{i(\mathbf{k}\cdot\mathbf{r}-\omega t)} \quad \text{or} \quad \Psi(\mathbf{r},t) = B_0 \sin(\mathbf{k}\cdot\mathbf{r} - \omega t + \delta) \qquad (1.10)$$

where: $\mathbf{k}\cdot\mathbf{r} = k_x x + k_y y + k_z z = |k||r|\cos\theta$, and θ is the angle between $\mathbf{k}$ and $\mathbf{r}$ (Fig. 1.12). Defining an arbitrary $\mathbf{x}$-coordinate along the plane-wave propagation direction vector $\mathbf{k}$ then returns to the one-dimensional equation.

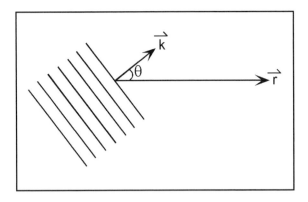

Figure 1.12: Propagation of uniform phase fronts for a plane wave traveling in an arbitrary direction.

One may alternately describe the propagation of spherical or cylindrical waves by solving the wave equation in spherical or cylindrical coordinates. The solution involves a transformation of coordinates as described next.

In spherical coordinates (r, θ, ϕ), the differential Laplacian operator is written as

$$\nabla^2 = \frac{1}{r^2}[\frac{\partial}{\partial r}(r^2 \frac{\partial}{\partial r})] + \frac{1}{r^2 \sin \theta}\frac{\partial}{\partial \theta}(\sin \theta \frac{\partial}{\partial \theta}) + \frac{1}{r^2 \sin^2 \theta}\frac{\partial^2}{\partial \phi^2} \qquad (1.11)$$

and the coordinates are as follows:

$$x = r \sin \theta \cos \phi$$

$$y = r \sin \theta \sin \phi$$

$$z = r \cos \theta$$

The solution in spherical coordinates can be found by the method of separation of variables (see Boas 1983, for example) or by transform methods (see Boas 1983). Separation of variables yields

$$\Psi(\mathbf{r}, \theta, \phi, t) = \sum_{n,\ell,m} A_{n\ell m} \sqrt{r} Z_{\ell + 1/2}(k_{\ell n}r) Y_{\ell m}(\theta, \phi) e^{i\omega t} \qquad (1.12)$$

where:

$r^{-1/2}Z_{\ell+1/2}(kr)$ = spherical Bessel functions of $1/2$ integer

$Z_{\ell+1/2}(kr) = J_{\ell+1/2}(kr)$ for standing waves with nodes at the

zeroes of the Bessel function

$Z_{\ell+1/2}(kr) = H^{(1)}{}_{\ell+1/2}(kr)$ for outward-traveling waves

$Z_{\ell+1/2}(kr) = H^{(2)}{}_{\ell+1/2}(kr)$ for inward-traveling waves

$Y_{\ell m}(\theta, \phi)$ = spherical harmonics = $P_\ell^{(m)}(\cos\theta)e^{\pm im\phi}$

$P_l^{(m)}(\cos\theta)$ = associated Legendre polynomials

It is interesting to note that Huygens's principle, which states that plane waves can always be treated as a sum of spherical wavelets and which is used to explain optical diffraction, is a direct result of the relationship between the mathematical expressions for the plane and spherical waves:

$$\Psi(\mathbf{r}) = A_0 e^{i\mathbf{k}\cdot\mathbf{r}} = A_0 e^{ikr\cos\theta} = A_0 \sum_{l=0}^{\infty}(2l+1)i^l J_l(kr)P_l(\cos\theta) \qquad (1.13)$$

Plane wave

Sum of spherical wavelets

1.5   Electromagnetic Waves

Since light is an electromagnetic wave, its behavior is determined by Maxwell's Equations. An examination of the fundamental physics behind these equations gives us insight into the principles underlying the interaction of light with matter. In the following treatment, **E** represents the electric field (volts/meter), **B** is the magnetic induction (teslas), $d\mathbf{S}$ indicates integration over a surface, $d\mathbf{V}$ indicates integration over a volume, $d\ell$ indicates integration over a line, Q is the total charge (coulombs) contained in a volume, and ρ is the electric charge density (coulombs/meter^3).

Gauss's law:

$$\oint\oint \mathbf{E}\cdot d\mathbf{S} = \int\int\int \rho dV = \frac{Q}{\epsilon} \qquad (1.14)$$

Gauss's law tells us that the outward flux of the electric field through any closed surface is equal to $1/\varepsilon$ times the net charge inside the volume, where ε is the electric premittivity of the medium. This means that the electric field over a surface results from the charge enclosed by that surface. Charges of the same sign will repel each other. Therefore, all charges will migrate to the surface in materials, such as metals, that have a reasonably high conductivity and allow charges to diffuse. Consequently, from Gauss's law, there is no static electric field inside a conductor, since the inside charge is zero. Note that for dielectrics, the total charge Q must include both the bound and the free charges.

Magnetic monopole equation:

$$\oint \oint \mathbf{B} \cdot d\mathbf{S} = 0 \tag{1.15}$$

The magnetic monopole equation is the magnetic analogue to Gauss's law and states simply that the integrated **B**-flux over any closed surface is zero; e.g., there are no magnetic monopoles.

Faraday's law

$$\oint \mathbf{E} \cdot d\ell = - \int \int \frac{\partial \mathbf{B}}{\partial t} \cdot d\mathbf{S} = -\frac{d\Phi}{dt} \tag{1.16}$$

Faraday's law states that a time-varying magnetic flux will induce an electric field in the loop defining the perimeter of the area enclosing the flux. In this equation, Φ is the magnetic flux.

Ampere's law (with Maxwell's modification):

$$\oint \mathbf{B} \cdot d\ell = \mu \int \int \left(\mathbf{J} + \varepsilon \frac{\partial \mathbf{E}}{\partial t} \right) \cdot d\mathbf{S} \tag{1.17}$$

Similar to Faraday's law, Ampere's law indicates that a time-varying electric flux will induce a magnetic field in the loop defining the perimeter of the area enclosing the flux. In this equation, **J** represents the real current density (A/m^2), and the displacement current, $\partial \mathbf{E}/\partial t$, was added later as Maxwell's modification.

In addition to Maxwell's equations, it is useful to consider an empirical equation that relates the electric field, **E**, to a current, **J**, through the conductivity, σ, of a material (Ohm's Law).

Ohm's Law:

$$\mathbf{E} = \mathbf{J}/\sigma \tag{1.18}$$

From Faraday's Law, a time-varying magnetic flux will induce a current of opposite sign. Combined, Faraday's and Ampere's laws suggest that time-varying electric and magnetic fields will induce real surface currents. These currents will tend to oppose the external fields and their penetration into a conductor. In other words, time-varying electric and magnetic fields will be partially shielded from the interior of conducting materials by induced currents and charges. This shielding effect is complete in perfect conductors, and the induced currents act to reflect the external electromagnetic fields completely. This is why superconductors make perfect mirrors.

Maxwell's equations can be written in differential form using the divergence and Stokes' theorems:

$$\int_V \mathbf{\nabla} \cdot \mathbf{A} \, dV = \oint_\mathbf{S} \mathbf{A} \cdot d\mathbf{S}; \qquad \int_\mathbf{S} \mathbf{\nabla} \times \mathbf{A} \cdot d\mathbf{S} = \oint_L \mathbf{A} \cdot d\ell \qquad (1.19)$$

Maxwell's equations in differential form are:

$$\textit{Free space}: \qquad \mathbf{\nabla} \cdot \mathbf{B} = 0 \qquad \mathbf{\nabla} \times \mathbf{E} = -\frac{\partial \mathbf{B}}{\partial t}$$

$$\textit{Free space}: \qquad \mathbf{\nabla} \cdot \mathbf{E} = \frac{\rho}{\epsilon_0} \qquad \mathbf{\nabla} \times \mathbf{B} = \mu_0 \left(\mathbf{J} + \epsilon_0 \frac{\partial \mathbf{E}}{\partial t} \right) \qquad (1.20)$$

$$\textit{In a medium}: \qquad \mathbf{\nabla} \cdot \mathbf{D} = \frac{\rho_{\text{free}}}{\epsilon} \qquad \mathbf{\nabla} \times \mathbf{H} = \mathbf{J} + \frac{\partial \mathbf{D}}{\partial t}$$

Additional equations of interest include:

$$\mathbf{D} = \epsilon \mathbf{E} = \epsilon_0 \mathbf{E} + \mathbf{P}$$

$$\mathbf{P} = \mathbf{E}(\epsilon - \epsilon_0) = \epsilon_0 \chi_P \mathbf{E}$$

$$\mathbf{B} = \mu \mathbf{H} = \mu_0 (\mathbf{H} + \mathbf{M}) = \mu_0 (1 + \chi_\mathbf{M}) \mathbf{H}$$

$$\rho = \rho_{\text{total}} = \rho_{\text{free}} + \rho_{\text{bound}} \ \text{ and } \ \rho_{\text{bound}} = -\mathbf{\nabla} \cdot \mathbf{P} \qquad (1.21)$$

$$\mathbf{J} = \sigma \mathbf{E}$$

$$\mathbf{\nabla} \cdot \mathbf{J} = -\frac{\partial \rho_{\text{total}}}{\partial t}$$

$$\textit{Lorentz force}: \quad \mathbf{F} = q(\mathbf{E} + \mathbf{v} \times \mathbf{B})$$

E = electric field vector (associated with all charges) (Joules/Coulomb = volts/meter)

D = electric field displacement vector (associated with free charges only)(Coulombs/meter^2)

P = polarization of the material (associated with bound charges only) (Coulombs/meter^2)

B = magnetic induction (associated with all currents)(Teslas = Webers/meter^2 = Newtons/(Amperes-meter))

H = magnetic field intensity (associated with true currents only) (Amperes/meter)

M = magnetization of the material (associated with bound magnetic moments) (Amperes/meter)

ϵ = electric permittivity (farads/meter)

ϵ_0 = electric permittivity of free space (8.85×10^{-12}farads/meter)

$\epsilon_D = \epsilon/\epsilon_0$ = dielectric constant (unitless)

μ = magnetic permittivity(henrys/meter = Tesla-meters/ampere)

μ_0 = magnetic permittivity of free space(1.26×10^{-6}Henrys/meter = Tesla-meters/Ampere)

$\chi_P = (\epsilon_D - 1)$ = electric susceptibility

χ_M = magnetic susceptibility

σ = conductivity (mhos/meter)

q = real charge

v = velocity of charged particle

The electric and magnetic susceptibilities are complex, and they depend on frequency (dispersion). Consequently, the electric field and polarization and the magnetic field and polarization have phase differences. Combining the preceding equations leads to the wave equation for electromagnetic fields:

$$\nabla \times \mathbf{E} = -\frac{\partial \mathbf{B}}{\partial t}$$

$$\nabla \times (\nabla \times \mathbf{E}) = \nabla(\nabla \cdot \mathbf{E}) - \nabla^2 \mathbf{E}$$

$$= -\nabla \times \frac{\partial \mathbf{B}}{\partial t} = -\frac{\partial}{\partial t}(\nabla \times \mathbf{B}) = -\frac{\partial}{\partial t}(\mu \mathbf{J} + \mu\epsilon \frac{\partial \mathbf{E}}{\partial t}) \qquad (1.22a)$$

$$= \nabla^2 \mathbf{E} = \mu\epsilon \frac{\partial^2 \mathbf{E}}{\partial t^2} + \mu\sigma \frac{\partial \mathbf{E}}{\partial t}$$

With the assumptions that there are no bound or free charges ($\nabla \cdot \mathbf{E} = 0$) and no real currents ($\partial \mathbf{J}/\partial t = 0$) and that the permittivities are not time dependent, this equation simplifies to the familiar wave equation:

$$\nabla^2 \mathbf{E} = \mu\epsilon \frac{\partial^2 \mathbf{E}}{\partial t^2} \qquad (1.22b)$$

By comparison to Eq. (1.3), the wave equation yields a definition for the velocity of the electromagnetic wave, v:

$$v = \frac{1}{\sqrt{\mu\epsilon}}$$

$$v_{\text{free space}} = \frac{1}{\sqrt{\mu_0 \epsilon_0}} = c = (2.9979 \times 10^8 \text{m/sec}) \qquad (1.23)$$

Using the equations starting with Faraday's law yields the wave equation for the magnetic field:

$$\nabla^2 \mathbf{B} = \mu\epsilon \frac{\partial^2 \mathbf{B}}{\partial t^2} \qquad (1.24)$$

These equations hold only for free space and insulators. Later, we will examine the result of allowing for a nonzero conductivity. In that case, we must use Eq. (1.22a) instead of Eq. (1.22b).

These equations are easily solved for various boundary conditions using either separation of variables or Fourier/Laplace transform methods. (See Appendix 1A).

If we now examine the behavior of plane-polarized waves with linear polarization in the y and z directions and propagating in free space along the x direction, we can see that Maxwell's Equations, using $\mathbf{E}$ and $\mathbf{H}$ vectors, will reduce to:

$$\nabla \cdot \mathbf{H} = 0 \qquad \nabla \times \mathbf{H} = -i\omega\epsilon_0\mathbf{E}$$
$$\nabla \cdot \mathbf{E} = 0 \qquad \nabla \times \mathbf{E} = i\omega\mu_0\mathbf{H} \tag{1.25}$$

The equations on the right show that $\mathbf{E}$ and $\mathbf{H}$ are perpendicular vectors. The wave equations for $\mathbf{E}$ and $\mathbf{H}$ can be found by taking the *curl* of *curl* of $\mathbf{E}$ and $\mathbf{H}$ in the preceding expressions:

$$\nabla^2\mathbf{E} + \epsilon_0\mu_0\omega^2\mathbf{E} = 0$$
$$\nabla^2\mathbf{H} + \epsilon_0\mu_0\omega^2\mathbf{H} = 0 \tag{1.26}$$

In this case, the field vectors, which are transverse to the propagation direction, are written as follows:

$$\mathbf{E} = u_y\mathbf{E}_{0y}e^{i(k_x x - \omega t)}$$
$$\mathbf{H} = u_z\mathbf{H}_{0z}e^{i(kx - \omega t)} \tag{1.27}$$

with

$$\mathbf{E}_{0y} = c\mu\mathbf{H}_{0z}$$

where u_y and u_z are the unit vectors in the y and z directions. If the three preceding expressions are combined, it can be seen that the electric and magnetic fields are related through the following expression:

$$\mathbf{E}/\mathbf{H} = \mu c = \omega\mu/k \tag{1.28}$$

The vector fields $\mathbf{E}$ and $\mathbf{H}$ are perpendicular to each other and in phase at all points, as shown earlier in Fig. 1.11. The magnetic field intensity is very much smaller than the electric field intensity; consequently, except for a few materials that have a large magnetic susceptibility, the electric field interaction dominates the behavior of most materials. In addition, since most materials are only weakly magnetic, the magnetic permittivity is not much different from that of free space ($\mu \approx \mu_0$).

1.6   Propagation Characteristics

The refractive index of a material is defined by the ratio of the velocity of light in free space to the velocity of light as it passes through the material. Hence, a material's index of refraction, n, may be written as

$$n = \frac{c}{v} = \sqrt{\frac{\epsilon\mu}{\epsilon_0\mu_0}}$$

Since $\mu \approx \mu_0$,

$$n \approx \sqrt{\frac{\epsilon}{\epsilon_0}} = \sqrt{\epsilon_D} \qquad (1.29)$$

The velocity of light traveling through a material is less than the velocity in free space (e.g., light travels more slowly through water, glass, and oil than it would in a vacuum). This means that the refractive index of a solid material, a gas, or a liquid will always be greater than 1. It is also true, however, that different wavelengths of light travel at different speeds through the same material. This leads to the fact that the refractive index of any matter, including air, has some wavelength dispersion, or a variation in value as a function of wavelength or frequency:

$$n = n(\omega) \quad \text{and} \quad \epsilon_D = \epsilon_D(\omega) \qquad (1.30)$$

Since materials undergo a certain amount of induced polarization, there is a loss associated with the propagation of light. This loss is the same as would have resulted from considering a nonzero conductivity in the solution to the wave equation. In insulators, the loss is small far from resonances where the induced polarization is negligible. It is substantial in conductors.

Let's look at the result of allowing for conductivity in Maxwell's equations:

$$\frac{\partial^2 \mathbf{E}}{\partial x^2} = \mu\sigma \frac{\partial \mathbf{E}}{\partial t} + \mu\epsilon \frac{\partial^2 \mathbf{E}}{\partial t^2} \qquad (1.31)$$

Assuming that the electric field solution is $\mathbf{E} = \mathbf{E}_0 e^{i(kx - \omega t)}$ and substituting into the differential equation leads to the following relationship:

$$k^2 = \mu\epsilon\omega^2 + i\mu\sigma\omega \qquad (1.32)$$

For a nonconductor, far from resonances in polarization, $\sigma = 0$ and $k = n\omega/c$. In a conductor and near polarization resonances, the conductivity is nonzero, which, therefore, causes the propagation vector, k, to be complex:

$$k = k_1 + ik_2 \qquad (1.33)$$

In a weak conductor, $k_2 \ll k_1$. Therefore, the two components of the propagation vector can be calculated to be:

$$k_1 = \frac{n\omega}{c} \quad (1/meter)$$

$$k_2 = \frac{\mu\sigma c}{2n} = \frac{\sigma}{2nc\epsilon_0} \quad (1/meter) \qquad (1.34)$$

As the conductivity increases, k_2 approaches the magnitude of k_1 and the more general solution must be used. Appendix 1B shows the derivation of the general expression. When the conductivity is very large, as in metals, $k_2 \approx k_1$, so this simplifies to the following expression:

$$k_1 = k_2 = \left[\frac{\mu\sigma\omega}{2}\right]^{1/2} \qquad (1.35)$$

Often the propagation constants are expressed in terms of a complex refractive index n^*, which is written in terms of a real refractive index, $n(\omega)$, and a real loss factor, $\kappa(\omega)$, which are unitless:

$$n^*(\omega) = n(\omega) + i\kappa(\omega) = \frac{c}{\omega[k_1(\omega) + ik_2(\omega)]}$$

$$n(\omega) = \frac{ck_1}{\omega}$$

$$\kappa(\omega) = \frac{ck_2}{\omega} \qquad (1.36)$$

$$\kappa(\omega) = \frac{\sigma}{2n\omega\epsilon_0} \quad \text{(poor conductors)}$$

$$\kappa(\omega) = c\sqrt{\frac{\mu\sigma}{2\omega}} \quad \text{(good conductors)}$$

Thus, the electric field strength can be rewritten as follows:

$$\mathbf{E} = \mathbf{E}_0 e^{i[(n\omega x/c) + (i\omega\kappa x/c) - \omega t]} = \mathbf{E}_0 e^{i\{[(n\omega x/c) - \omega t] - \omega\kappa x/c\}} \qquad (1.37)$$

The power density in an electromagnetic wave is calculated from the Poynting vector, $\mathbf{S}$. For a plane wave, the average value of $\mathbf{S}$ gives the

wave's intensity as a function of propagation distance. It can be noted that
S points in the direction of propagation of the electromagnetic wave:

$$\mathbf{S} = \frac{1}{\mu}\mathbf{E}\times\mathbf{B} = \mathbf{E}\times\mathbf{H} = \frac{\text{energy}}{\text{time}\times\text{area}}$$

$$I = \langle\mathbf{S}\rangle = \frac{c\epsilon_0}{2}\mathbf{E}_0{}^2 e^{-2\omega\kappa x/c} = I_0 e^{-\alpha x}$$

(1.38)

where the attenuation coefficient, α (in nepers/m), can be calculated as
follows:

$$\alpha = \frac{2\omega\kappa}{c} \quad\text{and}\quad I_0 = \frac{c\epsilon_0}{2}\mathbf{E}^2$$

(1.39)

The attenuation results from the fact that, since n and k are complex in a
medium, the **E** and **B** fields are not in phase with each other. The mean
penetration depth (in meters), often called the skin depth, defines the
distance into a material that a wave travels before its amplitude is
attenuated by a factor of $1/e$. It is the inverse of the attenuation coefficient:

$$\delta = \frac{1}{\alpha} = \frac{c}{2\omega\kappa}$$

(1.40)

The expression for skin depth is found to be:

$$\delta = \frac{nc\epsilon_0}{\sigma} \quad\text{(poor conductors)}$$

$$\delta = \left(\frac{2}{\omega\sigma\mu}\right)^{1/2} \quad\text{(good conductors)}$$

(1.41)

There are numerous ways to express loss. A few important ones are:

$$\frac{d\mathbf{B}}{m} = \frac{10}{2.303}\alpha\ (\text{nepers/m}) = \frac{10\log_{10}(I_0/I)}{x(\text{meters})}$$

$$\text{Absorbance} = \log_{10}(I_0/I) = \text{Optical Density}$$

$$\frac{d\mathbf{B}}{m} = \frac{10\times\text{OD}}{x(\text{meters})}$$

(1.42)

Beer-Lambert law : $I = I_0 e^{\epsilon_c[\text{M}]x}$

$$\text{OD} = \log_{10}(I_0/I) = \epsilon_c[\text{M}]x$$

where ϵ_c is the extinction coefficient (liters/mole·m), [M] is the concentration of the absorbing species (moles/liter), and x is the sample thickness in meters.

1.7 Dispersion

The three major effects of the presence of a medium are: (1) the wave is attenuated, (2) the electric and magnetic fields are no longer in phase, and (3) the wave velocity and propagation constants (refractive index and wave vector) are no longer constant and depend on frequency or wavelength. The strong dependence of the velocity of light traveling through a material (refractive index) on its wavelength or frequency is one of the important characteristics of the optical behavior of materials. This is a phenomenon called *dispersion*. Stated simply, it means that different frequencies of light will experience different indices of refraction in a given material. The molecular-level reason for this dispersion and the resulting dispersion behavior are different in each type of material, so this will be treated in the next chapters. One result of dispersion is that white light, which is composed of all of the colors of the visible spectrum, can be refracted into a rainbow of separated colors by a prism or by water droplets. This happens because a light wave incident on an interface will refract a given amount that is determined by its angle of incidence at the interface and by the refractive index it encounters on both sides of the interface. Consequently, the color components of white light or solar radiation, while incident at the same angle on the surface of a material (prism or rain droplet) will be refracted at different angles due to the dispersion of the material. Such behavior is described by Snell's law, which states that for each frequency:

$$n_1(\omega) \sin \theta_1 = n_2(\omega) \sin \theta_2. \tag{1.43}$$

The angles θ_1 and θ_2 are shown schematically in Fig. 1.13, and ω corresponds to each color of interest. In normal dispersion, $n(\omega_B) > n(\omega_R)$.

The combined effect of dispersion and Snell's law can be used to analyze the formation of rainbows in the sky. As shown in Figs. 1.14 and 1.15, sunlight is refracted and then reflected in a raindrop to form a visible image in which the colors have been separated. In both raindrops, the red light is refracted above the blue light at the air–water interface because of

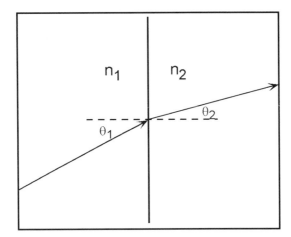

Figure 1.13: Schematic illustrating Snell's law.

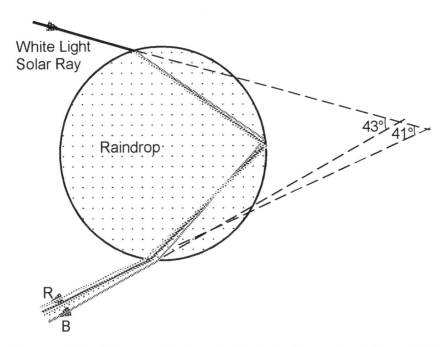

Figure 1.14: Color separation in a rain drop (after Nassau 1983). The red light ends on top after a single reflection, so the red arc in the primary rainbow has the lowest curvature.

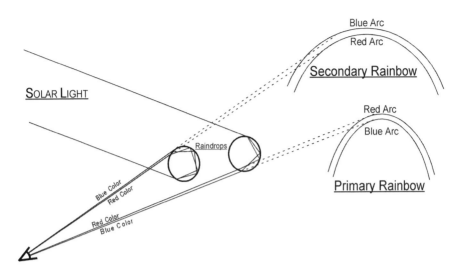

Figure 1.15: Primary rainbow formation at the 42° cone, with the red color on the outside. The reversed secondary rainbow is at a 52° cone. The second internal reflection in the water drop reverses the order of the colors and puts the blue arc on the outside of the rainbow.

the lower refractive index of the red wavelength. In the left raindrop, two internal reflections result in a reflection of the blue light above the red. In the right raindrop, a single internal reflection results in a reflection of the red light above the blue.

In the right raindrop, the reflected red light appears on the outside arc of the rainbow while the reflected blue light is on the inside. This is the primary rainbow. In the spherical raindrop, the reflected beam is scattered at 41–43° from the incident beam. This is the observation angle for the primary rainbow. The secondary rainbow, coming from the left raindrop results from two reflections within each rain drop. Its colors are reversed from the primary rainbow and the angle of observation is 52°. The secondary rainbow is much fainter than the primary. The difference in refractive index of water between blue and red wavelengths is only 0.0071 at 20 °C (0.5%), yet this produces the striking display of color in the sky.

1.8   Kramers–Kronig Relations

The relationship between the incident electric field and the induced polarization in a material is expressed by Eq. (1.21) through the electric susceptibility. Kramers–Kronig relations are equations that relate the real part to the imaginary part of the susceptibility. These relations arise from a mathematical constraint on complex functions that represent physical processes. Kramers–Kronig relations are a special case of Hilbert transforms. The complex susceptibility relates two real functions, the electric field, $E(t)$, and the polarization, $P(t)$; therefore, it must satisfy symmetry requirements between itself and its complex conjugate:

$$\chi(-\omega) = \chi_{\text{complex conjugate}}(\omega) = \chi^*(\omega) \qquad (1.44)$$

This forces a set of equality requirements between the real and imaginary parts of the complex frequency-dependent susceptibility. Causality requirements, which state that the polarizability can result only from an electric field applied in the present or past but not in the future, determine the form of the relationship between the complex parts of the frequency-dependent susceptibility.

In the following equations we show that the real and imaginary parts of the complex susceptibility are related to the refractive index and absorption coefficient:

$$\chi^*(\omega) = \chi_1(\omega) + i\chi_2(\omega) = (\frac{\epsilon^*(\omega)}{\epsilon_0} - 1)$$

$$\epsilon_D^*(\omega) = 1 + \chi^*(\omega)$$

$$\qquad (1.45)$$

$$n^*(\omega) = n(\omega) + i\kappa(\omega) = n(\omega) + i\frac{c\alpha(\omega)}{2\omega} = \epsilon_D^{*2}(\omega)$$

$$n^*(\omega) = [1 + \chi_1(\omega) + i\chi_2(\omega)]^2$$

Consequently, the frequency dependencies of these two properties are also related. Physically, what we see is that while an electromagnetic wave travels through a medium, its velocity is affected by the polarization behavior of that medium. However, it is that same polarization behavior that also causes absorption of the energy of the wave. Therefore, the absorption component and the velocity component of the propagation vector are related. The frequency-dependent changes in velocity (dispersion) and the absorption peaks occur near resonances in the induced polarization. One can visualize such an effect by assuming that polarization occurs through some vibration of charge under a springlike restoring force. Such a harmonic system would have a resonant frequency, ω_0. If the incident wave (electromagnetic field) is at a frequency ω that is much less than the resonant frequency $(\omega \ll \omega_0)$, then the charge can follow the electric field through its entire oscillation, thus remaining in phase with it. Under this condition, there is no absorption of energy from the wave and little change in the wave velocity through the material. If the incident field is at a much higher frequency $(\omega \gg \omega_0)$, then the polarization charge has no chance to respond to the applied field and effectively does not see it. This again leads to no absorption or change in wave velocity. At resonance, however, when the applied field is near the resonant frequency of the polarization process $(\omega \approx \omega_0)$, the charge reacts, but by the time it has reacted the field has changed direction. In this case, the external field and the polarization are out of phase with each other and energy is extracted from the field. (It's like trying to push a swing. It does not work if you push during the return cycle.) As energy is extracted from the wave, the interaction between the field and the induced polarization reaches a maximum and absorption becomes large. Due to the same interaction, the wave velocity in the material is also affected. [For example, the wave velocity is significantly reduced at $\omega < \omega_0$, for charges can follow the field oscillations. The reduction in velocity is much less at $\omega > \omega_0$, where the charges cannot follow the field oscillations, so $n(\omega < \omega_0) > n(\omega > \omega_0)$.]

The mathematical formalism that leads to the Kramers–Kronig equations is important, though complex, and it is essential to understanding the conditions of applicability of the equations. It is discussed in Appendix 1C. For most applications it is only necessary to know that the Kramers–Kronig relations hold for all linear processes. They also hold for a large variety of nonlinear processes, though not all. What breaks down is not the physical requirement of the Hilbert transform conditions but, rather, the simplifying mathematical method used to calculate the Kramers–Kronig equations. We refer the reader to Appendix 1C for a more complete discussion and the mathematical treatment.

The Kramers–Kronig equations, because they relate the frequency dependence of two physical functions, also have consequences on time dependence. Here, they pertain to the requirement of causality, which states that the time dependence of the polarization, $P(t)$, must follow that of the electric field, $E(t')$ for all $t' \leq t$ (causality principle).

Without proof, the form of the relationship, known as the Kramers–Kronig relations may be written as follows:

$$\chi_1(\omega) = \frac{2}{\pi} \int_0^{\infty} \frac{\omega' \chi_2(\omega')}{\omega'^2 - \omega^2} d\omega'$$

$$\chi_2(\omega) = \frac{2\omega}{\pi} \int_0^{\infty} \frac{\chi_1(\omega')}{\omega^2 - \omega'^2} d\omega'$$

(1.46)

In terms of refractive index and absorption coefficient, the equations become:

$$n(\omega) = \frac{c}{\pi} \int_0^{\infty} \frac{\alpha(\omega')}{\omega'^2 - \omega^2} d\omega'$$

$$\alpha(\omega) = \frac{c}{\pi} \int_0^{\infty} \frac{n(\omega')}{\omega'^2 - \omega^2} d\omega'$$

(1.47)

In all these equations, only the principal value of the integral is calculated (e.g., the singular point at $\omega' = \omega$ is omitted from the integral).

The implications of these relations are profound. They state that if the refractive index shows a dispersion (wavelength dependence), then the absorption coefficient must also show dispersion. The magnitude of the absorption is related to the dispersion of the refractive index, and vice versa. The form of the relations indicates that in regions of the spectrum where the propagating medium exhibits very little index dispersion, the absorption will be low. Conversely, regions of the spectrum with a strong dispersion in refractive index will exhibit a large absorption. This is consistent with our earlier discussion on the basis of the material's polarization response to an electromagnetic wave.

Often, materials do not lend themselves to measurements of both refractive index and absorption coefficient. Kramers–Kronig relations, in principle, allow calculating one from the other. There is a practical difficulty, in that the relations involve integration over the entire frequency or wavelength spectrum. This is generally not possible experimentally. However, the denominator of the argument in the integral is proportional to the difference between the squares of the frequency of interest, ω, and the integration variable ω'. This indicates that if measurements are conducted in a range of frequencies or wavelengths spanning the region of interest, and if no large variations are expected outside that region, then the error introduced by leaving out the unmeasured region may be small. But measurements must cover *all regions of high dispersion* that are near the range of interest! Note that, while the Kramers–Kronig relations are universal for stationary systems, their application will require some care in dynamic systems and in those with nonlinear processes.

1.9 Wave–Particle Duality

It is well known that there are some conditions under which electromagnetic waves must be treated as particles (photons) rather than as waves. These conditions arise, for example, when describing electron–photon interactions and when performing low intensity photon counting detection techniques. Such examples will be discussed in later chapters. Let's look at the fundamental equations that relate photons to wave properties.

The energy density of a wave can be expressed in terms of photon energies as follows:

$$I = \frac{c\epsilon_0}{2}\mathbf{E}_0^2 = Nh\nu \tag{1.48}$$

where $h\nu$ is the photon energy and N is the number of photons.

Now let's consider the relativistic equation for the energy, ξ, of a single particle:

$$(\xi)^2 = (pc)^2 + (m_0 c^2)^2 \tag{1.49}$$

where the first term is the kinetic energy in terms of the momentum, p, and the second term is the rest mass (m_0) energy. Photons have no rest mass $(m_0 = 0)$, so this gives the following relation between momentum and energy using Eqs. (1.48) and (1.49):

$$\xi = h\nu = pc \tag{1.50}$$

This leads to the following well-known equations for photon momentum:

$$p = \frac{h\nu}{c} = \frac{h}{\lambda} = \hbar k \tag{1.51}$$

1.10   Phonons

In processes involving light–matter interactions, momentum may be transferred to or absorbed from the atoms in a material. If the structure of the material has any regular order, the vibrational modes of the atoms are quantized into phonons of energy, $\hbar\omega$. Such acoustic vibrations in a medium will interact with the electrons in the medium and with photons transmitted by the medium. There are two kinds of phonons: acoustic and optical. Acoustic phonons are lower in energy than optical phonons, and the two types form different branches in the phonon dispersion relation. Figure 1.16 shows the vibrations corresponding to the optical and acoustic branches for a diatomic crystal. In the acoustic modes, adjacent atoms of different kinds vibrate in essentially the same direction with a periodically varying amplitude. In the optical modes, adjacent atoms of different kinds travel in opposite directions with amplitudes that are inversely proportional to the atom masses. Monoatomic materials and lattices exhibit acoustic phonons only, unless the lattice has atoms in two different sites. Materials made from more than one atom per unit cell exhibit both kinds of phonons: acoustic and optical. In three-dimensional structures, these branches are then divided into longitudinal and transverse modes, as shown in Fig. 1.17. Longitudinal waves are formed by atoms that are displaced along the direction of propagation of the wave; transverse waves are formed by atoms that are displaced

Motion associated with the acoustic phonon branch

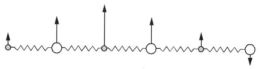

Motion associated with the optical phonon branch

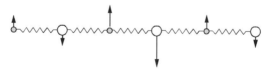

Figure 1.16: Sketch of atomic motions associated with phonon branches.

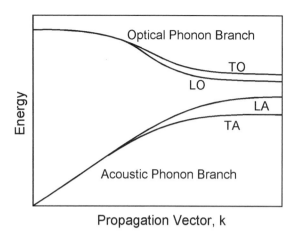

Propagation Vector, k

Figure 1.17: Energy dispersion curve for phonons in solids with at least two different kinds of atoms, so optical phonons are formed.

normal to the wave propagation direction. Cubic crystals have one longitudinal and two transverse waves. As the symmetry is decreased, more modes are present. Glasses are essentially isotropic in all directions, so they have one longitudinal and one transverse wave. One often uses the concept of phonons in glasses, but their quantization conditions are relaxed. The formation of phonons is possible in glasses, because, despite their lack of long range order, most glasses possess reasonable molecular or unit cell order, which gives them well-defined, though broad, modes of vibrations and phonon resonances. Liquids and gases support acoustic waves but not phonons, since they lack structural order.

A phonon of wave vector **k** will interact with photons and other particles as if it had momentum, $\hbar \mathbf{k}$. In reality an acoustic wave itself has no physical momentum, except at $\mathbf{k} = 0$, where the entire lattice undergoes a uniform translation. This is somewhat like standing in water at the beach and being hit by a wave. The wave itself carries no net momentum, but the impact you feel comes from the water (lattice) that is being moved by the wave at a localized position on its trajectory. Photon–phonon interactions occur in situations in which the value of the propagation vector of the photon is changed along with the energy. An example of this is found in indirect transitions in semiconductors, a topic to be discussed in more depth in Chapter 5.

1.11   Measurements

Key measurements of the propagation characteristics of light often involve the evaluation of: (1) the absorption or loss in intensity as a function of wavelength, (2) the scattering coefficient, and (3) the refractive index dispersion. In this chapter, we initiate the custom of describing measurement methods that produce the essential characteristics of the materials covered in each chapter. We begin by discussing ruled gratings and a variety of spectrometers.

1.11.1   Ruled gratings

The most fundamental form of diffraction grating consists of multiple parallel, equidistant slits in an opaque screen. If d is the distance between the centers of any two adjacent slits, β_m is the angle through which the light is diffracted, m is the order of the diffracted beam, and λ is the wavelength, then the diffraction of light from a grating is described by

$$d \sin \beta_m = m\lambda \qquad (1.52)$$

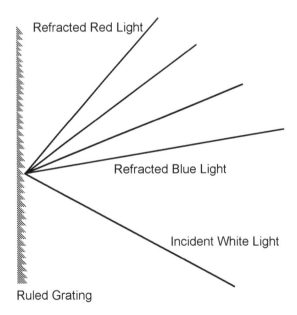

Figure 1.18: Color separation from a ruled grating.

where m is an integer. In the case of a ruled grating, color separation is made possible by capitalizing on the effect of interference of light at different angles. Ruled gratings are not made up of slits or lines; rather, they are composed of reflective surfaces that have been machined into a metallic surface. Such gratings are typically "blazed" at an angle, as shown in Fig. 1.18. The blaze angle chosen increases the amount of light diffracted into a selected order over a similar grating with flat blazing. This has the advantage of decreasing losses in the grating. In the case of a reflective grating with light incident at an oblique angle, α, the general grating equation becomes

$$d(\sin \alpha - \sin \beta_m) = m\lambda \tag{1.53}$$

where d is the center-to-center separation of the facets or grating lines, α is the angle of incidence (from normal), and β_m is the angle of the mth diffracted order (from normal). As will be seen later in this chapter, gratings play a major role in spectrometers for optical measurements.

Ruled gratings reflect light from closely spaced surfaces, so the optical path differs with angle, as shown in Fig. 1.19. When the incident light is passed through an entrance slit and the reflected light is viewed through an exit slit, different colors are formed as the grating is rotated about its

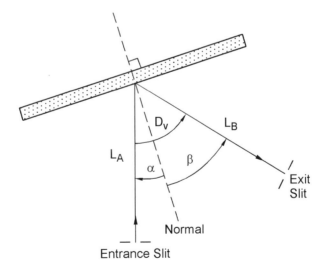

Figure 1.19: Sketch of the operation of a typical grating monochrometer with an entrance slit and an exit slit.

axis. The optical path difference for light of differing wavelengths is expressed as

$$\delta = \frac{2\pi}{\lambda} N\Delta x \quad \text{and} \quad \sin\beta_m = \sin\alpha - Nm\lambda \qquad (1.54)$$

If $\Delta x = m\lambda$, then the interference is constructive; if $\Delta x = (m - 1/2)\lambda$, then the interference is destructive.

The symbols used in Eq. (1.54) and in Fig. 1.19 have the following meanings:

δ = optical path difference

N = groove density in number/mm

Δx = dispersion of light at exit slit

α = angle of incidence at entrance slit

β = angle of diffraction of grating

m = diffraction order

λ = wavelength in nm

L_B = effective exit focal length

A general equation for the angular dispersion can be used to examine the effect of groove density of the grating on the range of wavelengths that can be scanned by a detector of fixed size.

$$\frac{d\beta}{d\lambda} = \frac{mN \times 10^{-6}}{\cos\beta} \quad \text{(radians/nm)}$$

$$\frac{d\lambda}{dx} = \frac{10^6 \cos\beta}{mNL_B} \quad \text{(nm/mm)} \qquad (1.55)$$

Thus a 640 mm spectrometer with a detector size of $\Delta x = 25$ mm containing 1000 detectors (with each detector covering 25 µm) will have the following resolution:

Groove density	150 g/mm	300 g/mm	1200 g/mm
5-pixel resolution	13 Å	6.5 Å	1.6 Å
dλ/pixel	2.6 Å	1.3 Å	0.3 Å

Diffraction gratings appear in nature:

- in opal stones, which have a structure of SiO_2 colloids aligned in streaks in a 3-dimensional grating
- in the indigo or gopher snake, whose skin is a 2-dimensional grating
- in the *Serica-serica* beatle and many other insects

As a source of biological color, diffraction gratings produce a strong iridescence in bright light, but the color disappears in low light.

1.11.2 The grating spectrometer

Grating spectrometers (Fig. 1.20) are used to break up white light into its wavelength components. They may be used in either of two modes: (1) to analyze broad-band spectral output from a light source or an emitting material, and (2) to filter and/or analyze narrow-wavelength components of a signal or of a source. In the first configuration, spectrometers are commonly referred to as spectrographs. In this case, the light from a source or the light signal from a luminescent or glowing material is focused into the spectrometer, dispersed using a diffractive grating, and then imaged onto a detector array (typically a CCD array or photodiode array). Because the light has been dispersed into its wavelength components by the grating, each pixel, or group of pixels, in the detector array records the intensity of light at a specific wavelength. The wavelength corresponding to a particular pixel may be calibrated using a calibration source with well-known emission characteristics (mercury lamp, for example). In this way, the intensity of light given off by the

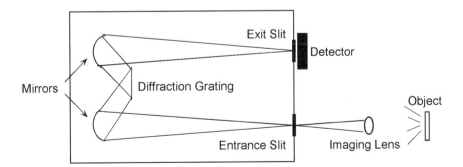

Figure 1.20: Schematic of a typical grating spectrometer.

object under study may be evaluated as a function of wavelength for a band of wavelengths simultaneously. If an exit slit is placed at the output of the spectrometer instead of a detector array, the spectrometer is now called a monochromator. In this case, only the single component (wavelength) of the dispersed light that falls on the exit slit is allowed to exit the spectrometer, hence the name monochromator. A detector placed at the output slit enables one to measure the intensity of light at the selected wavelength, or the monochromatic light that exits the slit may be used as a source for further optical tests. The particular wavelength that exits the spectrometer may be tuned by rotating the diffractive grating so that the dispersed light scans across the exit slit. Again the angle of the grating corresponding to a particular wavelength is determined using a calibration source. Monochromators may be used to filter a source, so a narrow wavelength of light can be used to excite a sample optically, or they may be used to analyze the spectral output of a luminescent sample or source.

Numerical aperture is very important to the design and use of spectrometers. Numerical aperture is related to the acceptance angle of the spectrometer and defines the angle of incidence of light rays that will fill the entire surface of the grating. Such a condition is desirable because it maximizes the sensitivity of the instrument. The F-number, $F\#$, focal length, f, grating size, D, numerical aperture, NA, and angle of incidence, θ, are related as follows:

$$F\# = \frac{f}{D} = \frac{2}{\text{NA}} = \frac{1}{2\tan\theta} \tag{1.56}$$

In order to fill the grating, the focusing lens at the spectrometer entrance must match the numerical aperture of the spectrometer. This allows the entrance beam to be focused into the entrance slit with an angle that matches the acceptance angle of the grating (see Fig. 1.21).

Grating spectrometers can be used to measure the absorption and/or reflection of light. Many spectrometers are built to operate with two separate beams (double-beam spectrometers) for increased accuracy of measurement by allowing a comparison of the intensity of the measuring beam with that of a standard beam. Briefly, a light source produces a broad range of wavelengths diffracted through a grating to form beams with a continuous variation in wavelength depending on the angle of the grating in the spectrometer. Sources for a UV-visible spectrometer consist of xenon or argon lamps. Sources for IR spectrometers consist of a glow

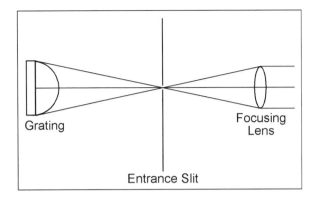

Figure 1.21: Importance of numerical aperture matching. The light from the collecting lens is incident on the spectrometer focusing lens. If the numerical aperture of the lens is properly matched to the numerical aperture of the instrument, the light fills the grating.

bar, which is simply a hot filament that emits light that approximates the black-body radiation spectrum. As shown in the illustration for a typical spectrometer in Fig. 1.22, the light output from the grating is split into two beams, one for the sample chamber and one for the reference

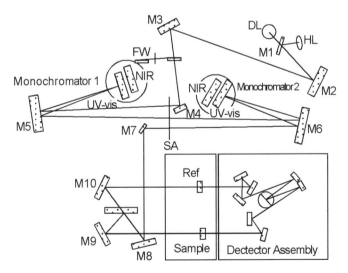

Figure 1.22: Sketch of a typical double-beam UV-visible spectrometer (after Perkin Elmer).

chamber. It is then mechanically chopped by a spinning wheel with regular openings positioned so that the sample beam is transmitted when the openings are present and the reference beam is reflected by the blocking parts of the wheel, and then transmitted through a different chamber. This produces the two beams with a chopping modulation out of phase with each other. After passing through the sample and reference chambers, the beams are then combined and detected by a phase-sensitive system that can measure the difference between the reference and sample beam intensities.

The spectrometer is first calibrated with the reference and sample chambers empty, to balance the transmitted amplitudes of the two beams. Then two kinds of measurements are possible: sample vs air, which gives the total loss, or sample vs another sample, which gives differential absorbance. The sample vs air measurement gives the total loss due to the presence of the sample, and this includes reflection (R) from both sample faces and (α) absorption and scattering from the sample of thickness Δx:

$$\frac{I_T}{I_0} = (1 - R^2)e^{-\alpha\Delta x} \qquad (1.57)$$

This type of measurement is limited to samples that transmit a sufficient amount of light to be measured by the detector system. There are other techniques, such as photoacoustic spectroscopy, for measuring absorption characteristics of highly absorbing systems. These are described in Chapter 2, "Optical Properties of Conductors".

Reflection measurements are also possible in this apparatus with the use of an M-shaped set of mirrors, as illustrated in Fig. 1.23. Here the incident light is reflected toward the sample by mirror M1. After reflection

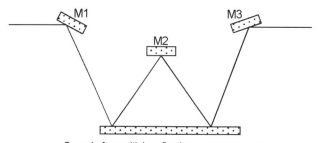

Sample for multiple reflection measurements

Figure 1.23: M-mirrors arranged for multiple reflections.

from the sample, depending on the angle of M1, the incident light may be reflected many times between the sample surface and mirror M2. Finally, the light is sent back to the spectrometer by mirror M3.

In a study of the amount of veiling glare in an automotive vehicle windshield that results from the reflection of sunlight from the dashboard of the vehicle, we designed a reflection apparatus for measuring reflection and scattering at various angles. A sketch of this apparatus is shown in Fig. 1.24. It consists of a light source and a grating monochrometer as in the instrument just described. The light is similarly chopped; then it is polarized and reflected from a sample target that is mounted on a rotating table. The detector is placed behind the analyzer, and both are mounted on an arm that can be freely rotated to obtain not just the light reflected at the angle of incidence on the sample (specular reflection) but also the light scattered at other angles (diffuse scattering).

Differential reflectivity measurements are often conducted on metal surfaces and will be described in Chapter 2, "Optical Properties of Conductors".

1.11.3 Fast Fourier transform spectrometers

Fourier transform (FT) spectrometers have an advantage over the dispersive spectrometers just described, because of the speed with which they can acquire data over a broad range of wavelengths. The dispersive spectrometers make measurements at each discrete wave-

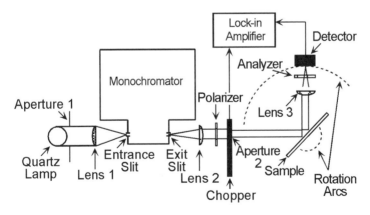

Figure 1.24: Reflection spectrometer.

length set by the operator. In order to scan a range of wavelengths, the grating must be adjusted to a precise angle; a measurement must be made, and the grating must then be moved. Fourier transform spectrometers, in contrast, obtain a broad range of wavelengths in a single measurement. Consequently, they are much faster than dispersive spectrometers. This speed allows them to make repeated measurements on the same sample and thus to detect a very weak transmission; they are very sensitive by comparison. However, they do not have the wavelength precision of the dispersive spectrometers.

Fourier transform spectrometers usually consist of Michelson interferometers, in which the source beam is split into two equal beams (Fig. 1.25). One beam goes to a fixed mirror; the second beam is reflected from a movable mirror. The two beams are recombined to form a wavelength-dependent interference pattern due to their optical path length difference, and they pass through a sample to a detector or are reflected from a sample to a detector. By rapid scanning of the movable mirror, a time-dependent optical pulse is received by the detector that is then Fourier transformed into a frequency spectrum. Limits on mirror speed have constrained the wavelength band typically measurable using FT spectrometers to the infrared (IR), generally from 1 to 15 μm.

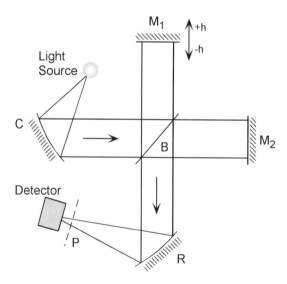

Figure 1.25: Sketch of a Fourier transform spectrometer.

1.11.4 Microscopes

Often, microscopes are used for light gathering in an optical measurement. Microscopes consist of two lenses separated by a long tube. The lens near the sample is the objective lens; that near the observer is the eyepiece. The separation between the eyepiece lens and the objective lens is the sum of the respective focal lengths of the lenses and a tube length, L, that separates the two focal points:

$$\text{lens} - \text{lens distance } = f_o + L + f_e \qquad (1.58)$$

The tube length is usually fixed at 160 mm, as shown in Fig. 1.26. The object is placed outside the focal point of the objective lens, which focuses the image at the field stop, which is also the focal point of the eyepiece. The final image is focused at infinity by the eyepiece. (In the calculation of lens focal length from magnifying power, this is assumed to be 254 mm.) This is a good viewing condition for optical microscopy, since the eye and the camera have their own lens that refocuses the magnified image onto the retina or the film. This is also a good viewing condition for using the microscope as a light-gathering instrument in front of a spectrometer, in order either to study very small samples or to receive light from only the surface of some object. Having the eyepiece focus at infinity allows varying the distance between the eyepiece and a final focusing lens that is ahead of the entrance slit of the spectrometer.

The magnification of a lens may be calculated from the ratio of its image distance to its focal length:

$$M_{\text{obj}} = \frac{160 \text{ mm}}{f_0}; \quad M_{\text{eyep}} = \frac{254 \text{ mm}}{f_e} \qquad (1.59)$$

The following table shows typical values for an objective and an eyepiece lens with a tube length of 160 mm; this leads to a lens-to-lens distance of 188.6 mm.

The focal spot diameter of a lens is also very pertinent to measurements made using a microscope. The spot diameter, d, is related to the numerical aperture, NA, of the lens as follows:

Lens	Power	Focal length
Objective	50 X	3.2 mm
Eyepiece	10 X	25.4 mm

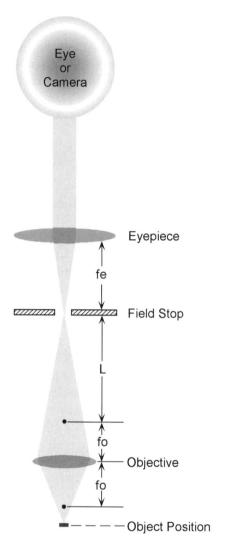

Figure 1.26: Sketch of an optical microscope (after Hecht 1990).

$$d = \frac{1.22\lambda_0}{\text{NA}} \qquad (1.60)$$

The lens working distance is greater than its focal length. Thus it is important to calculate the appropriate working distances, magnifications, and spot diameters when designing a microscope light-collection apparatus.

Appendix 1A

Solution of the Wave Equation by Transform Methods

This is simply an example to highlight the Fourier–Laplace transforms and contour integration in preparation for the more complex appendices later in the book.

Consider the problem of an infinite string driven by an external oscillator at one end of the string. The problem is one-dimensional and the oscillator is at $x = 0$. The 1-D wave equation is:

$$\frac{\partial^2 \Psi}{\partial x^2} = \frac{1}{c^2} \frac{\partial^2 \Psi}{\partial t^2} \tag{1.61}$$

Initial conditions:

$$\Psi(x, 0) = 0 \quad \text{and} \quad \frac{\partial \Psi(x, 0)}{\partial t} = 0 \tag{1.62}$$

Boundary condition:

$$\Psi(0, t) = f(t) \tag{1.63}$$

We look for solutions at $x \geq 0$ and at $t \geq 0$.

Since the t variable has two initial conditions, we use a Laplace transform on t. Since the x variable has one boundary condition, we use the Fourier sine transform on x:

45

$$\mathscr{F}_{\text{sine}}\left\{\Psi_{xx} - \frac{1}{c^2}\frac{\partial^2 \Psi_{xx}}{\partial t^2} = 0\right\} = -s^2\Psi_s + s\Psi(0,t) - \frac{1}{c^2}\frac{\partial^2 \Psi_s}{\partial t^2} = 0$$

$$\mathscr{L}\left\{-s^2\Psi_s + sf(t) - \frac{1}{c^2}\frac{\partial^2 \Psi}{\partial t^2}\right\} = -s^2\Psi_{sp} + sf_p - \frac{p^2}{c^2}\Psi_{sp} = 0 \tag{1.64}$$

This yields the transformed solution:

$$\Psi_{sp} = \frac{sf_p}{s^2 + p^2/c^2} \tag{1.65}$$

First, we invert the Fourier transform:

$$\Psi_p = \sqrt{\frac{2}{\pi}} f_p \int_0^\infty \frac{s\,\sin sx}{s^2 + p^2/c^2}\,ds \tag{1.66}$$

and follow by contour integration to get

$$\Psi_p = \frac{f_p}{\sqrt{2\pi}} Im \oint \frac{se^{isx}}{s^2 + p^2/c^2}\,ds = \frac{f_s}{\sqrt{2\pi}} Im[2\pi i\, Res(ip/c)] \tag{1.67}$$

The result by contour integration according to the associated drawing in Fig. 1.27 is

$$\Psi_s = \frac{f_s}{\sqrt{2/\pi}} e^{-px/c} \quad \text{for } x>0 \text{ only} \tag{1.68}$$

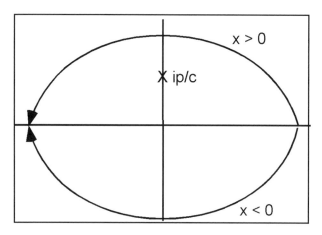

Figure 1.27: Schematic of the contour integration showing the pole in the upper half space.

Then we invert the Laplace transform to get:

$$\Psi(x,t) = \begin{cases} f(t - x/c) & \text{for } t - x/c > 0 \text{ (upper branch)} \\ 0 & \text{for } t - x/c < 0 \text{ (lower branch)} \end{cases} \tag{1.69}$$

This is a wave traveling to the right with the same shape as the wave produced by the oscillator.

Appendix 1B

General Solution for the Propagation Vectors

The general equation that results from the differential wave equation for either the electric or the magnetic field in the presence of a nonzero conductivity is written as

$$k^2 = \mu\epsilon\omega^2 + i\mu\sigma\omega \qquad (1.70)$$

The propagation constant, k, is complex and can be written as the sum of a real and an imaginary component:

$$k^2 = (k_1 + ik_2)^2 = k_1{}^2 - k_2{}^2 + 2ik_1k_2 \qquad (1.71)$$

a. For poor conductors, $k_2{}^2 \ll k_1{}^2$ and the functions simplify to:

$$k_1 = \frac{\omega n}{c}$$
$$k_2 = \frac{\sigma}{2nc\epsilon_0} \qquad (1.72)$$

b. For good conductors, we need to take the square root of the complex number in Eq. (1.70) by using polar coordinates. The general solution is:

$$k_1 = [(\mu\epsilon\omega^2)^2 + (\mu\sigma\omega)^2]^{1/4} \left[\frac{1}{2} \left(1 + \frac{1}{\sqrt{1 + \sigma^2/\epsilon^2\omega^2}} \right) \right]^{1/2}$$

$$k_2 = [(\mu\epsilon\omega^2)^2 + (\mu\sigma\omega)^2]^{1/4} \left[\frac{1}{2} \left(1 - \frac{1}{\sqrt{1 + \sigma^2/\epsilon^2\omega^2}} \right) \right]^{1/2}$$

(1.73)

c. For very good conductors, the general expression can be simplified using the assumption that $(\sigma/\epsilon\omega)^2 \gg 1$. Then the two components of the propagation vector become equal:

$$k_1 = k_2 = \frac{1}{2}[\mu\sigma\omega]^{1/2} \tag{1.74}$$

The skin depth, which is equal to $\delta = 1/(2k_2)$, is plotted in terms of the reduced conductivity $(\sigma\epsilon/\omega)$ in Fig. 1.28, which shows a wide variation in the penetration of the electromagnetic wave into the material. Nonconductors have skin depths of microns and larger, whereas good conductors have skin depths of only a few angstroms.

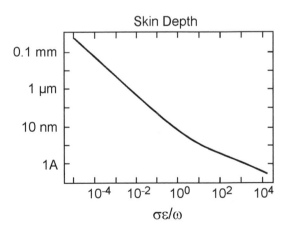

Figure 1.28: Variation in skin depth as a function of conductivity for a material with refractive index of 1.5.

Appendix 1C

Kramers–Kronig Relations

Kramers–Kronig relations, a special case of Hilbert transforms, are dispersion relations that relate the complex parts of the susceptibility of a material. Thus they relate the complex parts of the dielectric function and the complex parts of the propagation vectors (index of refraction and loss factor or absorption). The general behavior results from considering the effect of an external stimulus (electric field) on the response (polarization) of a material through a response function (the susceptibility). In a linear and local system, the polarization, $\mathbf{P}(t)$, and the electric field, $\mathbf{E}(t)$, are related as follows:

$$\mathbf{P}(t) = \epsilon_0 \int_{-\infty}^{t} \chi(t - t')\mathbf{E}(t')dt' \tag{1.75}$$

The response function or susceptibility can have a time-delayed response, so the polarization does not result just from the simultaneous electric field but may, by superposition, depend on the past values of the electric field. This explains the integral over time. The integral is limited to t in the upper limit, because the polarization at time t cannot be affected by an electric field in the future $(t' > t)$. This condition causes a problem when using Fourier transforms to represent the frequency-dependent relationship between the electric field and the polarization. This is shown next.

If the Fourier transforms are taken, we may write:

50

$$\mathbf{E}(\omega) = \int\limits_{-\infty}^{\infty} \mathbf{E}(t)e^{-i\omega t}dt$$

$$\mathbf{P}(\omega) = \int\limits_{-\infty}^{\infty} \mathbf{P}(t)e^{-i\omega t}dt \qquad (1.76)$$

$$\chi(\omega) = \int\limits_{-\infty}^{\infty} \chi(t)e^{-i\omega t}dt$$

If one considers the polarization to be a linear product of the susceptibility and the electric field:

$$\mathbf{P}(\omega) = \epsilon_0 \chi(\omega)\mathbf{E}(\omega) \qquad (1.77)$$

then the time-dependent polarization can be calculated by use of the convolution theorem to be

$$\mathbf{P}(t) = \epsilon_0 \int\limits_{-\infty}^{\infty} \chi(t - t')\mathbf{E}(t')dt' \qquad (1.78)$$

This expression is ergodic only if we define the susceptibility such that it satisfies the following condition:

$$\chi(t - t') = \begin{cases} 0 & \text{for } t' > t \\ \text{finite} & \text{for } t' \le t \end{cases} \qquad (1.79)$$

This condition is known as causality. Typically, if Eq. (1.75) had an integral over all times (e.g., from $-\infty$ to $+\infty$), then it could be shown that a material would exhibit polarization in advance of the arrival of a field.

The time-dependent functions must be real; however, all three components of the equation can be complex in Fourier transformed space (frequency space). The reality of the time-dependent relationship requires that the susceptibility satisfy a symmetry requirement expressed as:

$$\chi^*(\omega) = \chi(-\omega)$$

$$\chi_1(+\omega) - i\chi_2(+\omega) = \chi_1(-\omega) + i\chi_2(-\omega)$$

$$\text{Real terms} : \chi_1(+\omega) = \chi_1(-\omega) \qquad (1.80)$$

$$\text{Imaginary terms} : \chi_2(+\omega) = -\chi_2(-\omega)$$

However, the causality condition imposes a stricter limitation on the susceptibility, which combines with Eq. (1.78) to produce the Kramers–Kronig relations. This limitation is obtained by calculating the Fourier transform of the susceptibility subject to the causality condition.

Assume that the system has both propagation and loss components, so $\omega' \to \omega''$, with the latter as a complex variable $\omega'' = \omega' + i\zeta' = \rho' e^{i\theta'}$:

$$\chi(t - t') = \int\limits_{-\infty}^{\infty} \chi(\omega'') e^{i\omega'(t-t')} e^{-\zeta'(t-t')} d\omega'' \tag{1.81}$$

This expression is not causal, since it allows for t' to be larger than t. Therefore, conditions must be applied to the form of the complex susceptibility, $\chi^*(\omega)$, to force it to zero when $t' > t$.

This condition is derived by examining the behavior of the integrand as the imaginary frequency axis goes to infinity. If ζ' is positive, the integral decays when taken over the upper half-plane in complex space, since the function $e^{-\zeta'(t-t')}$ decays at $\zeta' = \infty$; but only for $(t - t') > 0$. This condition builds into the solution for the susceptibility the causality requirement that only the values of $t' < t$ are considered or $(t - t') > 0$ always.

The desired causal constraint on the Fourier component of the susceptibility is obtained from the contour integral of the function:

$$\oint \frac{\chi(\omega')}{\omega' - \omega} d\omega' \tag{1.82}$$

Since the susceptibility must be analytic, we solve for the contour integral of Fig. 1.29 using the Cauchy integral formula:

$$\chi(\omega) = \frac{1}{2\pi i} \oint \frac{\chi(\omega')}{\omega' - \omega} d\omega' = \frac{1}{2\pi i} \left[\int 1 + \int 2 + \int 3 + \int 4 \right] \tag{1.83}$$

The integral over the semicircle (4) vanishes because the complex function decays in the upper half-plane. Consequently, $\chi(\omega)$ reduces to the sum of the integrals over paths 1, 2, and 3. The integral over path 2 contains half a pole measured in the clockwise direction, so it can be written as follows:

$$\int 2 = {}^1\!/_2 [2\pi i \chi(\omega)] = i\pi \chi(\omega) \tag{1.84}$$

Consequently,

$$\chi(\omega) = \frac{1}{2i\pi} \left[\int 1 + \int 3 \right] + \frac{i\pi}{2i\pi} \chi(\omega) \Rightarrow \chi(\omega) = \frac{1}{i\pi} \left[\int 1 + \int 3 \right] \tag{1.85}$$

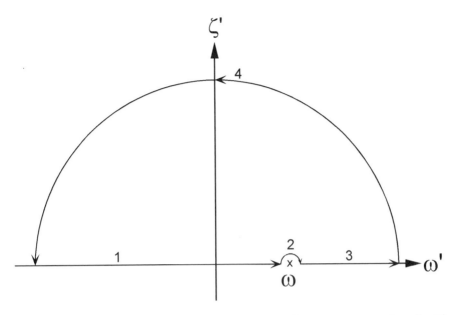

Figure 1.29: Contour integration in the complex frequency space $\omega'' = \omega' + i\zeta'$ for solution of the Kramers–Kronig condition on the real and imaginary parts of the susceptibility with a pole at $\omega' = \omega$.

and the equation reduces to

$$\int 1 + \int 3 = \lim_{\epsilon \to 0} \left[\int_{-\infty}^{\omega-\epsilon} \frac{\chi(\omega')}{\omega' - \omega} d\omega' + \int_{\omega+\epsilon}^{\infty} \frac{\chi(\omega')}{\omega' - \omega} d\omega' \right] \equiv \mathscr{P} \int_{-\infty}^{\infty} \frac{\chi(\omega')}{\omega' - \omega} d\omega'$$

$$(1.86)$$

where $\mathscr{P}$ indicates the principal value of the integral (e.g., the singular point at $\omega' = \omega$ is omitted.)

$$\chi(\omega) = \frac{1}{i\pi} \mathscr{P} \int_{-\infty}^{\infty} \frac{\chi(\omega')}{\omega' - \omega} d\omega' \qquad (1.87)$$

Equation (1.77) suggests that a monochromatic stimulus, $E(\omega)$, will cause a response, $P(\omega)$, of the same frequency. Equation (1.87) shows that if $\chi(\omega')$ has a pole at ω, then there can be a finite response even in the absence of a stimulus (e.g., normal modes of vibration of the system). The result of this equation, and in particular the presence of i on the right side, is that the real and imaginary parts of complex susceptibility $[\chi(\omega) = \chi_1(\omega) + i\chi_2(\omega)]$ are related to each other as follows:

$$\chi_1(\omega) = \frac{1}{\pi}\mathscr{P}\int_{-\infty}^{\infty}\frac{\chi_2(\omega')}{\omega' - \omega}d\omega'$$

$$\chi_2(\omega) = -\frac{1}{\pi}\mathscr{P}\int_{-\infty}^{\infty}\frac{\chi_1(\omega')}{\omega' - \omega}d\omega'$$

(1.88)

Application of the symmetry condition on the complex susceptibility $\chi^*(\omega) = \chi(-\omega)$ simplifies the integral through the following steps:

$$\chi_1(\omega) = \frac{1}{\pi}\mathscr{P}\int_{-\infty}^{\infty}\frac{\chi_2(\omega')}{\omega' - \omega}d\omega' = \frac{1}{\pi}\left[\mathscr{P}\int_{\infty}^{0}\frac{\chi_2(\omega')}{\omega' - \omega}d\omega' + \mathscr{P}\int_{0}^{\infty}\frac{\chi_2(\omega')}{\omega' - \omega}d\omega'\right] \quad (1.89)$$

By letting ω go to $-\omega$ in the first integral, the limits of integration can be reversed:

$$\chi_1(\omega) = \frac{1}{\pi}\mathscr{P}\int_{0}^{\infty}\frac{\chi_2(-\omega')}{-\omega' - \omega}d\omega' + \frac{1}{\pi}\mathscr{P}\int_{0}^{\infty}\frac{\chi_2(\omega')}{\omega' - \omega}d\omega'$$

$$= \frac{1}{\pi}\mathscr{P}\int_{0}^{\infty}\chi_2(\omega')\left[\frac{1}{\omega' + \omega} + \frac{1}{\omega' - \omega}\right]d\omega'$$

(1.90)

since $\chi_2(-\omega') = -\chi_2(\omega')$. Combining the two fractions leads to the simpler expression below. The expression for $\chi_2(\omega)$ is resolved similarly while remembering that $\chi_1(-\omega') = +\chi_1(\omega')$ to give the second expression in Eq. (1.91):

$$\chi_1(\omega) = \frac{2}{\pi}\mathscr{P}\int_{0}^{\infty}\frac{\omega'\chi_2(\omega')}{\omega'^2 - \omega^2}d\omega'$$

$$\chi_2(\omega) = -\frac{2\omega}{\pi}\mathscr{P}\int_{0}^{\infty}\frac{\chi_1(\omega')}{\omega'^2 - \omega^2}d\omega'$$

(1.91)

Since the dielectric function is related to the susceptibility through $\epsilon = 1 + \chi$, its components are also constrained by similar conditions:

$$\epsilon_1(\omega) - 1 = \frac{2}{\pi}\mathscr{P}\int_0^\infty \frac{\omega'\epsilon_2(\omega')}{\omega'^2 - \omega^2}d\omega'$$

(1.92)

$$\epsilon_2(\omega) = -\frac{2\omega}{\pi}\mathscr{P}\int_0^\infty \frac{\epsilon_1(\omega') - 1}{\omega'^2 - \omega^2}d\omega'$$

The results show that dispersion and absorption are intimately connected because they are the real and imaginary parts of the propagation vector. The results of the calculation of the real and imaginary parts of the dielectric function near a typical resonance are shown in Fig. 1.30.

The Kramers–Kronig relations hold if the system is stationary and you can use the convolution theorem. There is a breakdown in metals because

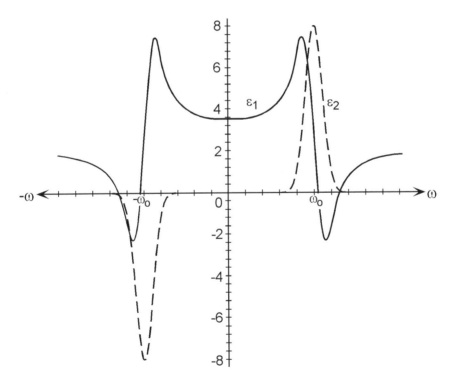

Figure 1.30: Plot of the causal real (ϵ_1) and imaginary (ϵ_2) components of the dielectric function as a function of frequency for a system with a resonance at $\omega = \omega_0$.

of a pole at $\omega = 0$ due to DC conduction. In this case, the pole is separated into an extra term in the imaginary part of the dielectric constant as follows:

$$\epsilon_2(\omega) = \frac{4\pi\sigma_0}{\omega} - \frac{2\omega}{\pi}\mathscr{P}\int_0^\infty \frac{\epsilon_1(\omega') - 1}{\omega'^2 - \omega^2}d\omega' \qquad (1.93)$$

References

Boas, M. L. 1983. *Mathematical Methods in the Physical Sciences*. New York: Wiley.

Born, M., and E. Wolf. 1980. *Principles of Optics*. Oxford: Pergamon Press.

Halliday, D., and R. Resnick. 1977 and later editions. *Physics*. New York:Wiley.

Hecht, E., 1990. *Optics*. Reading, MA: Addison-Wesley.

Instruments S. A. 1990. *Guide for Spectroscopy*. Edison, NJ: ISA.

Jackson, J. D. 1995 and later editions. *Classical Electrodynamics*. New York: Wiley.

Minneart, M. G. J. 1974. *Light and Color in the Outdoors*. New York: Springer-Verlag.

Nassau, K. 1983. *The Physics of Color*. New York: Wiley.

Perkin, Elmer. UV-Visible Lambda-9 Spectrometer, 761 Main Avenue, Norwalk, CT 06859.

Subramanian, S. 1996. A study of instrument panel reflectivity and its impact on veiling glare. Master's thesis, University of Florida, Gainesville.

Toll, J. S. 1956. Causality and the dispersion relation. *Phys. Rev.* 104:1760.

Chapter 2

Optical Properties of Conductors

2.1 Introduction

As will be seen, the optical properties of conductors are dominated by their high reflectivity. In optical systems, conducting materials are often used to reflect optical signals or to shield regions of space from electromagnetic waves. Conductive coatings commonly serve as mirror surfaces for moderate-energy infrared and visible light. X-ray optical systems frequently utilize gold films for mirrors and gratings. A familiar example of shielding is the metal sheeting and coating surrounding and inside a microwave oven that protects your kitchen from the microwave energy. Conductors can also be used to detect or emit electromagnetic signals of lower frequencies (e.g., antennae for AM and FM radios, TVs, microwaves, and radar) or to transport microwaves and radio frequency (RF) waves by confining them in rectangular waveguides.

In all materials, the energy levels of the electrons may be described by a set of energy bands that comprise the *band structure* of the solid (or liquid or plasma). Electrons bound to the energy states of an atom's outer shell are said to be valence electrons, which occupy the *valence bands* of the material. By contrast, the *conduction bands* are defined by the energy levels available to the "free" electrons. The separation between the top of the uppermost valence band and bottom of the lowermost conduction band is known as the *bandgap* of the material. There are four classes of materials that are generally distinguished by their conductivity. They are insulators, semiconductors, semimetals, and conductors. Insulators have such a large bandgap energy that the conduction bands are unoccupied. Semiconductors have a lower bandgap energy, and their conduction band states are partially occupied by a few "free" electrons at room temperature. Semimetals and metals have many "free" electrons, and their conduction bands have high occupancy. Figure 2.1 depicts the differences between the various classes of conductors and nonconductors in a two-band system consisting of bonding states and free states. The occupied states are shown with dashed lines. The Fermi energy, ξ_F, denotes the highest occupied electronic energy level at absolute zero. Hence, bands and states below the Fermi energy will be filled; bands that cross the Fermi energy will be partially filled; and bands and states above the Fermi energy will be empty. It can be seen that metals and semimetals have regions within wave vector space in which electrons may easily move from filled states to empty states within a single band. This is the source of

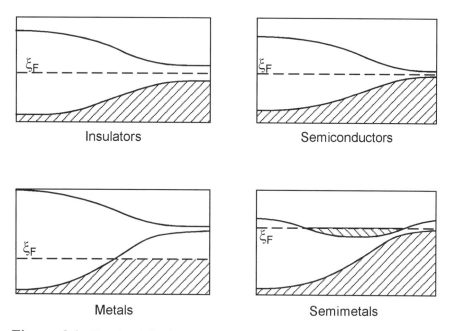

Figure 2.1: Sketch of the band structure for various materials in terms of energy vs wave vector. The occupied states are shown with dashed lines and define the Fermi Energy level ξ_F.

conductivity in these materials and the source of the sharp contrast between these materials and insulators, since electronic conduction is due only to electrons that exist in partially filled bands. A comparison of free carrier concentrations is given in Table 2.1.

Conductors are materials with free carriers and a reasonable conductivity. Generally, these materials have one or more *partially* filled bands in their electronic band structure. This allows charge to be transported via a field-driven diffusion process. Materials that exhibit these properties include the alkali metals, with partially filled s states, the transition metals, with partially filled s and d states, the rare earths, with partially filled f states, and the trivalent metals with partially filled p states. The divalent metals would appear to be insulators, since their electrons fill the s states. But it turns out that there is some overlap in the bands, leaving a few holes in the Brillouin zone, which gives them a poor but modest conductivity compared to "true" insulators. Column IVA elements like carbon, silicon, and germanium have four valence electrons, and, with the exception of β-tin and lead, they crystallize in the diamond cubic structure with tetrahedral covalent bonds that are fully saturated. This happens because their two electrons in the p states form hybrid σ orbitals by mixing with the two electrons in the s states to become the valence band (bonding electrons). This is accomplished through the excitation of one (s) electron to the third (p) state, as expressed by the following electronic configurations:

Material type	Free carriers (number/cm^3)	Examples
Metals	10^{22}–10^{23}	K, Na, Cu
Semimetals	10^{17}–10^{22}	Bi, graphite, Sb, As
Heavily doped semiconductors	10^{17}–10^{19}	n-ZnSe, n/p GaAs
Semiconductors	10^{13}–10^{17}	Si, Ge, undoped GaAs
Insulators	$<10^{13}$	Oxides, fluorides

Table 2.1: Carrier concentrations at room temperature for various types of materials.

$$s^2 p^2 \Rightarrow sp^3$$

$$[\text{filled}(n-1)\text{shell}](ns)^2(np)^2 \Rightarrow [\text{filled}(n-1)\text{shell}](ns)^1(np_x)^1(np_y)^1(np_z)^1$$

$$(2.1)$$

This sp^3 configuration has four bonding states per atom in the hybrid σ orbital, with four branches having tetrahedral symmetry and an occupation number of two electrons per branch. This leaves four empty hybrid states that become π orbitals to make up the conduction band. There is an energy separation between the filled sp^3 σ-bonding states and the empty π-bonding states. Consequently, these elements act as either insulators or semiconductors, depending on the energy difference between the filled and empty states (bandgap). Carbon (diamond) has $n = 2$; Si has $n = 3$; Ge has $n = 4$. For the heavier elements, Sn has $n = 5$ and Pb has $n = 6$, but they do not follow the same behavior.

In contrast, the graphite structure of carbon is formed by sp^2 bonding states that offer only planar trigonal symmetry for the covalent bonds and a weak, stretched bond out of the plane. This imparts to graphite a sheetlike structure that determines its electrical, optical, and mechanical properties (graphite is a good conductor, black in color, and easily friable). In graphite, the high electrical conductivity results because the σ and π orbitals touch, and there is a continuous energy transition between the filled and the empty states. Its behavior is described as that of a semimetal (see Fig. 2.1). The column VA and VIA elements of the periodic table, below the first row, have free electrons and act as either semimetals or semiconductors with very low bandgap energies, depending upon the relative energies of their π and σ orbitals. (Sulfur has six-fold bonding symmetry, resulting from hybrid $d^2 sp^3$ states.)

In the first chapter, we saw how the propagation characteristics of an electromagnetic wave may be described in terms of the complex wave vector, $k^*(\omega)$, the complex refractive index, $n^*(\omega)$, or the complex dielectric constant, $\epsilon^*(\omega)$. Historically, the study of metals has emphasized the use of dielectric constants. The relationships are as follows:

$$k^*(\omega)^2 = \frac{[n(\omega) + i\kappa(\omega)]^2 \omega^2}{c^2} = \frac{\epsilon_D^*(\omega)\omega^2}{c^2} \qquad (2.2)$$

In metals:

If $\kappa(\omega)$ is large and $n(\omega)$ is close to 1 $\rightarrow$ reflection
If $\kappa(\omega)$ is large and $n(\omega)$ is also large $\rightarrow$ absorption

The complex dielectric constant is written as:

$$\epsilon_D^*(\omega) = \epsilon_{D1}(\omega) + i\epsilon_{D2}(\omega) = [n(\omega) + i\kappa(\omega)]^2$$
$$\epsilon_{D1}(\omega) = n^2(\omega) - \kappa^2(\omega) \tag{2.3}$$
$$\epsilon_{D2}(\omega) = 2n(\omega)\kappa(\omega) = \frac{\sigma(\omega)}{\epsilon_0 \omega}$$

Combining Eqs. (1.40) and (1.41), the attenuation coefficient and skin depth can be written as:

$$\alpha(\omega) = \frac{2\omega\kappa(\omega)}{c} = \frac{1}{\delta(\omega)}$$

$$\delta(\omega) = \frac{n\epsilon_0 c}{\sigma(\omega)} \quad \text{(poor conductors)} \tag{2.4}$$

$$\delta(\omega) = \left[\frac{2}{\omega\sigma(\omega)\mu}\right]^{1/2} \quad \text{(good conductors)}$$

It is clear that a higher conductivity, $\sigma(\omega)$, leads to a greater loss and a smaller skin depth. As noted in Chapter 1, the free carriers that can support the induced current will act to shield the interior of the conductor from the incident electromagnetic wave.

2.2 Atomistic view: Drude model

Let us look at the electron density in a metal. Assume that each atom in general contributes all its valence electrons to the free-electron gas that bonds the metal. Let's calculate the density of free electrons in the metal. There are 6×10^{23} atoms/mole. Knowing the atomic weight, A, and the density, ρ, gives the number of valence electrons per unit volume, n, for a metal of valence Z_v, as follows:

$$n = 6 \times 10^{23} \left(\frac{Z_v \rho}{A}\right) \tag{2.5}$$

For Cu, $Z_v = 1$, $\rho = 9\text{g/cm}^3$ and $A = 64$ g/mole. The free-electron concentration is calculated to be: $n = 10^{23}$electrons/cm^3, giving a free-electron density of 0.1 electrons/Å^3. Since the electrons repulse each other, and assuming that they are uniformly distributed in the metal (in the absence of external fields), each electron occupies a volume of radius 1Å, or $2a_B$

(where a_B is the Bohr radius $\approx 0.529\text{Å}$) so the electron–electron separation is essentially 2Å ($4a_B$).

So the electrons are very close to each other and to the ionic cores. The latter is understandable, since it is the strength of the interaction between the ionic cores and the free-electron gas that provides the bonding energy of a metal. The former, however, would suggest very strong electron–electron interactions, yet the simple Drude model, which treats electrons as noninteracting particles, does a reasonable job of describing the gross characteristics of the behavior of metals. This is because the uniform mutual repulsion between the electrons reduces their interaction. This is somewhat like a crowd of people with a strong aversion to bumping or touching one another. They all stay out of each other's way and minimize collisions even in dense crowds. Thus the electron mean free path in a conductor is several orders of magnitude larger than the calculated separation.

Electron–electron interactions do occur, and they are responsible for the electrical resistance of metals at low temperatures. Electron–phonon interactions also occur, but they are weak in metals and only begin to contribute at higher temperatures.

A very effective way of treating the behavior of charges in an electromagnetic field is by modeling the charges as coupled harmonic oscillators for which the field creates the driving force. In the case of insulators, the charges are the bound electrons; therefore, the restoring force is the Coulomb attraction from the ionic cores. In metals, the charges are the free electrons; therefore, there is no restoring force. The Drude model uses this approach to give a good fundamental understanding of the behavior of free electrons in a metal under optical illumination, and allows the calculation of the induced polarization.

Let us look at a set of N_f free electrons per unit volume, each with mass m and charge q_e, exposed to an oscillatory electric field in the x direction, E_x. The differential equation of motion for the free electrons, written from Newton's second law, and its solution are:

$$m\frac{d^2x}{dt^2} = q_e E_x = q_e E_0 e^{i\omega t}$$

Letting $x(t) = x_0 e^{i\omega t}$ yields

$$x_0 = -\frac{q_e E_0}{m\omega^2} \tag{2.6}$$

Since $x(t)$ is the displacement of the electrons from equilibrium, we assume that a dipole moment is formed with the underlying charge of the ionic cores. This dipole moment, p, is equal to the product of the charge and the displacement:

$$p(t) = ql = q_e x(t) = -\frac{q_e^2 E(t)}{m\omega^2} \tag{2.7}$$

The polarization, P, of the material is simply the sum of dipole moments over all free electrons. In a first approximation, we neglect electron–electron interactions. All the free electrons are subject to the same force, so the polarization becomes

$$P(t) = \sum p_i = -\frac{N_f q_e^2 E(t)}{m\omega^2} \tag{2.8}$$

Recalling the relation between the electric permittivity, the electric displacement vector, the applied electric field, and the polarization ($\mathbf{D} = \epsilon \mathbf{E} = \epsilon_0 \mathbf{E} + \mathbf{P}$), we can write

$$\epsilon_D = 1 + \chi_P = n^{*2} = 1 + \frac{P(t)}{\epsilon_0 E(t)} = 1 - \frac{N_f q_e^2}{\epsilon_0 m \omega^2} \tag{2.9}$$

The dielectric constant is therefore negative at low frequencies, goes through zero at a point defined as the plasma frequency $[\epsilon_D(\omega_p) = 0]$, and then becomes positive at higher frequencies. The complex refractive index is essentially imaginary below the plasma frequency; thus the electromagnetic wave cannot propagate in the metal and is reflected. The plasma frequency, ω_p, is expressed as follows:

$$\omega_p^2 = \frac{N_f q_e^2}{m \epsilon_0} \tag{2.10}$$

This simplifies the expression for the dielectric constant to

$$\epsilon_D = 1 - \frac{\omega_p^2}{\omega^2} \tag{2.11}$$

Note that the value of the plasma frequency is dependent only on the number of free electrons and not on the conductivity. If the positive ion background has a dielectric constant ϵ_∞ up to high frequencies ($\omega > \omega_p$), then the dielectric constant must be written as

$$\epsilon_D = \epsilon_\infty \left[1 - \frac{\tilde{\omega}_p^2}{\omega^2} \right] \tag{2.12}$$

where $\tilde{\omega}_p^2 = N_f q_e^2 / m \epsilon_\infty \epsilon_0$. The dispersion can be represented by a relation between the frequency dependence of the dielectric constant and the wave vector as follows: $\epsilon_D \omega^2 = c^2 k^2$. Using Eq. (2.12) this becomes

$$\epsilon_D \omega^2 = \epsilon_\infty [\omega^2 - \tilde{\omega}_p^2] = c^2 k^2 \tag{2.13}$$

For $\omega^2 < \tilde{\omega}_p^2$, the propagation wave vector, k^*, is fully imaginary; consequently, the transverse wave does not propagate in the medium and it is reflected at the interface.

Alkali metals are very shiny (reflective) at frequencies below ω_p and transparent above it. The drop in reflectivity at the plasma frequency is quite pronounced (see Fig. 2.2). Assuming that the various metals have a number of free electrons per atom equal to their valence, it is instructive to compare the calculated plasma frequency with that measured (see Table 2.2). $N_{\text{effective}}$ is the ratio of the number of free electrons that gives the measured plasma frequency to the number of valence electrons. A ratio of 1 indicates that all free electrons are participating in the induced current; and a ratio of less than 1 indicates electron–electron interactions.

When the dielectric constant vanishes, the electric displacement vector also vanishes ($\mathbf{D} = 0$) and the electric field equals the polarization: $\mathbf{E} = -\mathbf{P}/\epsilon_0$. This represents a longitudinal wave that consists of free

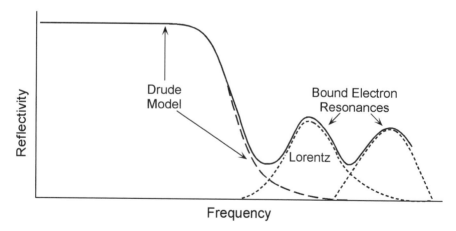

Figure 2.2: Plot of Drude model for reflectivity of metals (from Hummel 1993).

Metal element	Measured plasma energy (eV)	Measured v_p	Calculated v_p	$N_{effective}$
Li	8.0	19.4×10^{14}Hz	19.4×10^{14}Hz	1
Na	5.9	14.3	14.3	1
K	3.9	9.5	10.4	0.8
Rb	3.6	8.8	9.4	0.9
Cs	3.4	8.3	8.3	1
Mg	4.4	10.6	10.9	0.9
Al	6.3	15.3	15.8	0.9
Si	6.8	16.4–16.9	16	1
Ge	6.7	16.0–16.4	16	1
InSb	5.2	12.0–13.0	12	1

Table 2.2: Comparison of calculated and measured plasma frequencies.

longitudinal oscillations of the electrons in the material at the plasma frequency with a zero wave vector.

Equation (2.11) shows that the dielectric constant has a pole at $\omega = 0$. This divergence in the dielectric constant causes the electric field to vanish regardless of the electric displacement vector (external field). This pole requires a modification of the Kramers–Kronig relations in metals as shown in Eq. (1.93), Appendix 1C.

The relation between the dielectric constant and the wave propagation vector gives insight into the behavior of the material for different values of the dielectric constant:

$\epsilon > 0$ For real ω, k is real and the transverse electromagnetic wave propagates with velocity c/n.

$\epsilon < 0$ For real ω, k is imaginary and the wave is damped with a characteristic length of k^{-1}.

$\epsilon = \infty$ The system has a finite response in the absence of a force. These points define the frequencies of free oscillation of the medium and occur at $\omega = 0$ in metals.

$\epsilon = 0$ Plasma resonance. Transverse waves cannot propagate, but a longitudinally polarized resonance occurs through coherent longitudinal motions of the electrons.

$\epsilon = $ complex k is complex and the wave propagates with a damping constant.

The conductivity is important in the reflection process and comes in if a loss factor is introduced in the Drude model. In this case, the conductivity will be involved in the loss associated with the propagation of the wave in the metal and directly affects the skin depth, as shown in Eq. (2.4). Thus, we see that in the case of very high-conductivity metals, the penetration depth is very shallow and the light is completely reflected by the material (e.g., superconductors are perfect mirrors). As the conductivity decreases, more penetration is allowed into the material, and an associated absorption develops; thus only a part of the light is actually reflected.

According to the Drude model, the reflection coefficient of a metal behaves as shown in Fig. 2.2. The reflection coefficient for light incident normal to the metal surface can be calculated from the refractive index, $n(\omega)$, and the loss factor, $\kappa(\omega)$:

$$R = \frac{[n(\omega) - 1]^2 + \kappa^2(\omega)}{[n(\omega) + 1]^2 + \kappa^2(\omega)} \qquad (2.14)$$

For example, silver at about 580 nm has a refractive index of 0.18 and a loss factor of 3.6. This gives silver a reflection coefficient of 0.95 at the specified wavelength. (See Fig. 2.3 for the reflectance of Ag.) Sodium metal has respective values of 0.044 and 2.42, and aluminum has 1.44 and 5.23 for reflection coefficients of 0.97 and 0.83, respectively.

Some metals may also have a contribution from bound electrons that provide a weak oscillatory behavior in the frequency dependence of the reflection coefficient, as shown in Fig. 2.2. The effect of bound electrons (Lorentz model) will be treated mathematically in the next chapter, on insulators. This effect is quite important for transition metals and is discussed later.

The effect of finite conduction in the metal may be added to the Drude equation by looking at the harmonic oscillator differential equation and adding a damping term as follows:

$$m\frac{d^2x}{dt^2} + \gamma\frac{dx}{dt} = q_e E_0 e^{i\omega t} \qquad (2.15)$$

where $\gamma = N_f q_e^2/\sigma$. Using the same substitution as in Eq. (2.5), we get a modified expression for the induced dipole polarization, p_0, and the dielectric constant, ϵ_D:

$$p_0 = -\frac{q_e^2 E_0}{m\omega^2 - i\omega\gamma}$$

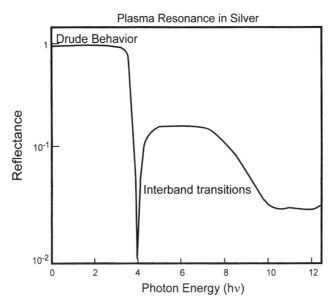

Figure 2.3: Reflectivity of silver showing Drude-like behavior in the IR and visible (from Irani, Huen, and Wooten 1971).

$$\overset{*}{\epsilon}_D = 1 - \frac{N_f q_e^2}{\epsilon_0} \frac{1}{m\omega^2 - i\omega\gamma} = 1 - \omega_p^2 \left[\frac{1}{\omega^2 - i\omega\gamma/m} \right]$$

$$\overset{*}{\epsilon}_D = \epsilon_1 + i\epsilon_2 = 1 - \omega_p^2 \left[\frac{\omega^2 + i\omega\gamma/m}{\omega^4 + \omega^2\gamma^2/m^2} \right] \tag{2.16}$$

$$= 1 - \frac{\omega_p^2}{\omega^2} \left[\frac{1}{1 + \gamma^2/m^2\omega^2} \right] + i \frac{\omega_p^2}{\omega^2} \left[\frac{\gamma/m\omega}{1 + \gamma^2/m^2\omega^2} \right]$$

Now let's look at the dielectric constant rewritten in terms of the plasma frequency and modified to include the background dielectric constant from the positive ionic cores as in Eq. (2.12).

$$\overset{*}{\epsilon}_D(\omega) = 1 - \frac{\omega_p^2}{\omega^2 - i\omega\gamma/m}$$

If the background dielectric constant is $\epsilon_B = \epsilon_\infty/\epsilon_0$ the total dielectric constant is:

$$\overset{*}{\epsilon}_D = \frac{\epsilon_\infty}{\epsilon_0} - \frac{N_f q_e^2}{\epsilon_0 m} \left[\frac{1}{\omega^2 - i\omega\gamma/m} \right] = \frac{\epsilon_\infty}{\epsilon_0} \left(1 - \frac{\tilde{\omega}_p^2}{\omega^2 - i\omega\gamma/m} \right) \tag{2.17}$$

2.3   Plasma frequency

At the plasma frequency, the propagation vector vanishes, so the electrons in the metal must interact collectively to ban the electromagnetic wave from the interior of the material. The electrons oscillate in phase, but there is no polarization charge. Consequently, this is a longitudinal oscillation. Recall that, unlike transverse waves, in which the atomic motions are *normal* to the direction of propagation of the wave, longitudinal waves are characterized by atomic motions *along* the propagation direction. These large longitudinal oscillations are called *plasmons*. The changes in the dielectric function and refractive index with frequency are shown in Figs. 2.4 and 2.5 for a typical high-conductivity metal with a plasma frequency of $\hbar\omega_p = 9\,\text{eV}$.

According to the Drude Model, both terms of the complex dielectric constant vanish at the plasma frequency. While this is approached quite well in some metals with very high conductivity (silver), electronic motion in real materials suffers from resistance, even for free electrons. This resistance comes from electron–electron and electron–phonon interactions. Consequently, the imaginary component of the dielectric constant does not vanish. The real part of the dielectric constant still vanishes, but the imaginary component decays slowly with increasing frequency. The loss factor, defined as the imaginary component divided by the modulus of

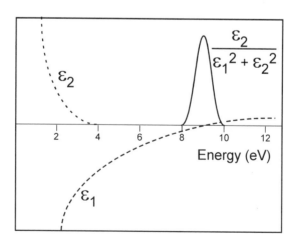

Figure 2.4: Behavior of complex dielectric constant through the plasma frequency.

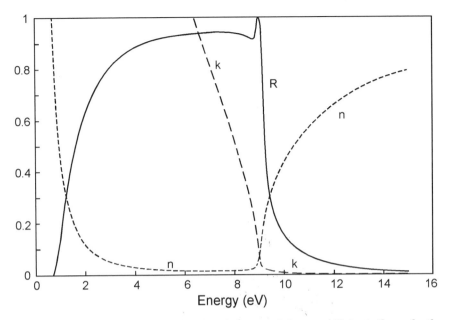

Figure 2.5: Variation of refractive index and loss coefficient through the plasma frequency.

the total complex dielectric constant, $\epsilon_2/(\epsilon_1^2 + \epsilon_2^2)$, has a maximum at the plasma frequency.

2.4 Band structures in metals

The electronic band structure of materials gives a powerful insight into their optical behavior, since it can show the various allowed carrier excitations (electron and hole transitions). The band structure of materials represents the dispersion relation between energy and momentum (frequency and wave vector). Typical plots of the electronic band structure of an alkali metal and a transition metal are shown in Fig. 2.6, with the Brillouin zones around the central point in the crystal (Γ point). Just for reminder, the Brillouin zones correspond to the minimum free-electron wavelength (maximum k vector) that can be propagated in any direction of the crystal. For example, it can be easily visualized in the case of phonon waves. If you have a chain of similar atoms that are a

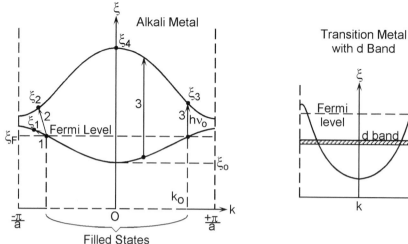

Figure 2.6: Band structure for an alkali metal and a transition metal with a d band, showing intraband transitions (1), indirect interband transitions (2), and direct interband transitions (3) (after Wooten, 1972).

distance a apart, then the shortest wavelength that can be propagated through the chain is that in which adjacent atoms are moving in opposite directions. Shorter wavelengths cannot be supported because there are no atoms any closer than a. This defines the minimum wavelength of $\lambda = 2a$, which corresponds to a maximum $k = \pi/a$. This is similar to the Brillouin zone boundary. (More is said in Chapter 5, "Optical Properties of Semiconductors.")

In a metal, the electrons fill only partially the valence band up to the Fermi energy ξ_F; thus excitations are possible at all energies. Since electrons obey the Pauli exclusion principle, all states below the Fermi energy are filled at $0\,\mathrm{K}$, and all states above it are empty. At higher temperatures, a distribution of energies will be occupied following the Fermi–Dirac distribution function for the occupation probability, $n(\xi)$:

$$n(\xi) = \frac{1}{1 + e^{(\xi-\xi_F)/k_B T}} \tag{2.18}$$

In the graph of Fig 2.6, a variety of electronic transitions are depicted in the band structure for an ideal metal. The Brillouin zones are at $\pm\pi/a$, where a is the lattice spacing. The graph depicts an alkali metal with an s valence band and a noble transition metal with a filled d band. There are four kinds of excitations possible in the structures shown:

Intraband transitions: transitions between two states in the same band. An electron at ξ_F goes to $\xi_F + \Delta\xi$. This transition requires momentum transfer ($\Delta k \neq 0$), which is possible only through a lattice interaction, since the photon momentum is negligible ($p_{photon} = \xi_{photon}/c$).

Indirect interband transitions: transitions between two states in different bands. An electron at ξ_F goes to ξ_2 in the conduction band. Since the transition is indirect, there is a change in momentum ($\Delta k \neq 0$), which requires a lattice interaction (phonon).

Direct interband transitions: transitions between two states in different bands but with the same momentum. Here ($\Delta k = 0$), and the transition does not require a phonon. An electron at ξ_F goes to ξ_3 and ξ_4. The ξ_3 transition defines the minimum, or threshold, direct transition energy. The maximum direct transition energy comes from the bottom of the valence band and corresponds to a transition between ξ_0 and ξ_4.

Underlying filled bands: If there is an underlying filled d band (noble metals: Cu, Ag, Au), then there is an additional possible interband transition from ξ_d to ξ_F.

Direct interband transitions dominate in metals, because the electron–phonon interaction is weak. The range of possible direct transition energies is: $\xi_3 - \xi_F < \xi < \xi_4 - \xi_0$, so the optical absorption dominates in that range.

2.4.1 Density of states

It is interesting to compare the density of states of an alkali metal, Li, and two transition metals, Fe and Cu. Their electronic structures are as follows: $n(\text{Li}) = [\text{He}](2s)^1, n(\text{Fe}) = [\text{Ar}](3d)^6(4s)^2$, and $n(\text{Cu}) = [\text{Ar}](3d)^{10}(4s)^1$. The density of states for these materials is shown in Fig. 2.7. In lithium, the free electrons fill only half of the $(2s)$ band, so we have the expected Drude behavior shown in the last section. In iron, the $3d$ and $4s$ electrons can each be continuously excited in their individual bands. Iron, like the other transition metals with an unfilled d-band, does show a decrease in reflectivity but exhibits no sharp plasma frequency–associated drop as seen in the alkali metals. The noble metals, like copper, silver, and gold, show a strong contribution from the filled d bands and the unfilled s band. In copper, the $3d$ electrons are bound in a filled band and

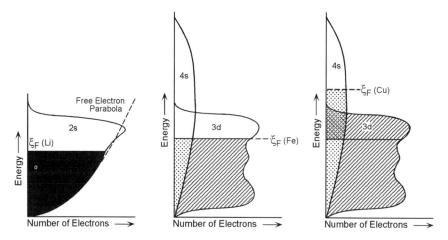

Figure 2.7: Electronic band structure of Li, Fe, and Cu, showing the overlap of d bands when present (after Slater 1951 and Nassau 1983).

only the $4s$ electrons can be continuously excited. The top of the filled $3d$ band occurs 2.2 eV below the Fermi level in the $4s$ band. Consequently, the $3d$ bound electrons can be excited with photons of energy equal to or greater than 2.2 eV. The dielectric constant then consists of contributions from the free electrons and from the bound electrons: $\epsilon_D = \epsilon_{free} + \epsilon_{bound}$. Similar behavior is found in silver and gold, for which $n(\text{Ag}) = [\text{Kr}](4d)^{10}(5s)^1$ and $n(\text{Au}) = [\text{Xe}](5d)^{10}(6s)^1$. Silver has an energy difference of 4 eV between the filled d band and the Fermi energy. The energy difference for Au is 2.5 eV.

An interesting comparison can be made between copper and silver (Fig. 2.8) to examine the effects of the bound electron excitation and the plasma resonance on the optical properties of the two metals. Copper has a d-band excitation threshold at 2.2 eV, while silver's is at 4.0 eV. Their free-electron gas contribution to the dielectric constant is about the same for each and would cause the dielectric constant to change sign at about 8.7 eV for copper and 9.2 eV for silver. The contributions from both the free-electron gas and the bound electrons to the dielectric constant are displayed for both metals in Fig. 2.8. The interesting effect is that the actual plasma frequencies for the two metals are quite different and that for copper is higher than that for silver, yet their reflection behavior follows more closely the bound electron excitation energies and has copper's reflection coefficient decreasing at lower energies than silver.

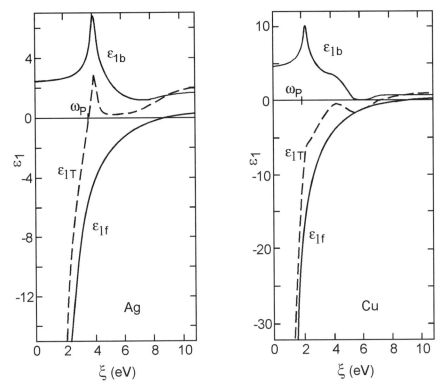

Figure 2.8: Dielectric constant behavior for Ag and Cu, showing the difference that arises from the binding energy of the d band. Silver, with a higher d band energy, has a clear plasma frequency behavior; Cu, with lower d band energy, does not (from Ehrenreich and Philipp 1962).

What we see is that in silver, since the bound electron contribution to the dielectric constant is at 4.0 eV, it is large enough to overcome the small negative free-electron gas contribution and to drive the total dielectric constant across zero to positive values. This places the plasma frequency near 3.9 eV. In copper, however, the bound electron excitation contributes to the dielectric constant at 2.2 eV, where the free-electron gas term is still strongly negative. Thus, adding the positive bound electron contribution does not raise the total dielectric constant sufficiently to cross zero. Consequently, copper does not have the abrupt change in reflection coefficient that silver has, and does not show a clear plasma resonance at either 2.2 or 7.2 eV, where the total dielectric constant eventually crosses zero. Between these two energies, the real part of the dielectric constant remains small. Consequently, the optical properties of copper between 2.2 and 7.2 eV are essentially determined by the imaginary component of the dielectric constant.

Nickel behaves much like iron. Its reflection coefficient also does not show an abrupt Drude-like decrease, because its Fermi energy intersects the unfilled d band. Here interband transitions between the s and d bands begin at about 0.3 eV, and a gradual drop in reflectance follows (Fig. 2.9). Note that most of the transition metals with an unfilled d band will exhibit a behavior similar to that of Ni and Fe. Those metals will still appear shiny, due to the presence of a reasonably high reflectivity in the visible, but they won't have the glitter of silver.

The rare earths are treated by assuming that the conduction band contains a number of free electrons equal to the chemical valence of each ion (nominally 3), and the dielectric constant is affected by both the free electrons and the bound electrons from the filled $5d$ band. The $4f$ orbitals are much more localized than the d orbitals; consequently, the $4f$ band exerts little influence on the behavior of the dielectric constant.

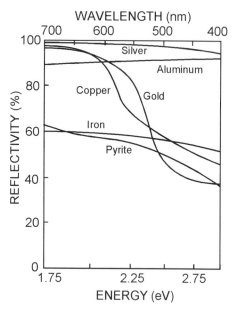

Figure 2.9: Reflectivity from various metals (from Nassau 1983).

2.5 Coloration in metals

The abrupt change in sign of the dielectric constant at the plasma frequency determines the reflected color of various metals. For example, silver, with a plasma wavelength at 310 nm, reflects over most of the visible. Silver also has a very high conductivity, which gives it a very high reflection coefficient at 95%. However the plasma resonance at 310 nm is lower that for than aluminum. In fact, the reflection of silver begins to decrease in the violet, while aluminum does not show any decrease. This gives silver a warmer white color, since its reflection spectrum has less blue and violet than aluminum, which reflects a cooler white.

In copper, the dielectric constant does not cross zero at 2.2 eV, but it changes abruptly due to the excitation of bound $3d$ electrons. This causes a drop in reflection coefficient, so wavelengths longer than 564 nm are reflected while shorter wavelengths are absorbed by the excitation of bound $3d$ electrons into the $4s$ band. This gives copper its reddish color. The bound-electron excitation threshold corresponds to the difference between the $3d$ energy level and the Fermi energy. The Fermi energy is the maximum value of energy occupied by the electrons. Since the latter are fermions, they obey the exclusion principle. Consequently if a substitutional impurity is added to Cu with extra electrons per atom, such as Zn, then, the Fermi surface is increased correspondingly. The $3d$ band itself remains essentially unchanged at low dopant concentrations; thus the difference between the Fermi energy and the top of the $3d$ band increases linearly with Zn content. Thus the color of the new alloy shifts to shorter wavelengths and begins to acquire a yellow tone. This is precisely the reason for the color of brass (an alloy of Zn and Cu). The alloying of copper with Ga(+ 3), Si(+ 4), Ge(+ 4) and Sn(+ 4) has the same result except that the threshold energy for Ga increases faster than that for Zn as a function of concentration due to its addition of two excess electrons per atom. Si, Ge, and Sn each add three excess electrons per atom in substitution for Cu, and their threshold energies increase correspondingly faster with dopant concentration. This is true only at very low alloying concentrations. At higher alloy concentrations, the band structure becomes distorted and the observed linear increase in threshold energy ceases.

Gold has a plasma resonance at 2.5 eV. Here the difference between the Fermi energy and the top of the $6d$ band reflects light at wavelengths longer than 496 nm and absorbs shorter wavelengths, giving the metal its characteristic yellow color. Interestingly, since the density of states for the $6s$ band of gold decreases at increased energies (as already shown for Cu), the probability of photoexcitation of electrons decreases, and a very thin sheet of gold transmits slightly in the blue-green.

Alloys of gold can be adjusted to give a wide range of colors:

Au + 25% Ag	green gold
Au + 25% Cu	red gold
Au + Fe	blue gold
Au + Al	purple gold
Au + Pt	white gold
Au + Pd	"
Au + (Cu + Ni + Zn)	"

Thus we see that two major factors affect the color of transition metals:

1. a large change in reflection coefficient due to an abrupt change in the real part of the dielectric constant due to either a plasma resonance or photoexcitation of bound d band electrons,
2. absorption associated with the photoexcitation of d band electrons.

The transition metal ions can also be incorporated into insulators by forming oxide compounds. In this case they provide the insulators with a broad range of interesting colors, which vary with host composition. This effect will be discussed in detail in the next two chapters.

2.6   Coloration by means of small metal particles

The deep brilliant color of "struck" metal-containing glasses is well known. Glasses with low concentrations (0.01% or less) of noble metals exhibit strong colors (ruby-red for gold, yellow for silver, and pink for platinum) when the metal is precipitated from solution in the glass to form small colloids (5–30 nm). The color is complementary to the region of light absorption in the metallic state. The same principles discussed earlier are at work, except the optical properties of the colloidal particles are determined by surface effects—for example, they exhibit a strong surface plasma resonance. The behavior of small metal particles is determined by simply solving the wave equation in a medium under an external field. The solution for spheres and ellipsoids was derived by Mie in 1908 and gives the relative magnitudes of the absorption and scattering coefficients as a function of wavelength. The theory fits the scattering measurements well.

Gold colloids are formed by including less than 0.1% gold in a soda-lime-silica or borosilicate glass. This ensures solubility at the melting point of the host glass. The colloids precipitate from solution with heat treatments at 600–700°C, where the solubility of the metal in the glass is decreased sufficiently. The particles grow with heat-treatment time. If their diameter is less than 10 μm, there is a weak Rayleigh scattering. Generally, spheres of 100-Å diameter give the well-known ruby-red color. The plasma resonance is due to bound electron plasma oscillations. The absorption of particles with sizes between 26 and 100 Å is sharp and independent of size. Samples with Au particles in this size range exhibit the well-known ruby-red color because of strong absorption of blue and green wavelengths (see Fig. 2.10). For colloids greater than 100 Å, the

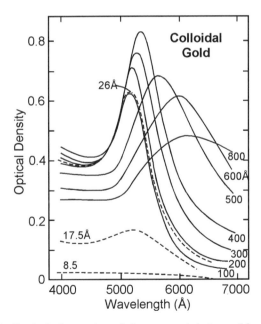

Figure 2.10: Optical absorption of glasses containing gold colloids of different sizes (from Doremus 1964).

absorption peak broadens and shifts to longer wavelengths; consequently, the color shifts from ruby-red to purple, then to violet, and then to pale blue (by about 800 Å) as the net absorption decreases at short wavelengths. Reflection from the particles, leading to scattering, is weak at small sizes and increases for particles greater than 500 Å, where it begins to add a reddish brown color.

In the solution of the wave equation for a small sphere in an external field (fully worked out by Mie), the electric field inside the sphere has the form described in Eq. (1.12) using the Bessel functions of half integer. Applying the condition of continuity of the tangential components of the internal and external fields across the boundary leads to a calculation of the scattering and absorption terms. The results show that the particles have only a weak scattering component and a strong absorption. Following the trend shown in Fig. 2.10, it can be seen that for very large sizes (several microns), the scattering component becomes stronger and wavelength independent. This is often called Mie scattering. (The reader is encouraged to follow the derivation in Appendix 2A.) Computer programs have been developed to obtain the dependence of the scattering and absorption cross sections on particle size and incident wavelength.

In small spheres, surface plasmons occur at $\epsilon_c = -2\epsilon_m$, where ϵ_c is the particle dielectric constant and ϵ_m is that of the matrix. This shifts the plasma frequency of the metal, ω_p, by

$$\omega_c = \frac{\omega_p}{(3\epsilon_m)^{1/2}} \tag{2.19}$$

This point is often referred to as the Frohlich frequency and corresponds to the vanishing of the dielectric constant of the particle. The plasma oscillation is a surface wave, hence its description as a surface plasmon.

2.7 Optical properties of superconductors

Superconductors have an energy gap that corresponds to the energy of the electron–electron interactions that order the electrons in **k**-space. Photons of energy less than the gap energy and at temperatures below the critical temperature are not absorbed by the superconductor. The conductivity component of the propagation vector diverges at absolute zero for frequencies below the gap energy. Photons with energy greater than the gap energy behave as in the normal metallic state, since they can cause transitions to "normal" free electron states.

As the temperature is lowered from the critical temperature, the energy gap increases and the conductivity term for photons with frequency below the gap energy increases. Due to the increased conductivity of the superconducting state, the reflection coefficient is increased and the skin depth is reduced. At zero frequency (DC field), the superconducting electrons short out the "normal" electrons, excited above the gap. (Here one may think of a two-fluid model of superconductivity in which one fluid is composed of paired superconducting electrons and the other is composed of "normal" electrons.)

High-T_c superconductors have a more complex behavior, and the presence of a well-defined energy gap is not established. High-T_c superconductors exhibit a broad absorption peak at zero frequency and one in the mid-IR. Their behavior is consistent with a two-fluid model in which the fluid populations are temperature dependent. Reflectance shows a steady drop through the IR region. In yttrium–barium–copper–oxide (YBCO), a broad peak appears in the reflectivity near $400\,\mathrm{cm}^{-1}$, most likely associated with localized transitions in the internal structure of the superconductor (chains), and a minimum is observed at about $800\,\mathrm{cm}^{-1}$ in the superconducting state.

2.8   Measurement techniques

2.8.1   Photoacoustic absorption spectroscopy

Often, a material absorbs too strongly to allow transmission measurements to yield any useful information, even in very thin layers. However, the details of the dielectric constant are of great interest, even in the absorption region. This is particularly true for conducting materials.

Photoacoustic spectroscopy relies on the heating of the gas above the sample surface; consequently, it is well suited for highly absorbing media. This method may also be used to measure the properties of films deposited on an absorbing substrate without interference from the substrate. The method uses a simple modification of the grating spectrometer. A beam of monochromatic light is incident upon the sample. This beam is chopped at acoustic frequencies. Depending on the degree of optical absorption at the incident wavelength, the sample will absorb light and expand or contract at the chopping frequency. This produces an acoustic signal that can be picked up by a receiver in the sample chamber. The intensity of the acoustic signal can then be related to the absorption by the material.

A related technique relies on heating the air adjacent to the material surface and probes the absorption by bouncing a laser beam from the spot where the incident chopped monochromatic light hits the sample. Using grazing incidence, the laser beam is then detected past a sharp edge by means of phase sensitive techniques (e.g., a lock-in amplifier). Since the air adjacent to the surface is heated by the incident monochromatic beam from the spectrometer when there is absorption, its refractive index will change accordingly and will refract the probe laser at different angles. This method is very precise and can be used to determine the *position* of absorption features in the spectrum of nontransparent materials.

2.8.2 Differential reflection spectroscopy

This technique is described in detail in Hummel (1993). Briefly, it consists of using chopped monochromatic light from a grating spectrometer. The light is cycled between two targets, and the reflected beams are mixed and detected using a phase-sensitive method to subtract their intensities. The difference in reflection coefficient will clearly identify any change in either of the two targets. This method is very sensitive to plasma resonance, and, since it is a remote technique when a telescope is used to gather the reflected light, it can be used over a wide range of temperatures, both hotter and colder than ambient.

Appendix 2A

Solution of the Mie Theory Equations

The Mie theory treats the problem of solving the vector wave equation for the electric and magnetic fields inside and outside a homogeneous isotropic sphere. The problem is algebraically complicated by the need to keep straight all the vector components, but the principle is essentially simple. One solves the wave equation inside and outside the sphere and applies boundary conditions of continuity of the tangential components of the electric and magnetic fields (no currents). The actual method uses the solution of the *scalar* wave equation in spherical coordinates and transforms the resulting functions into the vector fields by a method discovered by Hertz and Debye. Here, we outline the procedure and present only the essential results. For reference, we recommend two books: Bohren and Huffman (1983) and Van de Hulst (1981).

We write the wave equations of Chapter 1 for the $\mathbf{E}$ and $\mathbf{H}$ vectors, letting the effect of conductivity and absorption be included in the complex propagation vector, $k = k_1 + ik_2$. If one assumes that the form of the solution will follow

$$\mathbf{E}(x,t) = \mathbf{E}(x)e^{-i\omega_0 t} \tag{2.20}$$

then the wave equations simplify to the Helmholtz equation:

$$\nabla^2 \mathbf{E} + k^2 \mathbf{E} = 0$$
$$\nabla^2 \mathbf{H} + k^2 \mathbf{H} = 0 \tag{2.21}$$

with $k^2 = \omega^2 \epsilon \mu$ and $\nabla \times \mathbf{E} = i\omega\mu\mathbf{H}$ and $\nabla \times \mathbf{H} = -i\omega\epsilon\mathbf{E}$ with the following boundary conditions:

$$(\mathbf{E}_2 - \mathbf{E}_1) \times \hat{e}_n = 0$$
$$(\mathbf{H}_2 - \mathbf{H}_1) \times \hat{e}_n = J_s$$
$$(\mathbf{B}_2 - \mathbf{B}_1) \bullet \hat{e}_n = 0 \qquad\qquad (2.22)$$
$$(\mathbf{D}_2 - \mathbf{D}_1) \bullet \hat{e}_n = \sigma_s$$

where $\hat{e}_n$ is the unit vector normal to the interface, J_s is the surface current density, and σ_s is the surface charge density.

Rather than try to solve the *vector* wave equation, a *scalar* generating function, Φ, is defined from which the vector fields may be constructed as follows:

$$\nabla^2 \Phi + k^2 \Phi = 0$$

with

$$\mathbf{M} = \nabla \times (\mathbf{r}\Phi) \qquad \text{and} \qquad \mathbf{N} = \frac{\nabla \times \mathbf{M}}{k}$$

leading to

$$\mathbf{E} = \mathbf{M}_o + i\mathbf{N}_e \qquad \text{and} \qquad \mathbf{H} = n(-\mathbf{M}_e + i\mathbf{N}_o) \qquad (2.23)$$

where the subscripts o and e stand for odd and even components of the functions and n is the index of refraction.

The solution to the Helmholtz equation by separation of variables is

$$\Phi(r, \theta, \phi) = \sum_{m=0}^{\infty} \sum_{\ell=m}^{\infty} A_{\ell m} Z_{\ell+1/2}(kr) P_{\ell}^{(m)}(cos\theta)e^{im\phi} \qquad (2.24)$$

where $P_{\ell}^{(m)}(\cos\theta)$ are the associated Legendre polynomials and $Z_{l+1/2}(kr)$ are spherical Bessel functions of order ℓ. If we assume a plane wave incident on the particle, we see immediately that there are several restrictions that must be applied to the solution. For example, the incident wave can be written as a plane wave polarized in the x direction:

$$\mathbf{E}(r, t) = E_0\hat{e}_x e^{ikr\cos\theta}e^{-i\omega t} = \mathbf{E}(r)e^{-i\omega t} \qquad (2.25)$$

The presence of only $(\cos\theta)$ in the exponent dictates that only the $m = 1$ (m equals one) terms are nonzero. Inside the sphere, only the first-order Bessel function is finite at the origin, so $Z_{\ell+1/2}(kr)^{\text{inside}} = j_{\ell+1/2}(kr)$. Outside the sphere, the wave propagates outward, so we use the first-order Hankel function: $Z_{\ell+1/2}(kr)^{\text{scattered}} = h_{\ell+1/2}^{(1)}(kr)$. Finally, if we divide the waves into an incident field, an inside field, and a scattered field, and if we use Huygen's principle, we may write the incident plane wave as a sum of wavelets, or spherical harmonics, as shown in Chapter 1:

$$\mathbf{E}^{(i)}(r) = \hat{e}_x \sum_{\ell=1}^{\infty} E_\ell P_\ell^1(\cos\theta) j_{\ell+1/2}(kr) e^{i\phi}$$

$$\mathbf{H}^{(i)}(r) = -\frac{k\hat{e}_y}{\mu\omega} \sum_{\ell=1}^{\infty} E_\ell P_\ell^1(\cos\theta) j_{\ell+1/2}(kr) e^{i\phi}$$

(2.26)

with

$$E_\ell = i^\ell E_0 \frac{2\ell+1}{\ell(\ell+1)}$$

New definitions and use of the Ricatti–Bessel functions as defined here simplify all the solutions with no loss of generality. Let $\rho = kr$, $x = ka$, $n = \frac{n_{\text{sphere}}}{n_{\text{matrix}}}$

$$\psi_\ell(\rho) = \rho j_{\ell+1/2}(\rho), \qquad \xi_\ell(\rho) = \rho h_{\ell+1/2}^{(1)}(\rho)$$

$$\pi_\ell(\cos\theta) = \frac{P_\ell^{(1)}(\cos\theta)}{\sin\theta}, \qquad \tau_\ell(\cos\theta) = \frac{dP_\ell^{(1)}(\cos\theta)}{d\theta}$$

(2.27)

The six field components become:

Incident EM wave

$$E_\theta^{(i)} = \frac{\cos\phi}{\rho} \sum_{\ell=1}^{\infty} E_\ell(\psi_\ell \pi_\ell - i\psi_\ell' \tau_\ell), \qquad H_\theta^{(i)} = \frac{k}{\mu\omega}(\tan\phi) E_\theta^{(i)}$$

$$E_\phi^{(i)} = \frac{\sin\phi}{\rho} \sum_{\ell=1}^{\infty} E_\ell(-\psi_\ell \tau_\ell + i\psi_\ell' \pi_\ell), \qquad H_\phi^{(i)} = -\frac{k}{\mu\omega}(\cot\phi) E_\phi^{(i)}$$

Scattered EM wave

$$E_\theta^{(s)} = \frac{\cos\phi}{\rho} \sum_{\ell=1}^{\infty} E_\ell(-b_\ell \xi_\ell \pi_\ell + ia_\ell \xi_\ell' \tau_\ell), \quad H_\theta^{(s)} = \frac{k}{\mu\omega}\frac{\sin\phi}{\rho} \sum_{\ell=1}^{\infty} E_\ell(-a_\ell \xi_\ell \pi_\ell + ib_\ell \xi_\ell' \tau_\ell)$$

$$E_\phi^{(s)} = \frac{\sin\phi}{\rho} \sum_{\ell=1}^{\infty} E_\ell(b_\ell \xi_\ell \pi_\ell - ia_\ell \xi_\ell' \tau_\ell), \quad H_\phi^{(s)} = \frac{k}{\mu\omega}\frac{\cos\phi}{\rho} \sum_{\ell=1}^{\infty} E_\ell(-a_\ell \xi_\ell \tau_\ell - ib_\ell \xi_\ell' \pi_\ell)$$

(2.28)

Inside EM wave

$$E_\theta^{(m)} = \frac{\cos\phi}{\rho} \sum_{\ell=1}^{\infty} E_\ell(c_\ell \psi_\ell \pi_\ell - id_\ell \psi_\ell' \tau_\ell), \qquad H_\theta^{(m)} = \frac{k_s}{\mu_s\omega}(\tan\phi) E_\theta^{(i)}$$

$$E_\phi^{(m)} = \frac{\sin\phi}{\rho} \sum_{\ell=1}^{\infty} E_\ell(-d_\ell \psi_\ell \tau_\ell + ic_\ell \psi_\ell' \pi_\ell), \qquad H_\phi^{(i)} = -\frac{k_s}{\mu_s\omega}(\cot\phi) E_\phi^{(i)}$$

The amplitude coefficients; a_ℓ, b_ℓ, c_ℓ and d_ℓ are calculated by applying the boundary conditions at $r = a$, to obtain the following relations for the scattered amplitude coefficients:

$$a_\ell = \frac{n\psi_\ell(nx)\psi'_\ell(x) - \psi_\ell(x)\psi'_\ell(nx)}{n\psi_\ell(nx)\xi'_\ell(x) - \psi'_\ell(nx)\xi_\ell(x)}$$

$$b_\ell = \frac{\psi_\ell(nx)\psi'_\ell(x) - n\psi_\ell(x)\psi'_\ell(nx)}{\psi_\ell(nx)\xi'_\ell(x) - n\psi'_\ell(nx)\xi_\ell(x)}$$

(2.29)

and the following amplitude coefficients for the wave inside the sphere:

$$c_\ell = \frac{\psi_\ell(x)\xi'_\ell(x) - \psi'_\ell(x)\xi_\ell(nx)}{n\psi_\ell(nx)\xi'_\ell(x) - \psi'_\ell(nx)\xi_\ell(x)}$$

$$d_\ell = \frac{\psi_\ell(x)\xi'_\ell(x) - \psi'_\ell(x)\xi_\ell(x)}{\psi_\ell(nx)\xi'_\ell(x) - n\psi'_\ell(nx)\xi_\ell(x)}$$

(2.30)

These equations lead to the scattering and extinction cross sections per particle:

$$C_{\text{scat}} = \frac{W^{(s)}}{I^{(i)}} = \frac{2\pi}{k^2} \sum_{\ell=1}^{\infty} (2\ell + 1)(|a_\ell|^2 + |b_\ell|^2)$$

$$C_{\text{ext}} = \frac{W^{(\text{ext})}}{I^{(i)}} = \frac{2\pi}{k^2} \sum_{\ell=1}^{\infty} (2\ell + 1)\text{Re}(a_\ell + b_\ell)$$

(2.31)

References

Bohren, C. F., and D. R. Huffman. 1983. *Absorption and Scattering of Light by Small Particles*. Wiley.

Doremus, R. H. 1964. Optical properties of gold particles. *J. Chem. Phys.* 40:2389–2396.

Ehrenreich, H., and H. R. Philipp. 1962. Optical Properties of Ag and Cu. *Phys. Rev.* 128:1622.

Hummel, R. E. 1993. *Electronic Properties of Materials*. 2nd ed. Berlin:Springer Verlag.

Irani, G. B., T. Huen, and F. Wooten. 1971. *Phys. Rev.* B3:2385.

Nassau, K. 1983. *The Physics of Color*. New York: Wiley.

Slater, J. 1951. *Quantum Theory of Matter*. New York: McGraw-Hill.

Van de Hulst. 1981. *Light Scattering by Small Particles*. New York: Dover.

Wooten, F. 1972. *Optical Properties of Solids*. New York: Academic Press.

Chapter 3

Optical Properties of Insulators — Fundamentals

3.1 Introduction

In contrast to conductors, insulating materials cannot support an induced current, so they cannot shield their interior from external, time-varying electromagnetic fields. Consequently, optical waves propagate through insulators, and do so with characteristics determined by the dielectric constant and refractive index of the material and by any absorptive or scattering processes present. As a result, insulators and particularly glasses (amorphous or noncrystalline insulators) have a broad window of transparency over some part of the optical spectrum. In this "transparency window," the dielectric constant generally has a weak dispersion; therefore, the materials transmit light with very little loss. (This is seen by applying Kramers–Kronig relations.) However, wavelengths outside the transparency region can induce strong polarization processes that cause a dispersion in the dielectric constant or refractive index, with an associated increase in propagation loss or absorption. Crystalline insulators have microstructure that can have an additional effect on transparency. If multiple phases are present, or in the presence of grain boundaries within a single phase, strong scattering processes may occur from any region with a sharp change in refractive index, and this scattering makes the insulator opaque. Another source of opacity in insulators can come from dopants that absorb light through bound-electron excitations. Transition-metal and rare-earth oxides are important examples of such absorbing species, and they add color to insulators. The optical properties of insulators are therefore dominated by their transparency and/or the color of their dopants. The ability to manipulate these two characteristics of insulators, and thus to control the characteristics of any light propagating through or reflecting from these materials, has given rise to numerous applications. Insulators are commonly used for optical lenses, optical waveguides (as in fiber communications), and host media with optically active ions for nonlinear optical switching. Colorants (transition metals) are added during the production of optical filters and sunglasses as well as various colored glasses for artistic applications. Finally, the addition of rare-earth ions with narrow-band luminescence centers has led to the development of a wide variety of lasers.

Let us start by examining the first class of insulators just mentioned, bulk glasses that are transparent at visible wavelengths. It is instructive to examine the source of transparency in these materials and of their ultraviolet (UV) and infrared (IR) absorptions. Glasses are highly transparent in the visible for three reasons:

(1) Their polarization processes are either too slow or too fast to keep up with the oscillations in electromagnetic fields associated with the visible optical wave; consequently, the refractive index is only weakly dependent on wavelength in that region of the electromagnetic spectrum.

(2) Their constituents do not have electronic states that allow free-electron or bound-electron transitions in the visible.

(3) Their microstructure is homogeneous and isotropic, and their refractive index is dependent on neither spatial position nor direction.

If we look at polarization processes, we see, as exhibited in Fig. 3.1, that insulators can undergo several major changes in polarization over the electromagnetic spectrum. All such changes will affect the dispersion in refractive index and cause absorption. These changes consist of polarization processes associated with the following structural excitations:

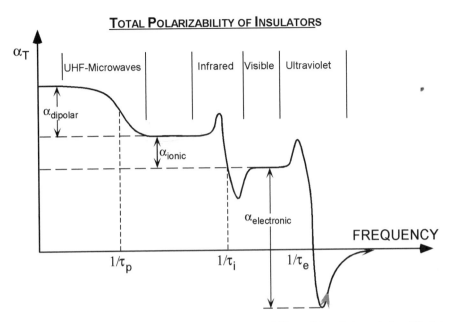

Figure 3.1: Components of the total polarizability of insulators (after Hecht 1998).

Orientational or dipolar polarization:
If polar molecules are present in the insulator, thermal agitation keeps the dipoles randomly oriented. Applied electric fields will align the dipoles. This alignment takes a certain time, τ_p, characteristic of the orientational motion of the polar molecule.

Ionic polarization:
The core electrons around each ion, and valence electrons involved in bonding, are polarizable through a slight shift in the electron cloud and the ionic core, leading to the formation of induced dipoles. The characteristic time for distortion of the electronic cloud, τ_i, is much shorter than the orientational polarization time, τ_p.

Electronic polarization:
Valence electrons can be photoexcited out of their bonding states, providing free electrons that contribute to the polarization, as in metals. This process is very fast, and its characteristic time, τ_e, is much shorter than τ_i or τ_p.

In between these regions of resonance, the polarization processes are either too slow or too fast for the incident electromagnetic wave. In the case where the polarization processes are too slow, the molecules or electrons do not have time to respond to the applied oscillating field, so they cannot extract energy from it. In the case where the polarization processes are too fast, the ions or electrons react immediately to the change in direction of the applied oscillating field and remain in phase with it. Again this does not lead to any exchange of energy (or absorption). Only when the period of the wave matches the reaction time of the polarization process is there an exchange of energy. In this case, as the field is applied, the dipoles begin to reorient; however, by the time they have aligned themselves with the field, it has changed direction. This maximizes the phase difference between the applied field and the induced polarization, leading to maximum energy extraction from the wave.

The refractive index and absorption coefficients are affected in a way similar to that for the induced polarization, as shown in Fig. 3.2. "Normal dispersion" corresponds to a decrease in refractive index with increasing wavelength (decreasing frequency). "Anomalous dispersion" refers to the opposite behavior. In other words, in the normal-dispersion regime, the higher-frequency (blue) light will travel more slowly (and thus have a larger refractive index) than the lower-frequency (red) light.

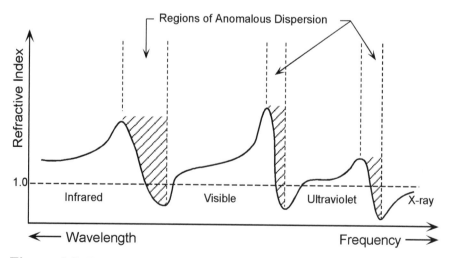

Figure 3.2: Dispersion in the refractive index showing the variation from the infrared to the x-ray region with increasing frequency (after Hecht 1998).

3.2 Harmonic oscillator theory

3.2.1 Classical model (Lorentz)

Lorentz oscillator theory provides a very good approach for estimating the optical behavior of insulators. The theory treats the polarization of insulators as arising from motions of simple harmonic oscillators. The success of this approach results from the fact that the potential energy of an electron in the bottom of a potential well (in equilibrium) can be approximated by a quadratic function that simplifies to the simple harmonic oscillator function. We will treat the simple harmonic oscillator classically and quantum mechanically. The reader may choose either or both paths. However, only the quantum mechanical treatment shows the source of quantized transitions and the existence of selection rules on optical transitions.

Consider an atom with a single electron bound to a nucleus. The electron cloud moves in the field of an electrostatic Coulomb attraction [second term Eq. (3.1)] and repulsion [first term Eq. (3.1)] from the nucleus and oscillates in the field of the applied electromagnetic wave. The classical force and potential are written as follows, with q_e the electron charge, r the distance from the nucleus, A the repulsion force constant, and ρ a hardness factor associated with the compressibility of the bond:

$$\mathbf{F} = Ae^{-r/\rho} - \frac{q_e{}^2}{4\pi\epsilon_0 r^2}$$

$$U = \rho Ae^{-r/\rho} - \frac{q_e{}^2}{4\pi\epsilon_0 r} \tag{3.1}$$

For a material in equilibrium, the potential energy has a minimum at some point r_0. The potential energy $U(r)$ near that point may be approximated by a Taylor series expansion about that point:

$$U(r - r_0) = U(r_0) + (r - r_0)\left(\frac{\partial U(r)}{\partial r}\right)_{r=r_0} + \frac{1}{2}(r - r_0)^2\left(\frac{\partial^2 U(r)}{\partial r^2}\right)_{r=r_0} + \cdots \tag{3.2}$$

The first term is a constant that can be discarded by proper choice of the zero point in potential energy. The linear term is discarded because this is a stable equilibrium (e.g., the potential energy is at a local minimum).

Thus, the Lorentz approach approximates the potential at equilibrium by the third term in the expansion, which corresponds to the quadratic spring equation by letting $k = (\partial^2 U(r)/\partial r^2)_{r=r_0}$:

$$U = \frac{1}{2}k(r - r_0)^2$$

$$F = -\nabla U = -k(r - r_0) = -kx$$

(3.3)

The electron cloud can be thought of as a spherical distribution of charge whose center is attached to the ion by a spring, as depicted in Fig. 3.3. A one-dimensional approximation just assumes that the spring stretches along its length without kinks. The corresponding one-dimensional differential equation of motion of the electron cloud under the oscillatory electric field is shown in Eq. (3.4), where the first term in the equation is the acceleration; the second term is the damping force due to radiative energy losses and/or collisions with phonons or other electrons, with $m\Gamma$ as the damping constant; the third term is the restoring force of the simple harmonic oscillator; and the right-hand term is the *local* applied field:

$$m\frac{d^2x(t)}{dt^2} + m\Gamma\frac{dx(t)}{dt} + kx(t) = +q_e E_0 e^{i\omega t}$$

(3.4)

Two assumptions are made here: (1) The mass of the nucleus is sufficiently large compared to the electron that we may neglect the

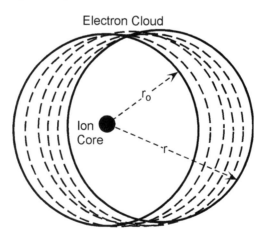

Electron Cloud

Figure 3.3: Illustration of the harmonic motion of the electron cloud with respect to the ion core. Intermediate positions are drawn with dashed lines.

motion of the nucleus. (2) The magnetic force term $[\mathbf{F}(t) = -q_e\, \boldsymbol{v} \times \mathbf{B}]$ is negligible, since the velocity of the electron, $\boldsymbol{v}$, is small and the magnetic susceptibility of the material, μ, is close to that of free space so that the local $\mathbf{B}$-field of the wave is small.

The differential equation of Eq. (3.4). has two solutions: a homogeneous or transient solution, $x_h(t)$:

$$x_h(t) = x_0 e^{-\frac{1}{2}\Gamma t} e^{\pm i\omega_0 t} \tag{3.5}$$

where: $\omega_0 = \sqrt{k/m - (\Gamma/2)^2} \approx \sqrt{k/m}$ is the natural frequency of the system, and an inhomogeneous or steady-state solution, $x_p(t)$, obtained by simple substitution:

$$x_p(t) = -\frac{q_e/m\; E_0}{(\omega_0{}^2 - \omega^2) - i\Gamma\omega} e^{i\omega_0 t} \tag{3.6}$$

Following the same polarization arguments with the local field as in the Drude equation, we can get an expression for the atomic polarizability, α:

$$\alpha(\omega) = \frac{q_e{}^2}{m} \frac{1}{(\omega_0{}^2 - \omega^2) - i\Gamma\omega} \tag{3.7}$$

Note that the electric field used here is the local field. Unlike the free-space condition, since there is strong local polarization, the local electric field strength will differ markedly from that of the applied field. This effect is discussed in Section 3.4.1.1, "Clausius-Mosotti Equation."

These equations need to be modified only by the number of oscillators available for the transitions and by the probability that they will take part in the optical absorption or emission. The latter term, called the *oscillator strength*, corresponds to the relative intensity of the transition for each resonance. The actual value of the oscillator strength can be obtained from the quantum mechanical treatment.

3.2.2 Quantum mechanical treatment

The Rayleigh–Jeans radiation law predicted that matter could absorb and emit radiation continuously over the electromagnetic spectrum. But experimental evidence from the photoelectric effect and black-body radiation spectrum showed that the interaction of radiation with matter is by discrete amounts of energy, namely, optical quanta. This was Planck's radiation law. The medium cannot be assumed to have freely adjustable oscillators. Thus oscillator energies must be quantized, and the emission or absorption of radiation is by quantized energy packets (photons). This means that at sufficiently low temperatures, when the thermal energy drops below the energy of a single energy quantum or packet, the system stops emitting light.

For the interested reader, the equations for a simple harmonic oscillator that lead to the quantization of energy are shown in Appendix 3A, giving the very elegant treatment with quantum mechanical raising and lowering operators. It is recommended that readers unfamiliar with that approach take the time to read the appendix to familiarize themselves with the powerful methods of quantum mechanics while examining one of its most interesting examples, one that is highly pertinent to this topic, since it shows how selection rules are obtained. The appendix will show that the energy levels of a simple harmonic oscillator, ξ_n, are quantized and written as follows:

$$\xi_n = \hbar\omega_0(n + \tfrac{1}{2}) \tag{3.8}$$

where ω_0 is the natural frequency of the oscillator $(\omega_0^2 \approx k/m)$ and n is an integer. Appendix 3A also shows that optical transitions in a simple harmonic oscillator are allowed only between states of adjacent energy levels: $\Delta n = \pm 1$. Thus, the emission or absorption of energy is done only by quanta between adjacent discrete levels of the harmonic oscillator, as shown in the following equation:

$$h\nu_{\text{photon}} = \hbar\omega_{\text{photon}} = \xi_f - \xi_0 = \hbar\omega_0(n_f + \tfrac{1}{2}) - \hbar\omega_0(n_0 + \tfrac{1}{2})$$
$$= \hbar\omega_0(n_f - n_0) = \pm\hbar\omega_0 \tag{3.9}$$

The equation shows that emitted or absorbed energy must be in the form of discrete quanta, each with a packet of energy $\hbar\omega_0$, and it explains the basis of Planck's radiation law.

We use a time-dependent perturbation approach to solve for the polarization of a simple harmonic oscillator. We consider the system in the ground state and allow only polarization and absorption but not emission. Before time $t = 0$, we have an unperturbed system. The perturbation, which consists of an interaction with the radiation, is turned on at $t = 0$. This leads to the time-dependent Schroedinger equation, using the Hamiltonian form:

$$-\frac{\hbar}{i}\frac{\partial\Psi}{\partial t} = (H_0 + H')\Psi \tag{3.10}$$

where the Hamiltonian operators are defined as:

$$H_0 = -\frac{\hbar^2}{2m}[\nabla^2 + V(r)] \quad \text{with } H_0\phi(\mathbf{r}) = \xi\phi(\mathbf{r})$$

$$H' = \frac{1}{2}(e^{i\omega t} + e^{-i\omega t})\mathbf{E_0} \cdot \mathbf{r}$$

(3.11)

H_0 gives the time-independent solution, $\phi(\mathbf{r})$; H' gives the time dependent perturbation. Solutions may be written as

$$\Psi(\mathbf{r},t) = \sum_n a_n(t)\phi_n(\mathbf{r})e^{-i\xi_n t/\hbar}$$

(3.12)

where $\phi_n(r)$ contains only the spatial dependence of the wave function. Using orthogonality of the solutions, it is possible to solve for the series factors, a_n, with the following results:

$$i\hbar\left(\frac{da_m}{dt}\right) = \sum_n a_n H_{mn}e^{i(\xi_m - \xi_n)t/\hbar}$$

(3.13)

where

$$H_{mn} = \int \phi_m^* H' \phi_n dV.$$

The dipole moment, p, is found through the spatial average of the oscillator displacement from equilibrium:

$$p = -q_e \sum_{k,\ell} a_k^* a_\ell e^{i(\xi_k - \xi_\ell)t/\hbar} x_{k,\ell}$$

(3.14)

where

$$x_{k,\ell} = \int \phi_k^* \, x \, \phi_\ell dV.$$

Here, $x_{k,\ell}$ is the dipole matrix element. Following a time average, and considering only transitions from the ground state, the polarizability can be written as

$$\alpha_{pol} = \sum_m \frac{2q_e^2|x_{m0}|^2}{\hbar} \frac{\omega_{m0}}{\omega_{m0}^2 - \omega^2}$$

(3.15)

where

$$\omega_{m0} \equiv (\xi_m - \xi_0)/\hbar.$$

If absorption is considered, the $i\Gamma\omega$ term is included as in the classical case, and the dielectric constant can be calculated to be

$$\epsilon_D^* = 1 + \frac{q_e^2}{m\epsilon_0} \sum_m \frac{N_m f_{m0}}{\omega_{m0}^2 - \omega^2 + i\Gamma_{m0}\omega} \qquad (3.16)$$

where

$$f_{m0} = \frac{2m\omega_{m0}|x_{m0}|^2}{\hbar}$$

The oscillator strength, f_{m0}, must obey the sum rule: $\sum_m f_{m0} = 1$.

N_m is the density of electrons bound to states associated with the ω_{m0} transition. The total complex dielectric constant is simply integrated over all possible transitions, each with its bound-electron transition and oscillator strength.

Note the similarity between the quantum mechanical result and the classical equation. The major difference is the quantization of the energy, as just shown and as in Appendix 3A. The real and imaginary parts of the dielectric constant may be easily calculated from Eq. (3.7) and are plotted in Fig. 3.4.

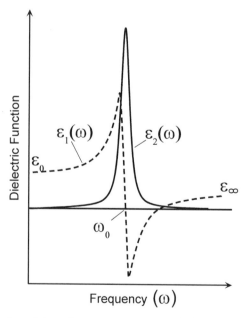

Figure 3.4: Real and imaginary components of the dielectric constant for a damped oscillator.

3.3 Selection rules for transitions between atomic levels

The value of the quantum mechanical treatment is seen when examining the implications of the oscillator strength expression [Eq. (3.16)] in terms of the transition dipole matrix element, x_{mn}, for transitions between atomic states. This term determines the allowed transitions that are optically active. For example, forbidden transitions will be associated with states whose dipole matrix elements and oscillator strengths are zero. Physically, we see that a certain symmetry must be maintained between the states of an allowed optical transition. This symmetry is reflected in the matrix element expression in upcoming Eq. (3.19). It can be expressed as a requirement to conserve angular momentum (orbital and spin angular momentum) in the creation or absorption of a photon. If the photon is assumed to have a spin of $S = 1$, then the total angular momentum difference between the initial and final states of the optical transition must differ by a quantum number: $\Delta j \pm 1$. In addition, since the optical wave has well-defined polarization directions, there are similar symmetry or conservation law restrictions on the planar projection of the angular momentum represented by the quantum number m. These symmetry and conservation law requirements restrict the transitions that can be effected by photons.

Calculation of the transition probability and selection rules requires a knowledge of the initial and final state wave functions. For a hydrogen atom, the wave functions can be written as

$$\phi_n(r, \theta, \phi) = \sum_{\ell, m} R_{n,\ell}(r) P_\ell^{(m)}(\cos \theta) e^{im\phi} \qquad (3.17)$$

where the functions $P_l^{(m)}(\cos \theta)$ are the associated Legendre polynomials and the radial dependence functions $R_{n\ell}(r)$ are composed of the associated Laguerre polynomials L_β, as follows:

$$R_{n\ell}(r) = -2\left(\frac{Z}{na_B}\right)^{3/2} \sqrt{\frac{(n-\ell-1)!}{n[(n+\ell)!]^3}} \left(\frac{2Zr}{na_B}\right)^\ell e^{-Zr/na_B} L_{n+\ell}^{2\ell+1}\left(\frac{2Zr}{na_B}\right) \qquad (3.18)$$

where a_B is the Bohr radius of hydrogen and the indices are as follows:

n (principal or total quantum numbers) = integer values $\geq \ell + 1$

ℓ (angular momentum quantum numbers) = integer values > 0

m (magnetic quantum numbers) = integer values, $-\ell \leq m \leq \ell$.

We will write the expression for a transition between state {n' ℓ' m'} and state {n ℓ m} of the hydrogen atom as an example. Other atoms will have more complex expressions; however, the implications of the symmetries are similar. The dipole moment direction is expected to be random with respect to the direction of polarization of the electromagnetic wave. So we take three projections parallel to the Cartesian axes. The dipole moment components will have the following respective forms with (x, y, z) and (r, θ, ϕ) as the coordinates of the electron with respect to the nucleus:

Component direction	Dipole moment
z	$r \cos \theta$
x	$r \sin \theta \cos \phi$
y	$r \sin \theta \sin \phi$

In the first transition probability calculation, we have aligned the dipole moment vector with the z component of the spherical coordinate axes to obtain the following expectation value equation:

$$z_{nn'} = \int \Phi_n^*(r, \theta, \phi) z\, \Phi_{n'}(r, \theta, \phi) dV$$

$$z_{nn'} = \int R_{n\ell}^*(r) P_\ell^{(m)}(\cos \theta) e^{-im\phi}(r \cos \theta) R_{n'\ell'}(r) P_{\ell'}^{(m')}(\cos \theta) e^{im'\phi} dV \tag{3.19}$$

The matrix element $z_{nn'}$ then becomes determined by the following integrals:

$$z_{nn'} = \int R_{n\ell}^*(r)\, r\, R_{n'\ell'}(r^2 dr) \int P_\ell^{(-m)}(\cos \theta) e^{-im\phi} \cos \theta\, P_{\ell'}^{(m')}(\cos \theta) e^{im'\phi}(\sin \theta d\theta d\phi) \tag{3.20}$$

The angle-dependent functions are orthogonal, but because of the $\cos \theta$ term from the $z_{nn'}$ matrix element, the angular indices ℓ and ℓ' must be different by 1 (e.g., the indices of the Legendre polynomials must satisfy $\ell' = \ell \pm 1$) while the magnetic quantum number is unchanged ($\Delta m = 0$). The orthogonality condition for the associated Legendre polynomials is expressed as

$$\int_{-1}^{1} P_{\ell'}^{(m')}(\cos \theta) P_\ell^{(m)}(\cos \theta) d(\cos \theta)$$

$$= \begin{cases} 0 & \text{for } \ell' \neq \ell \text{ and } m' \neq m \\ (2^\ell \ell!)^2 \frac{4\pi}{2\ell - 1} \frac{(\ell - m)!}{(\ell + m)!} & \text{for } \ell' = \ell \text{ and } m' = m \end{cases} \tag{3.21}$$

The selection rule comes from the following substitution:

$$\cos \theta \, P_\ell(\cos \theta) = \frac{\ell+1}{2\ell+1} P_{\ell+1}(\cos \theta) + \frac{\ell}{2\ell+1} P_{\ell-1}(\cos \theta) \qquad (3.22)$$

Equations (3.21) and (3.22) show that the integral of Eq. (3.20) is nonzero only when $\ell' = \ell + 1$ or $\ell' = \ell - 1$, and when $\Delta m = 0$.

For the other two orientations of the dipole moment components, there will be a contribution from the ϕ component of the integral due to the presence of the $\cos \phi$ term in x and the $\sin \phi$ term in y. These both require the condition $\Delta m = \pm 1$. The $\sin \theta$ term in both x and y components gives the same restriction as the $\cos \theta$ term in the z component, so $\Delta \ell = \pm 1$.

Together, these components impose the well-known selection rules for optical transitions. Thus, without spin the electric dipole selection rules are as follows:

$$\Delta \ell = \pm 1$$
$$\Delta m = 0, \pm 1 \qquad (3.23)$$

This selection rule is interesting because it requires that, in any dipole transition between two states, the total angular momentum must change by $\pm \hbar$. Since momentum must be conserved during the absorption or emission of photons, this mathematical result suggests that photons *must carry angular momentum*. This is indeed the case, and the angular momentum of photons comes into play when one considers the interaction of polarized light with matter (see Chapter 5, on semiconductors).

If spin is included, things get more complicated, because a new quantum number, j, must be introduced. It is defined as the vector sum of the orbital angular momentum and the spin angular momentum, represented by the quantum numbers ℓ and s, respectively, with the condition $\ell - s \leq j \leq \ell + s$. The electric dipole selection rules become:

$$\Delta j = 0, \pm 1; \quad \text{but } j = 0 \text{ cannot go to } j = 0$$
$$\Delta m = 0, \pm 1; \quad \text{but } m = 0 \text{ cannot go to } m = 0 \text{ when } \Delta j = 0 \qquad (3.24)$$

Electric quadrupole transitions:

$$\text{neglecting spin:} \quad \Delta \ell = 0, \pm 2 \quad \text{and} \quad \Delta m = 0, \pm 1, \pm 2 \qquad (3.25)$$

$$\text{including spin:} \quad \Delta j = 0; \pm 1, \pm 2 \quad \text{and} \quad \Delta m = 0, \pm 1, \pm 2 \qquad (3.26)$$

Magnetic dipole transitions:

$$\text{neglecting spin:} \quad \Delta\ell = 0 \quad \text{and} \quad \Delta m = 0, \pm 1 \qquad (3.27)$$

$$\text{including spin:} \quad \Delta j = 0 \quad \text{and} \quad \Delta m = 0, \pm 1 \qquad (3.28)$$

The more complex elements and the modifications imposed by neighboring bonds and electron–electron interactions will reduce the symmetry of the stationary-state wave functions. This reduces the limitations of the selection rules, and more transitions are allowed.

Unlike the simple harmonic oscillator, the coulomb potential yields atomic states whose r-dependence has no restrictions on optical transitions. Consequently, there is no restriction on Δn for atomic states. This means that hydrogen and other atoms (which are all governed by Coulomb potentials) can be ionized directly from the ground state without having to staircase up the energy ladder, as you would have to do for the simple harmonic oscillator (where Δn is restricted to ± 1).

3.4   Propagation of light through insulators

Light incident upon an insulator undergoes several processes:

(1) dispersion during propagation (variation in index with wavelength),
(2) reflection from any boundary that is associated with a change in refractive index,
(3) absorption,
(4) scattering (absorption and *elastic* re-emission in random directions),
(5) luminescence or fluorescence (absorption and *inelastic* re-emission in random directions),
(6) birefringence (variation in refractive index between different polarization directions).

First, using the results of Chapter 2, we can look at a transparent medium and examine its behavior according to the values of its refractive index and loss factor (Fig. 3.5). For an ionic crystal like KCl, with an intrinsic resonance at 4 eV, the refractive index and loss coefficients are as in the figure, and their behavior can be divided into 4 regions:

Region I: $n(\omega)$ is large and $\kappa(\omega)$ is small $\rightarrow$ transmission
Region II: $n(\omega)$ is large and $\kappa(\omega)$ peaks $\rightarrow$ absorption
Region III: $n(\omega)$ is small and $\kappa(\omega)$ is large $\rightarrow$ reflection

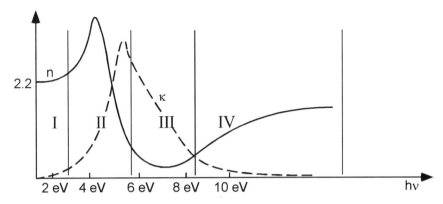

Figure 3.5: Plot of the behavior of the refractive index and loss coefficient for an ionic crystal with an oscillator at about 4 eV (after Kittel 1996).

Region IV: $n(\omega)$ is small and $\kappa(\omega)$ is small → transmission

Now let's look at the mechanisms underlying these processes.

3.4.1 Refractive index and dispersion

3.4.1.1 Clausius–Mosotti equation

The Lorentz equation derived earlier for the bound electrons calculates the polarization due to the action of an external electric field on the charges in the material to form dipoles. However, the local field acting on the interior region of an insulator is not equal to the applied field, due to the effect of the other local charges surrounding the region of interest. This problem is resolved by application of the Clausius–Mosotti Equation. This equation derives a relationship between the local polarizability, α_{pol}, and the dielectric constant that takes into account the local field. First, we write the induced polarization, **P** as a sum over all the j states, each having an occupancy of N_j:

$$\mathbf{P} = \sum_j N_j \alpha_{polj} \mathbf{E}_j(\text{local}) \tag{3.29}$$

It is possible to calculate the modification in the local field resulting from the induced polarization in a uniformly polarized medium. This is done by calculating the field due to an induced surface charge on a sphere surrounding the point of interest. The results show that the local field may be written as $\mathbf{E}(\text{local}) = \mathbf{E}_0 + (4\pi/3)\mathbf{P}$. This leads to expressions for the polarization and the susceptibility:

$$\mathbf{P} = \sum_j N_j \alpha_{\text{polj}} (\mathbf{E}_0 + \frac{4\pi\mathbf{P}}{3})$$

$$\chi = \frac{\mathbf{P}}{\mathbf{E}_0} = \frac{\sum N_j \alpha_{\text{polj}}}{1 - \frac{4\pi}{3} \sum N_j \alpha_{\text{polj}}}$$

(3.30)

This leads to the well-known Clausius–Mosotti equations for the dielectric constant and the refractive index:

$$\frac{\epsilon - 1}{\epsilon + 2} = \frac{n^2 - 1}{n^2 + 2} = \frac{4\pi}{3} \sum_j N_j \alpha_{\text{polj}}$$

(3.31)

This result applies equally well for either the dipole, the ionic or the electronic polarizations. At visible frequencies, the electronic polarization dominates. Pauling values of electronic polarizability for various ions are listed in Table 3.1. The higher values lead to higher-index dispersions in the UV. It is clear that the monovalent and divalent *anions* control the dispersion of the materials in which they are present. Among the *cations*, the heavier metals have the stronger effect.

3.4.1.2   Dispersion

When combined, the Clausius–Mosotti and Lorentz equations lead to the following dispersion relation for insulators with ω_{0j} representing the frequencies of the normal modes of vibration of the system:

						He(0.201)
Li^+(0.029)	Be^{+2}(0.008)	B^{+3}(0.003)	C^{+4}(0.001)	O^{-2}(3.88)	F^-(1.04)	Ne(0.39)
Na^+(0.179)	Mg^{+2}(0.094)	Al^{+3}(0.052)	Si^{+4}(0.016)	S^{-2}(10.2)	Cl^-(2.95)	Ar(1.62)
K^+(0.83)	Ca^{+2}(0.47)	Sc^{+3}(0.286)	Ti^{+4}(0.185)	Se^{-2}(10.5)	Br^-(4.77)	Kr(2.46)
Rb^+(1.40)	Sr^{+2}(0.86)	Y^{+3}(0.55)	Zr^{+4}(0.37)	Te^{-2}(14.0)	I^-(7.10)	Xe(3.99)
Cs^+(2.42)	Ba^{+2}(1.55)	La^{+3}(1.04)	Ce^{+4}(0.73)			

After Pauling 1927 and Tessman, Kaha, and Shockly 1953.

Table 3.1: Pauling electronic polarizabilities at the sodium D line in 10^{-30} m^3.

$$\frac{n^2 - 1}{n^2 + 2} = \frac{Nq_e^2}{3\epsilon_0 m} \sum_j \frac{f_j}{\omega_{0j}^2 - \omega^2 + i\Gamma_j\omega} \qquad (3.32)$$

Colorless, transparent media have their characteristic frequencies, ω_{0j} outside the visible spectrum. In glasses, the electronic polarization frequencies are in the UV, where they induce a strong absorption, and the molecular polarization frequencies are in the infrared, where they cause multiphonon absorption. In between, the refractive index has a weak frequency dependence, decreasing with increasing wavelength (normal dispersion) (Fig. 3.6). In the UV, with increasing frequency as the applied frequency approaches ω_{0j}, two things happen: The refractive index increases drastically (normal dispersion regime) and then decreases with increasing frequency above ω_{0j} (anomalous dispersion regime), and there is a large absorption.

Numerous expressions, which are based on differences between the optical refractive index of a glass at specified wavelengths, have been derived to define some measure of dispersion for the glass. A commonly

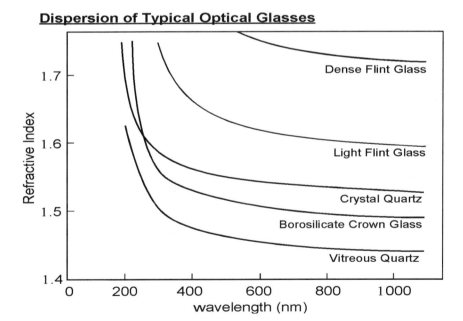

Figure 3.6: Typical dispersion for optical glasses of high, medium, and low index.

used measure of the optical dispersion is through the Abbe number, ν, as follows:

$$\nu = \frac{n_D - 1}{n_F - n_C} \qquad (3.33)$$

where:

$$n_D = n(589.3\,\text{nm}) \quad \text{Na lamp}$$

or

$$= n(587.56\,\text{nm}) \quad \text{He lamp}$$

$$n_F = n(486.13\,\text{nm}) \quad \text{H}_2 \text{ lamp}$$

$$n_C = n(656.27\,\text{nm}) \quad \text{H}_2 \text{ lamp}$$

The Abbe number is also frequently referred to as the V-value, the ν-value, or the reciprocal dispersive power. The refractivity of a glass is defined by $n_D - 1$ and the mean dispersion by $n_F - n_C$; the value $n_D - n_C$ refers to the partial dispersion. The partial dispersion ratio is given by the ratio of the partial dispersion to the mean dispersion. Often, glass manufacturers specify glass type using a designation based on the glass refractive index at the sodium D-line and on the Abbe number. A glass of type *abcdef* will have an index of $n_D = 1.abc$ and a ν-value of *de.f*. Various families of glasses fall within well-defined ranges of Abbe number and refractive index, as shown in Fig. 3.7.

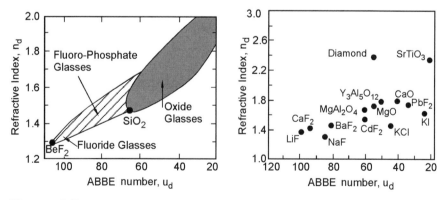

Figure 3.7: Graph of the relationship between index and dispersion for various families of glasses (after Weber, Milem and Smith 1978).

The dispersion equation [Eq. 3.32] for a single-component glass often reduces to the more commonly known Sellmeier form:

$$n^2(\lambda) = 1 + A + \frac{B}{\lambda^2} - C\lambda^2 \qquad (3.34)$$

The Sellmeier coefficients A, B, and C have been determined experimentally for a variety of different materials and may be found in the literature. Thus, given the proper coefficients, one may determine the refractive index dispersion of a material over any desired wavelength range. For fused silica, the constants are: $A = 1.099433$, $B = 10974.1$, $C = 9.5988 \times 10^{-9}$. The result is shown in Fig. 3.8.

Excellent fits to experimental dispersion data are commonly obtained using the three-term Sellmeier expression of Eq. (3.34). The general Sellmeier expression is as follows:

$$n^2(\lambda) - 1 = \sum_{i=1}^{3} \frac{A_i \lambda^2}{\lambda^2 - L_i^2} \qquad (3.35)$$

This form may be applied to multicomponent glasses, whose index dispersion may be expressed by modification of Eq. (3.35). This is known

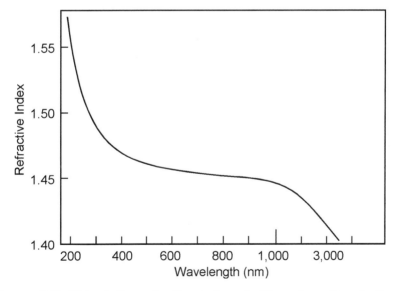

Figure 3.8: Plot of the refractive index of silica glass. For clarity, the wavelength scale beyond 1,000 has been compressed by a factor of 10.

as the mixed Sellmeier equation and represents a glass of composition x (composition 1): $(1-x)$(composition 2). An example from the $x\text{GeO}_2 : (1-x)\text{SiO}_2$ system (after Fleming 1984) is given by

$$n^2(\lambda) - 1 = \sum_{i=1}^{3} \frac{[SA_i + X(GA_i - SA_i)]\lambda^2}{\lambda^2 - [SL_i + X(GL_i - SL_i)]^2} \qquad (3.36)$$

In this equation, X represents the mol fraction of GeO_2 in the glass, SA, SL, GA, and GL are the Sellmeier coefficients A_i and L_i for SiO_2 and GeO_2 glasses, respectively, which are listed in Table 3.2.

The dispersion relation described by the Sellmeier expressions can be used to give information about the UV band edge of materials. Figures 3.6 and 3.8, in agreement with the expected dispersion behavior, show that the index of silicate glasses rises dramatically near 200 nm as the electron clouds of the ions approach resonance. In pure silica, the UV edge is near 160 nm (8 eV), and the bandgap (free-electron formation), which is the difference between the ionization potential and the electron affinity energies or the Si–O antibonding states, is around 120 nm (10.4 eV). The UV absorption edge corresponds to the frequency at which the bonds polarize. This is usually lower than the frequency of free-electron formation. Impurities lower the absorption edge further. Estimates of the energy bands of SiO_2 and GeO_2, based on UV photoelectron spectroscopy and electron loss spectroscopy by Rowe (1974) are shown in Fig. 3.9. The reflectance spectra of α-quartz and fused silica are also given in the figure.

Sellmeier coefficients	SiO$_2$	GeO$_2$
A_1	0.6961663	0.80686642
L_1	0.0684043	0.068972606
A_2	0.4079426	0.71815848
L_2	0.1162414	0.15396605
A_3	0.8974794	0.85416831
L_3	9.896161	11.841931

After Fleming 1984 and Potter 1994.

Table 3.2: Sellmeier coefficients for SiO_2 and GeO_2 glasses.

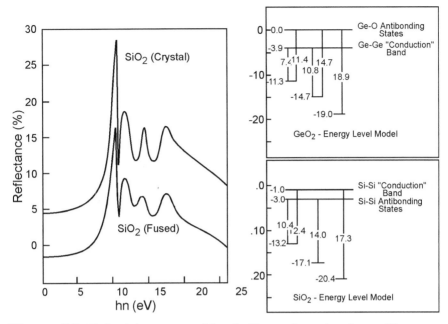

Figure 3.9: Reflectivity spectra of fused silica and quartz, along with energy band models for SiO_2 and GeO_2 (after Sigel 1977).

Reilly used a molecular orbital approach to calculate the energy states and concluded that the absorption edge position is associated with the lone-pair oxygen valence orbitals, and the 10.2-eV peak corresponds to a transition to an exciton state between the $2p_x$ hole orbital of oxygen and the $3s$ electron orbital of oxygen. Ruffa (1968), using a valence bond approach, calculated the same result with a Wannier exciton at 10.2 eV using an electron/hole reduced mass of 0.5 and a dielectric constant of 2.3.

Germanium dioxide has a predicted bandgap at 7.4 eV between the valence and the Ge–Ge band. However, the dominance of oxygen-deficient centers in germania induces strong absorption peaks at 242 and 202 nm (5.1 and 6.1 eV). These peaks can be photoexcited and saturated and are the basis for the photosensitivity shown by silica-germania fibers and thin films. More will be said on this subject in Chapter 7.

The UV edge has been measured in several network formers:

SiO_2 glass   155–160 nm (7.8 eV)
GeO_2 glass   200 nm (6.2 eV)

B_2O_3 glass	172 nm (7.2 eV)
P_2O_5 glass	145 nm (8.6 eV) (P_2O_5 has a tetrahedral structure with a double-bonded oxygen)
Al_2O_3 single-crystal alumina	145 nm (8.55 eV)
Al_2O_3 anodized films	182 nm (6.8 eV).

Of interest is the UV modification of silicate glasses by network modifiers. Structurally, alkali metals and alkaline-earth metals break up the silicate tetrahedral network. Oxygens that cannot link two Si atoms and instead link an Si atom to an alkali or an alkaline-earth atom are called nonbridging oxygen (NBO) atoms. This produces electrons with lower binding energies and a lower-frequency UV edge. Additions of Al_2O_3 and B_2O_3 improve the tetrahedral structure of the glass by reducing the NBO concentration, and they move the UV edge back up to higher frequencies. High-field-strength alkali ions, because they bind the oxygen ions more strongly, will cause a lesser reduction in the UV edge, even though they break up the corner- shared tetrahedral structure. UV edge values are as follows:

Li_2O-SiO_2 glass	188 nm (6.6 eV)
Na_2O-SiO_2 glass	207 nm (6.0 eV)
K_2O-SiO_2 glass	214 nm (5.8 eV)

Similarly, $MgO > CaO > SrO > BaO$. Also, following this trend, *PbO*, which is present in moderate concentrations in optical glasses, lowers the UV edge significantly.

In the infrared, the index decreases as the frequency is lowered to the resonances of the multiphonon vibrations of the constituent atoms or of the water or OH species often found in glass. In silica, the Si–O, the Si–OH, and the OH multiphonon vibrations contribute to the absorption spectrum in the IR, as shown in Fig. 3.10.

The fundamental absorption peaks occur further in the IR and are best measured by Raman spectroscopy. Raman-active, normal-mode resonances, corresponding to the structural motions of the SiO_4 cage, are shown in the sketches of Fig. 3.11. Infrared and Raman spectra for various inorganic materials can be found in a variety of references. The Raman spectra of SiO_2 and GeO_2 glasses are shown in Fig. 3.12. They show the gradual shifts in structure in the amorphous phase between SiO_2 and GeO_2.

1000
500
100
50
10
5
Attenuation (dB/km)
500 700 900 1100 1300 1500 1700
Wavelength (nm)

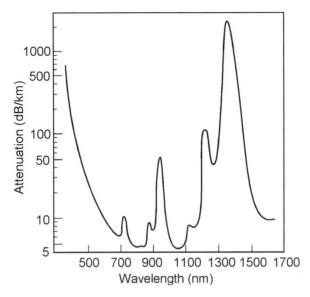

Figure 3.10: Infrared spectrum of silica (after Maurer 1973). The 900 and 700 cm^{-1} peaks are multiphonon vibrations of water or hydroxyl contaminants. The SiO_2 multiphonon peaks occur above 1,100 nm.

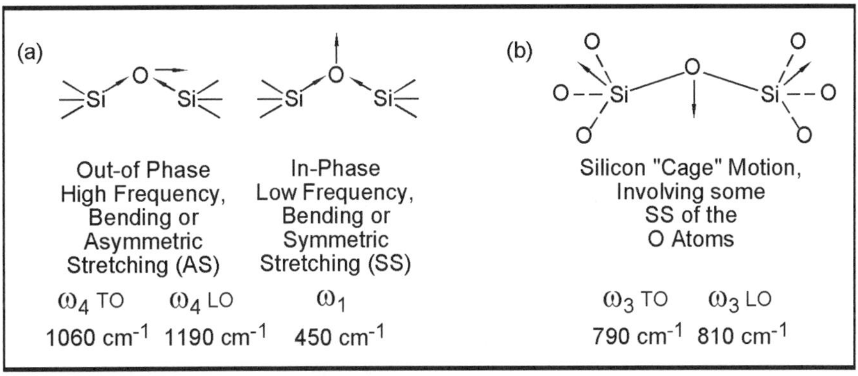

(a)
Out-of Phase High Frequency, Bending or Asymmetric Stretching (AS)
In-Phase Low Frequency, Bending or Symmetric Stretching (SS)
ω4 TO ω4 LO ω1
1060 cm-1 1190 cm-1 450 cm-1
(b)
Silicon "Cage" Motion, Involving some SS of the O Atoms
ω3 TO ω3 LO
790 cm-1 810 cm-1

Figure 3.11: Sketch of representative ionic motions for the phonon normal modes of silica, with estimated corresponding frequencies (after Wallace 1991).

For equations giving the refractive indices and dispersions of a wide variety of glasses and crystals, the reader is referred to Tropf, Thomas, and Harris (1995).

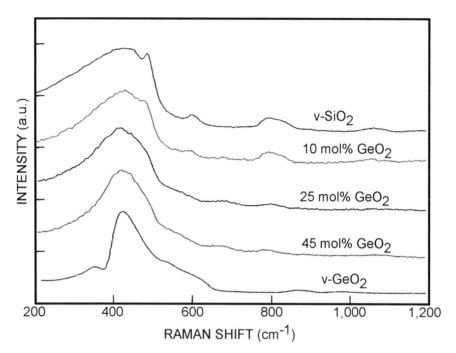

Figure 3.12: Raman spectra from SiO_2 and GeO_2 and some mixed glasses (after Chen 1994).

3.4.1.3   Composition dependence and calculations of the refractive index

There are numerous methods for calculating the refractive index of glasses. They all depend on calculated tables of coefficients, which often depend on the role of each ion in the glass structure. In general, the anions contribute the major part of the polarization, consequently their identity and concentration essentially determine the refractive index. Two major methods are described in Appendix 3B. Many databases now provide refractive indices for a wide selection of materials.

3.4.1.4   Temperature dependence of the refractive index

The effect of temperature on refractive index depends upon 2 counteracting effects:

(1) The growth in specific volume caused by the increase in temperature (positive coefficient of thermal expansion). This decreases the refractive index.
(2) The increase in polarizability caused by the rising temperature. This increases the refractive index.

The competition of these two effects determines the actual behavior of the refractive index with temperature. On the whole, materials with a low coefficient of thermal expansion (CTE) will have a positive change in index with increasing temperature, while materials with high CTE values will exhibit the opposite behavior. In some instances, as the CTE changes with temperature, the temperature derivative of the refractive index will actually change sign.

If we define the change in polarizability with temperature as

$$\Phi = \frac{1}{P}\left(\frac{dP}{dT}\right)$$

and the CTE through the molar volume, M_v, as

$$\text{CTE} = \frac{1}{M_v}\left(\frac{dM_v}{dT}\right)$$

then

$$\frac{dn}{dT} = \frac{(n^2-1)(n^2+2)}{6n}[\Phi - \text{CTE}] \tag{3.37}$$

Figure 3.13 presents three graphs showing the behavior of B_2O_3, SiO_2, and a borosilicate glass and exhibiting the kinds of temperature

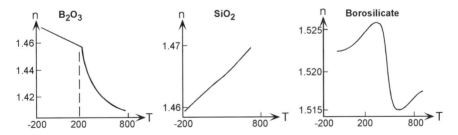

Figure 3.13: Temperature dependence of the refractive index of several glasses (after Fanderlik 1983).

dependences that are possible from this relationship. Three basic kinds of behavior have been observed:

(1) If $CTE > \Phi$, then the index decreases with increasing temperature. (This is commonly seen in B_2O_3 glass, borates, phosphates, organic glasses.)

(2) If $CTE < \Phi$, then the index increases with increasing temperature up through the transformation range. There, the liquid CTE grows by about a factor of 3 or more and can change the temperature dependence of the index. (Silica glass, with its low coefficient of expansion, exhibits an increase in index over the entire temperature range of the solid.)

(3) If $CTE \approx \Phi$, then, depending on the individual temperature dependences, the refractive index may either increase, decrease, or do both as the temperature is changed, as in borosilicate glasses (Fig. 3.13).

3.4.2   Reflection and transmission

Light incident upon the smooth surface of an insulator or upon the interface between two insulators with different refractive indices will be reflected and refracted. Figure 3.14 shows the trace of these three rays. The incident, reflected, and refracted rays lie in the same plane of incidence. The reflected beam from a flat, polished surface will propagate at an angle $\theta_r = -\theta_i$ equal to the angle of incidence (specular reflection). The law of reflection holds over any area sufficiently smooth to produce specular reflection (e.g., surface roughness less than 1/10 the wavelength of light). The refracted, or transmitted, beam will propagate at an angle θ_t that obeys Snell's Law:

$$n_i \sin \theta_i = n_t \sin \theta_t \tag{3.38}$$

The reflected and refracted beam intensities must satisfy the requirement that the parallel components of the total electric and magnetic fields be continuous across the insulator boundaries. These relationships lead to the Fresnel formulae [see upcoming Eqs.(3.45)].

At normal incidence $(\theta_i = \theta_t = 0)$, the amplitude reflection and transmission coefficients are:

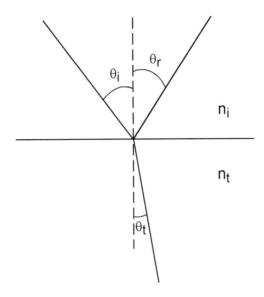

Figure 3.14: Geometry of light rays for specular reflection and transmission.

$$r_\parallel = r_\perp = \frac{n_i - n_t}{n_i + n_t}, \qquad t_\parallel = t_\perp = \frac{2n_i}{n_i + n_t} \tag{3.39}$$

The reflectance (or reflection coefficient), defined as the ratio of the intensities of the reflected and the incident beams, and the transmittance (or transmission coefficient), given by the ratio of the intensities of the transmitted and incident beams, are obtained from the amplitude reflection and transmission coefficients:

$$R = r^* r$$
$$T = \left(\frac{n_t \cos \theta_t}{n_i \cos \theta_i}\right) t^* t \tag{3.40}$$

For normal incidence, the reflectance (or reflectivity) R and transmittance T can be written as:

$$R = \frac{I_R}{I_0} = \left(\frac{n_i - n_t}{n_i + n_t}\right)^2, \quad T = \frac{I_T}{I_0} = \frac{4 n_i^2}{(n_i + n_t)^2}\left(\frac{n_t}{n_i}\right) \tag{3.41}$$

The transmittance has an extra term [in parentheses in Eq. (3.40)] due to the change in refractive index between the incidence medium and the transmission medium. This term is canceled out by multiplication with its inverse if another interface is present, and the optical beam returns to the medium of incidence.

For m surfaces, at normal incidence and with air as the medium of incidence and n as the refractive index of the medium, the reflectivity is written as

$$R_m = 1 - \left[1 - \left(\frac{n-1}{n+1}\right)\right]^m \qquad (3.42)$$

For light traversing a glass plate of index 1.5, the reflection from each surface is approximately 4%, giving a transmission of 92%. A rough approximation often used to estimate the reflection from two glass surfaces is

$$R_2 \approx 2n/(n^2+1) \approx 1.287 - 0.243n \qquad (3.43)$$

If a material has a non-negligible absorption coefficient α, with loss factor $\kappa = \alpha\lambda/4\pi$, then the reflectivity and transmittance at normal incidence become:

$$R_1 = \frac{(n_i - n_t)^2 + \kappa^2}{(n_i + n_t)^2 + \kappa^2}, \quad T_1 = \frac{4n_i^{\,2}}{(n_t + n_i)^2 + \kappa^2}\left(\frac{n_t}{n_i}\right)^{+i\kappa} \qquad (3.44)$$

Because of the differences in the continuity requirements for optical beams with different polarizations, the amplitude reflection and transmission coefficients at angles away from normal incidence have different equations. In keeping track of polarization, it is necessary to distinguish polarizations with the electric field in the plane of incidence (parallel, | |, or p-polarizations) from polarizations with the field normal to the plane of incidence (normal, ⊥, or s-polarizations). Using n_i and n_t for the refractive indices of the incidence and transmission media, respectively, the amplitude reflection and transmission coefficients are written as follows (Fresnel equations):

$$r_s(\perp) = \frac{n_i \cos\theta_i - n_t \cos\theta_t}{n_i \cos\theta_i + n_t \cos\theta_t}$$

$$r_p(\|) = \frac{n_i \cos\theta_t - n_t \cos\theta_i}{n_t \cos\theta_i + n_i \cos\theta_t}$$

$$t_s(\perp) = \frac{2n_i \cos\theta_i}{n_i \cos\theta_i + n_t \cos\theta_t} \tag{3.45}$$

$$t_p(\|) = \frac{2n_i \cos\theta_i}{n_t \cos\theta_i + n_i \cos\theta_t}$$

The amplitude coefficients are shown in Fig. 3.15 as a function of angle of incidence for an air–glass ($n = 1.5$) interface in which $n_i < n_t$. The reflection coefficients, calculated from Eq. (3.40) are shown in Fig. 3.16. Of particular interest is the angle at which the R_p term vanishes. This is Brewster's angle, the angle at which only the component parallel to the

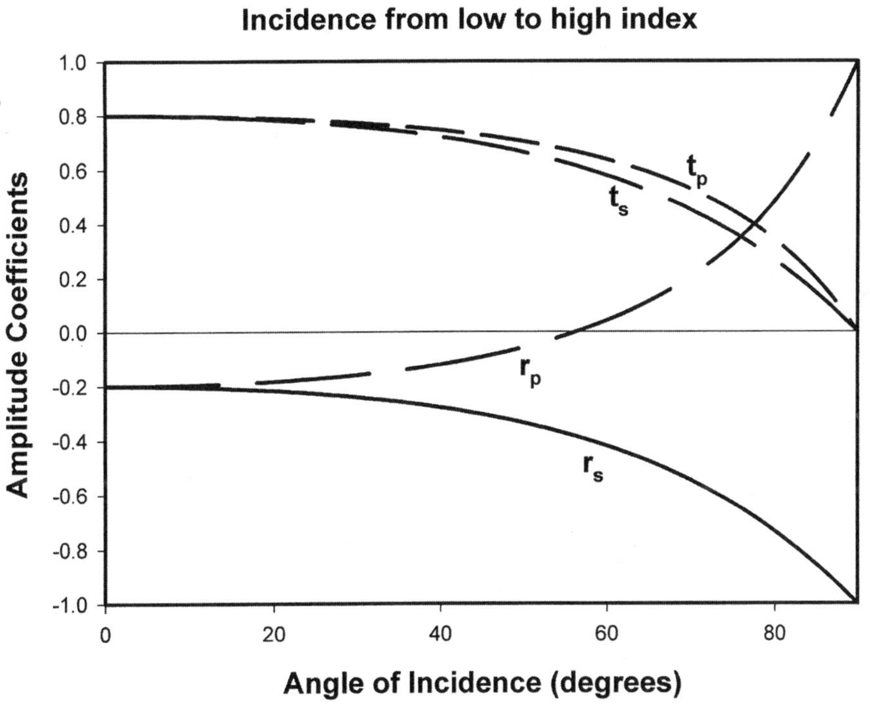

Incidence from low to high index

Figure 3.15: Reflection and transmission amplitudes for incidence from low ($n = 1$) to high ($n = 1.5$) refractive index.

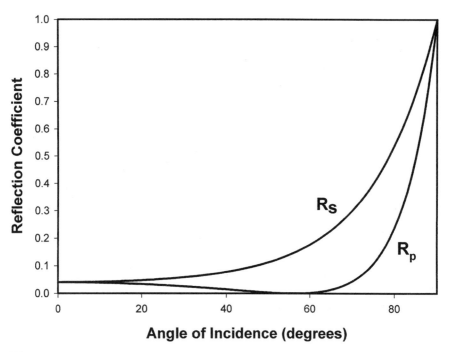

Angle of Incidence (degrees)

Figure 3.16: Reflection coefficients for s- and p-polarizations from an air–glass plane interface for incidence from low ($n = 1$) to high ($n = 1.5$) refractive index.

interface (s-polarization) is reflected and the nonparallel component (p-polarization) is totally transmitted ($R_p = 0$). This occurs when

$$n_t \cos \theta_B = n_i \cos \theta_t \qquad (3.46\text{a})$$

or when

$$\sin \theta_B = n_t / (n_i^2 + n_t^2)^{\frac{1}{2}} \qquad (3.46\text{b})$$

Now let's look at the light as a function of angle of incidence for each polarization direction reflected from a glass plate of index 1.5. We see a monotonic increase in reflectivity with increasing angle of incidence for the s-polarization and a decrease in reflectivity up to Brewster's angle followed by a sharp increase in p-polarization (Fig. 3.16).

If the incidence medium has a lower index than the transmission medium, then there is no problem with Snell's law, because the following expression is always less than 1:

$$\sin \theta_t = (n_i / n_t) \sin \theta_i, \qquad (3.47)$$

Incidence from high to low index

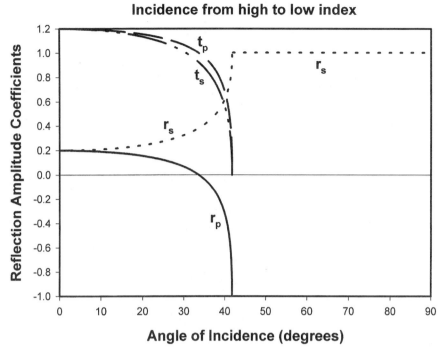

Angle of Incidence (degrees)

Figure 3.17: Reflection and transmission amplitudes for incidence from high $(n = 1.5)$ to low $(n = 1)$ refractive index.

thus, θ_t is always real

However, if the medium of incidence has a higher refractive index than the transmission medium, as shown by the amplitude coefficients in Fig. 3.17, then there will be an angle of incidence for which transmission (Fig. 3.18) is not possible for any polarization because $\sin \theta_t$ will be greater than 1, according to Snell's law. At this point,

$$\theta_t = \frac{\pi}{2} \tag{3.48a}$$

when

$$n_i > n_t,$$

and

$$\theta_i = \theta_c = \sin^{-1} \frac{n_i}{n_t} \tag{3.48b}$$

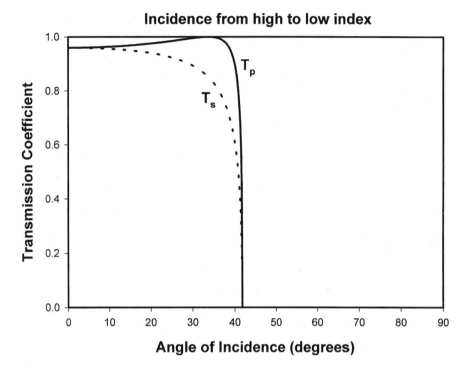

Incidence from high to low index

Angle of Incidence (degrees)

Figure 3.18: Transmission coefficients for s-and p-polarizations from an air–glass plane interface for incidence from high $(n = 1.5)$ to low $(n = 1)$ refractive index.

The angle of incidence at which the transmitted angle reaches 90° is called the angle of total internal reflection, θ_c, or the critical angle. Note that if you know either n_i or n_t accurately, a careful measurement of the critical angle, θ_c, gives the other refractive index.

Total internal reflection can be exploited in the use of prisms to guide light, as shown in Fig. 3.19. This is commonly practiced in microscopes and periscopes. Also shown in the figure is the use of "frustrated" internal reflection to guide light into thin films. This method is commonly practiced to inject or launch light into film optical waveguides. The optical energy couples from the prism across the air gap into the film:

$$E_t(\text{film}) = E_t(\text{prism})e^{-\beta y}e^{i(kx \sin \theta_i/n_i - \omega t)} \qquad (3.49)$$

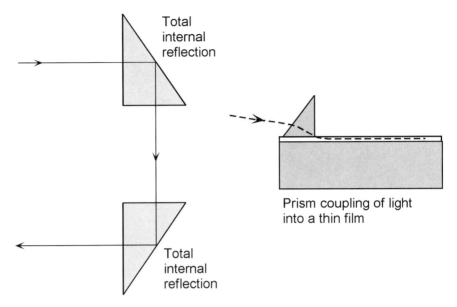

Figure 3.19: Total internal reflection in prisms and "frustrated" internal reflection for the injection of light into thin films.

3.4.3 Nonspecular reflection

If light is incident upon a rough surface whose roughness is of the order of (or less than) the wavelength of light, diffuse reflection, or scattering, occurs. In this case, the reflected beam is not simply a single beam that propagates at an angle equal to the angle of incidence of the beam. Diffuse reflection is more or less isotropic and often displays the form given in Fig. 3.20. The scattered light will form a plume around the specular reflection angle, as shown. The plume may exhibit interference patterns that depend on the ratio of the size of the microstructure to the incident wavelength.

If the scale of the roughness is greater than the wavelength of light, then the angle of diffuse reflection is related to the distribution of reflection angles in the rough surface, as specular scattering occurs for each angle present in the surface. This is shown in Fig. 3.21. Since the intensity of the reflected light increases with angle for all angles of incidence in s-polarization, the maximum intensity of the s-scattered light from a rough surface will occur in a forward direction from (i.e., at an

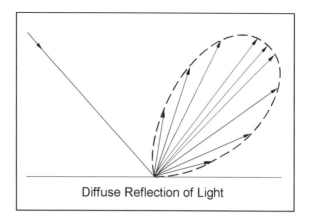

Diffuse Reflection of Light

Figure 3.20: Intensity pattern for diffuse reflection.

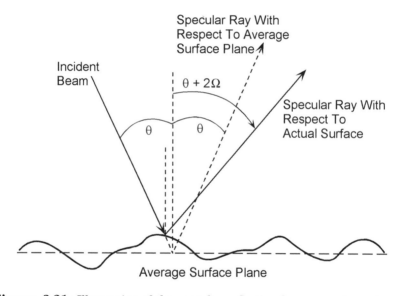

Average Surface Plane

Figure 3.21: Illustration of the specular reflection from a rough surface with roughness larger than the wavelength of light. The specular angle for the average surface plane is θ. The angle of inclination of the surface normal from the average surface plane is Ω.

angle greater than) specular. In p-polarization, the reflected light intensity decreases with angle of incidence until Brewster's angle, at which point it increases sharply. Therefore, p-scattered light from a rough surface with an average angle of incidence less than Brewster's angle will exhibit a small backward scattering from specular. If the average angle of incidence is larger than Brewster's angle, a large forward scattering will occur. Finally, if the angle of incidence matches Brewster's angle, the scattered light will occur in two lobes surrounding the specular angle, with negligible light at Brewster's angle. This problem is easily treated mathematically by assuming a distribution of angles of incidence resulting from the distribution of angles from the rough surface. The range of scattering angles is calculated directly by assuming a distribution of specular reflection processes based on the distribution of angles of incidence, as shown in Eq. (3.50). The angular distribution of intensities seen by the observer depends on the aperture of the detector, as shown in the equation:

$$I_R(\theta) = \int\limits_{-\phi/2}^{\phi/2} R_{s,p}(\theta + 2\Omega)G(\Omega)d\Omega \tag{3.50}$$

where $R_{s,p}$ is the reflection coefficient for s- or p-polarizations, $G(\Omega)$ is the distribution of surface angles (Ω) measured from an average plane through the surface, θ is the angle of incidence of the light measured from the normal to the average plane, and ϕ is the aperture of the detector.

3.4.4 Optical attenuation

The transparency window in insulators is determined by the UV edge at short wavelengths and by the multiphonon vibration spectrum at longer wavelengths. In between, absorption and scattering processes will reduce transparency. Figure 3.22 depicts this behavior for a silica glass and several silicate glass fibers; we can see the IR losses from the phonon absorption spectrum of water and hydroxyl in the glass and the effect of Rayleigh scattering at short wavelengths. The phonon vibration spectrum corresponds to a mixture of fundamental and overtones of the HOH and the OH vibrations and the Si–O stretch vibration. The peak positions and intensities are listed in Table 3.3, they reach from the deep IR well into the visible, although the higher overtones are very weak.

Rayleigh scattering causes the slowly increasing loss toward the blue. The fact that Rayleigh scattering (Fig. 3.23) decreases with increasing wavelength to the inverse fourth power (λ^{-4}) is a major reason for the need to conduct optical communications at the longest possible wavelength. Thus, silicate-fiber communications are conducted at 1.3

Wavelength (nm)	Intensity (dB/km)	Source of phonon
12 μm	–	$\nu_1 = $ Si–O stretch fundamental
2730	>40,000	$\nu_0 = $ OH stretch fundamental
1370	2900	$2\nu_0$
1230	150	$2\nu_0 + \nu_1$
1125	3.4	$2\nu_0 + 2\nu_1$
1030	0.4	$2\nu_0 + 3\nu_1$
950	72	$3\nu_0$
880	6.6	$3\nu_0 + \nu_1$
825	0.8	$3\nu_0 + 2\nu_1$
775	0.1	$3\nu_0 + 3\nu_1$
725	6.4	$4\nu_0$
685	0.9	$4\nu_0 + \nu_1$
585	0.5	$5\nu_0$

From Midwinter 1979.

Table 3.3: Fundamentals and overtones of the water-associated phonon absorption.

and 1.55 μm, two high-transmission regions around the large OH absorption band at 1.37 μm, and just before the onset of the multiphonon vibration absorption of the glass. Notably, the region around 1.3 μm is also an ideal operational wavelength regime, since it is generally the "zero-dispersion" wavelength region for silicate fibers. This is the wavelength region at which minimum pulse broadening (and, thus, signal distortion) occurs, since the group velocity dispersion of the material is minimum (see Agrawal 1989 for more details). Studies have been conducted to find glasses with a transmission window that extends further into the infrared. It is possible to alter the multiphonon frequencies by increasing the mass of the cation and anion atoms in the insulator composition. Fluoride and chalcogenide glasses have had great promise for moving the communications wavelengths to 3–6 μm. These materials replace the rather light oxygen anion with heavier fluoride or sulfur and selenium

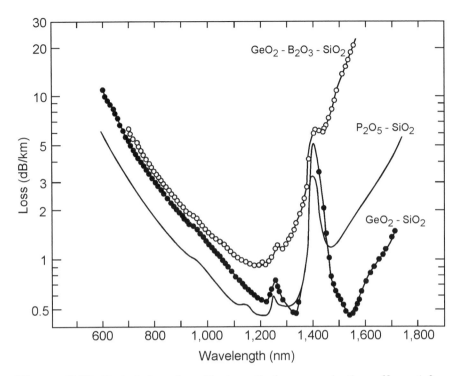

Figure 3.22: Typical loss for silicate optical communications fibers (after Osonai et al. 1976).

ions. This replacement pushes the multiphonon-vibration absorption lines further into the IR region, providing transparency at near- and mid-IR wavelengths. Impurity and scattering problems, however, have kept the losses in these materials above those of silica-based fibers. Chloride and bromide glasses, while exhibiting a multiphonon absorption at longer wavelengths, do not have sufficient chemical resistance to aqueous attack to be useful for long-distance-communication fibers. (Nonoxide glasses are discussed further in Chapter 4.)

Looking at the dependence of the upper end of the transmission window on composition, we have already observed the lowering of the UV edge in oxide insulators from the addition of ionic components with easily ionized electrons. Many nonoxide insulators exhibit much lower bandgap energies than the oxide counterparts, and their behavior bridges the difference between insulators and semiconductors. Thus materials like

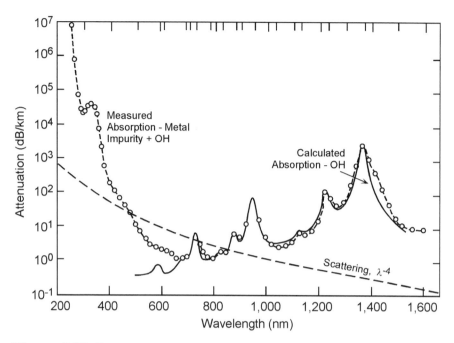

Figure 3.23: Loss processes in a silica fiber broken down into the various contributions from the OH overtones, Rayleigh scattering and a transition metal impurity [after Keck].

ZnS, CdS, CdSe, and CdTe have decreasing bandgap energies from the blue end of the spectrum through deep red to the near infrared. These materials are usually treated as semiconductors rather than insulators. Table 3.4 presents bandgap energies and UV absorption edges for a wide variety of insulator and semiconductor materials.

The replacement of oxygen by sulfur and selenium reduces the structural bonding energy of the material. Consequently, this replacement lowers the bandgap energy. And, although chalcogenide materials have been produced into commercial insulators with good transparency in the infrared, they absorb strongly in the visible and appear black to the eye. The fluoride bond strength, due to its high ionicity, is as high as that of oxygen, so fluorides have a much broader transparency window and reach from the mid IR to the UV. These materials are discussed in detail in the next chapter.

For some glasses and ceramics, Table 3.4 lists absorption edges only instead of bandgap energies, because these materials are susceptible to

Material	Bandgap energy (eV)	Absorption edge (nm)	Exciton Bohr diameter (Å)
Lithium fluoride (LiF)	12	120	
Calcium fluoride (CaF$_2$)	10	130	
Sapphire (Al$_2$O$_3$)	9.9	150	
Silica (SiO$_2$)	8.4	160	
Alumina (px*)(Al$_2$O$_3$)		200	
Diamond (C)	5.5	240	
Magnesium oxide (MgO)		250	
Zinc sulfide (ZnS)	3.8	325	
Copper chloride (CuCl)	3.40	365	13
Gallium nitride (GaN)	3.40	365	
Zinc oxide (ZnO)	3.2	390	
Silicon carbide (SiC)	2.8–3.2	390–440	
Titanium dioxide (TiO$_2$)	2.9	430	
Magnesium fluoride (MgF$_2$)		450	
Zinc selenide (ZnSe)	2.58	480	84
Cadmium sulfide (CdS)	2.53	490	56
Zinc telluride (ZnTe)	2.28	550	
Aluminum arsenide (AlAs)	2.14	580	
Cadmium selenide (CdSe)	1.74	710	106
Cadmium telluride (CdTe)	1.50	830	150
Gallium arsenide (GaAs)	1.43	870	280
Indium phosphide (InP)	1.28	970	
Silicon (Si)	1.11	1,130	40l, 90t
Germanium (Ge)	0.67	1,850	50l, 400t
Lead sulfide (PbS)	0.41	3,024	400

px* indicates polycrystalline.

Table 3.4: Bandgap energies and UV absorption edges.

high-energy defects that often move the absorption edge below the bandgap energy. The Bohr diameter, $2a_x$ the size of the correlated electron/hole pair (exciton) in the solid, defines the smallest structure that will exhibit bulk behavior $(D > 10a_x)$. Note that the subscripts ℓ and t for the Bohr diameters of silicon and germanium represent the longitudinal and transverse values and are so distinguished in these two materials because they are indirect gap semiconductors with anisotropic curvature of the conduction and valence bands in three-dimensional k-space. More will be said on this topic in Chapter 5.

Within the transparency window, absorption peaks may be caused by color centers and ions with internal electronic states that may be easily photoexcited at low energies. Such ions include the transition metals and the rare earths. The transition metals have unfilled $3d$ and $4d$ electronic shells; the rare earths have unfilled $4f$ shells that allow optically active transitions. Such transitions absorb in the visible and can impart a strong color to the insulator despite a low concentration. Table 3.5 lists various sources of absorptive color in insulators.

3.4.5   Ligand field theory

It is clear that the light absorbed or emitted (photoluminescence) by color-center ions depends on the environment of these ions. For example, transition-metal ions will change color depending on the host composition (Table 3.5) and are very sensitive to the composition of their nearest neighbor and next-nearest neighbor shells. This is because the optical transitions of transition-metal ions are caused by the energy levels of the d electrons, which are involved in interatomic bonding. Rare-earth ions are less susceptible to their environment, because the color results from transitions involving the $4f$ states, and the $4f$ electrons are kept closer to the core and do not participate in interatomic bonding. Yet measurements conducted on materials containing rare-earth ions do show some variation with host material composition and structure.

A startling example of the effect of local environment on the color of transition metal ions is exhibited by the difference between ruby and emerald gemstones. Both colors are formed by an absorption process involving the $3d$ electrons in the Cr^{+3} impurity that substitutionally replaces the octahedrally coordinated Al^{+3} ion. The nearest-neighbor oxygen shell about Cr has the same structure in both crystals. In ruby, the host crystal is corundum (Al_2O_3 α-phase); the host crystal of emerald is

Colorant	Host glass	Concentration (wt%) for 0.1 OD	Color or shade	Source of color
CoO	Silicates	0.0018	Cobalt blue	$Co^{+2}(3d^7)$
	Borates		Pink	
NiO	Silicates	0.0078	Neutral grey	$Ni^{+2}(3d^8)$
	Borates		Blue	
Se		0.0250	Pink	Se^{+4}
Cr_2O_3	Most compositions	0.0324	Serpentine green	$Cr^{+3}(3d^3)$
CuO	Silicates (oxy)	0.116	Blue	$Cu^{+2}(3d^9)$
Cu_2O	Silicates (red)		Red	$Cu^{+1}(3d^{10})$
MnO	Silicates	0.207	Violet	$Mn^{+3}(3d^4)$
	Silicates		Pale yellow	$Mn^{+2}(3d^5)$
U_3O_8		0.381	Pyrethrum (yellow with green fluorescence)	U^{+6}
Fe_2O_3	Silicates (oxy)	0.384	Yellow brown	$Fe^{3+}(3d^5)$
FeO	Silicates (red)		Blue green	$Fe^{+2}(3d^6)$
V_2O_3	Silicates	0.655	Serpentine green	$V^{+3}(3d^2)$
VO_2	Silicates		Blue	$V^{+4}(3d^1)$
Nd_2O_3	Silicates	0.92	Purple with red fluorescence	Nd^{+3}
Pr_2O_3	Silicates	2.38	Greyish green with yellow fluorescence	Pr^{+3}
Ti_2O_3	Phosphates		Violet-brown	$Ti^{+3}(3d^1)$
$TiO_2 + CeO_2$	Silicates	6.6	Yellow	$TiO_2 + CeO_2$

Table 3.5: Coloring agents in glasses.

Beryl ($3BeO \cdot Al_2O_3 \cdot 2SiO_2$). Consequently, there is a difference in the next-nearest-neighbor shells, Al ions in ruby and Si ions in emerald. Silicon bonds more strongly to oxygen than does aluminum. As a result, the nearest-neighbor Cr–O bonds in emerald are longer and weaker than in ruby. This difference shifts the absorption bands that center at violet and green in ruby to bands that center at blue and red in emerald. Thus, ruby is red with a weak blue component, and emerald is green. Good ruby gems have an additional sparkle from a red luminescence at the transmission band wavelength.

Ligand field theory (LFT) has been used to calculate the influence of the local environment on the energy levels of transition-metal and rare-earth ions. Ligand field theory is a modification of crystal field theory (CFT). In both approaches, the ligand atoms are represented as point charges, and the electrostatic field of the environment is calculated at the location of the ion of interest. The techniques make rigorous use of the symmetries of the local neighborhood. A major drawback of the CFT method is that it does not take into account any form of covalent bonding, and while it is possible to include partial charge transfer between the metal and the ligand in ionic bonding, the covalent bonding effects are ignored. The CFT method has been modified to the ligand field theory by the introduction of other parameters to introduce some effects of covalency, such as partial charge exchange. The fundamental equations for application of ligand field theory are given in Appendix 3C.

3.4.6   Optical scattering

When light propagates through a medium, it can undergo several kinds of scattering mechanisms. A scattering process is one that receives incident light of a given frequency with a specific propagation vector and that produces an emission of light with either a changed propagation vector (elastic scattering) or both a changed propagation vector and a changed frequency (inelastic scattering).

When only the propagation vector is altered, two processes are possible. The first consists simply of an apparent change in propagation direction, as would be observed in the reflection of light from a smooth or a rough surface. This is discussed in Section 3.4.3. The second process consists of the absorption of the incident light by a molecular process, followed by the re-emission of the same light packet without noticeable change in frequency. This process is called *Rayleigh scattering*. In the scattering process, the molecule or atom can exchange kinetic energy of motion with the incident photon, so the process is not strictly elastic and some small frequency shift is present in the scattered light. The shift is essentially unmeasurable in solids, since the atoms are confined to structural cages in which they undergo thermal vibrations only. Thus, the allowed atomic velocities are small. In liquids, however, diffusional motions are possible and the width of the Rayleigh line broadens. Rayleigh scattering line width measurements can be used to determine the diffusional properties of the liquid.

When the frequency is altered in addition to the propagation vector, again two processes may be occurring. Both consist of the absorption of the incident light by a molecular process, as in Rayleigh scattering, but the photon either loses or gains energy due to scattering from the phonon branches of the atomic vibration spectrum associated with the thermal motion of the lattice. The first process, called *Brillouin scattering,* is characterized by two scattering regions around the central Rayleigh line when the scattered light is viewed as a function of frequency or wavelength. The higher-energy region results from the *absorption* of an *acoustic* phonon; that below the Rayleigh peak results from the *emission* of an *acoustic* phonon. Both Brillouin scattering regions actually consist of several peaks. These are due to the dominant absorption/emission of the longitudinal acoustic phonon and a weaker coupling to one or several transverse acoustic modes. In the second inelastic scattering process, phonons from the *optical* dispersion branches of the atomic vibration spectrum are either absorbed or emitted while the photon is re-emitted, thus increasing or decreasing the energy of the "scattered" photon. This process is called *Raman scattering.* The difference between acoustic and optical phonon branches is discussed in Chapter 1. Rayleigh, Brillouin, and Raman scattering processes are described in detail shortly. Figure 3.24 shows the frequency ranges of the scattering mechanisms discussed, where ω_s and ω_i are the angular frequencies of scattered and incident radiation, respectively. Note that the Brillouin spectrum near zero frequency shift is not easily resolved due to the presence of the strong Rayleigh scattering peak.

3.4.6.1 Rayleigh and Brillouin scattering

An experimental curve of Rayleigh and Brillouin scattering from silica glass is shown in Fig. 3.25. Rayleigh scattering is caused by local

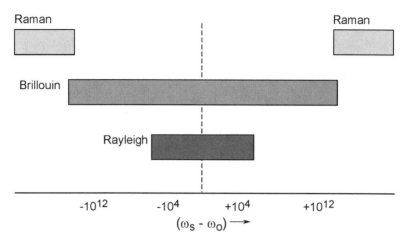

Figure 3.24: Ranges of frequencies corresponding to the processes of Rayleigh, Brillouin, and Raman scattering (after Hopf and Stegeman 1985).

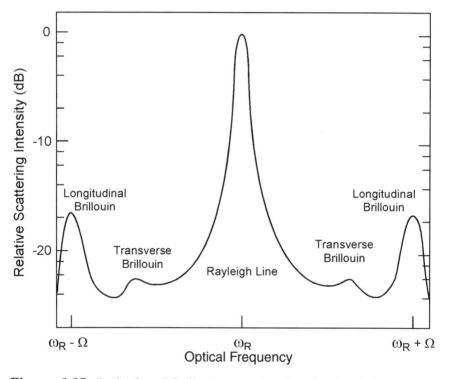

Figure 3.25: Rayleigh and Brillouin scattering from longitudinal and transverse acoustic modes of silica (from Rich and Pinnow 1976).

fluctuations (in density, structure, composition, or temperature, for example) that have no particular dynamic structure. Rayleigh scattering can occur in bulk media or from a single molecule if light is absorbed and re-emitted. In liquids, scattering results from nonpropagating diffusive fluctuations, whereas in solids, since the fluctuations are essentially frozen, scattering is from nonpropagating density, temperature, or frozen-in composition fluctuations. In a glass, density fluctuations are frozen near a viscosity of 10^{12} Pa·s, composition fluctuations are frozen near $10^4 - 10^5$ Pa·s. In condensed media, Rayleigh light scattered by neighboring centers is emitted nearly in phase. If the medium is, in addition, homogeneous, the scattered light appears only in the forward direction. The forward scattered light remains in phase with the incident light (i.e., the scattering process does not retard the phase of the light), so light emitted from the scattering medium appears to have simply traveled more slowly than expected in propagating through the medium. This condition is identical to the situation in which light propagates unscattered through a medium that has a refractive index greater than unity. For inhomogeneous media, the scattered waves do not cancel in the nonforward direction, and one observes Rayleigh scattering in all directions with an intensity that results from the presence of density, temperature, or composition fluctuations. However, even in this case the Rayleigh line width in solids is very narrow, usually unmeasurable.

Rayleigh scattering exhibits the well-known λ^{-4}-wavelength dependence (when light scatters from particles in the size range $\lambda/20$ to about λ). This is the source of Tyndall blue (blue eyes and blue sky). This scattering is also the source of the red color of the setting sun, by subtraction of the blue colors from the solar spectrum by means of Rayleigh scattering by the atmosphere. Phase-separated glasses with small microstructure (less than 5000 Å) also exhibit the characteristic Tyndall blue color when observed in reflection and the expected complementary reddish color when seen in transmission. Many biological colors owe their hue to Rayleigh scattering. For example, blue is usually a direct result of Rayleigh scattering. Green can result from the combination of Rayleigh scattering and a yellow pigment. Purple can be the combination of Rayleigh scattering and a red pigment.

For particles of radius a that is small compared to the wavelength of light, λ, the resulting Rayleigh scattered intensity, I_s, can be expressed as follows for unpolarized incident light, I_0:

$$\frac{I_s}{I_0} = \frac{8\pi^4 N a^6}{\lambda^4 r^2} \left|\frac{m^2 - 1}{m^2 + 1}\right|^2 (1 + \cos^2 \theta) = Q(1 + \cos^2 \theta) \qquad (3.51)$$

where N is the number of particles, θ is the scattering angle, r is the distance of the observation point from the scattering center, and m is the ratio of the refractive index of the scattering center to that of the medium ($m = n_1/n_0$). For a medium with density or composition fluctuations, the absorption due to Rayleigh scattering, α, can be written as:

$$\alpha = \frac{8\pi^3}{3\lambda^4} (m^2 - 1)^2 \beta k_B T \qquad (3.52)$$

where β is the isothermal compressibility of the medium, k_B is the Boltzmann constant, and T is the ambient temperature of the medium if it is a liquid. It the medium is a glass, then the ambient temperature must be replaced by the fictive temperature of the glass (e.g., the temperature at which the fluctuations are frozen).

For polarized light:

$$I_{\parallel} = I_0 Q \cos^2 \theta$$

$$I_{\perp} = I_0 Q \qquad (3.53)$$

$$I_s = \frac{1}{2}(I_{\parallel} + I_{\perp})$$

$I_{\parallel}$ is produced when the incident light is polarized parallel to the scattering plane (p-polarization). $I_{\perp}$ is produced when the incident light is polarized perpendicular to the scattering plane (s-polarization). Thus, if the incident light, I_0, is 100% polarized, the scattered light is similarly polarized. However, since the different polarizations scatter differently, unpolarized light will partially polarize upon scattering. The degree of polarization parallel to the plane of scattering, P_p, can be expressed as

$$P_p = \frac{1 - \cos^2 \theta}{1 + \cos^2 \theta} \qquad (3.54)$$

The maximum polarization occurs at normal incidence ($\theta = 0°$), where the scattering component parallel to the plane of incidence vanishes. Solar light scattered in Earth's atmosphere is polarized parallel to the ground ($I_{\perp}$, horizontally polarized). For this reason, polarized sunglasses are polarized in the vertical direction in order to remove the stronger Rayleigh scattered component of sunlight.

If the index of the scattering particle is approximately that of the medium, then one uses the Rayleigh–Gans approximation to Rayleigh scattering theory. Here, only the phase difference between scattering centers is considered, and the scattered intensity at a distance r from the medium is written as

$$\frac{I_s}{I_0} = \frac{4\pi^2 V^2}{r^2 \lambda^4} (n_0 - 1)^2 [\tfrac{1}{2}(1 + \cos^2 \theta)] |\Phi(\theta, \lambda)|^2 \qquad (3.55)$$

where V is the scattering volume, θ is the angle between the observer position and the direction of incidence of the light, and the scattering function, $\Phi(\theta, \lambda)$, is written as

$$\Phi(\theta, \lambda) = \frac{1}{V} \int e^{i\delta} dV \qquad (3.56)$$

The phase difference δ between the scattered rays is integrated over the sample volume. The scattering function $\Phi = 1$ in forward scattering ($\theta = 0°$), and $|\Phi| < 1$ in other directions.

3.4.6.2 Mie scattering

When the size of the scattering particle is equal to or larger than the wavelength of the incident light, the scattering centers can be considered as separate media, and the propagation of optical waves in the scattering media must be considered as well. This causes the incident light to be both absorbed and scattered. A general theory was first developed by Mie, who calculated the solutions for light scattering from any size spherical and elliptical particles. Essentially, the method solves the wave equation outside and inside the scattering object and uses the boundary conditions for Maxwell's equations to obtain the scattered light intensity, as shown in Chapter 2. The absorption coefficient α for N spheres per unit volume with complex dielectric constant $\varepsilon^* = \varepsilon_1 + i\varepsilon_2$ embedded in a medium of refractive index n_0 is written as:

$$\alpha = \frac{18\pi N n_0^3 \varepsilon_2}{\lambda(\varepsilon_1 + 2n_0^2) + \varepsilon_2{}^2} \qquad (3.57)$$

Mie scattering is the source of color in glasses containing colloids of noble metals (Ag, Au, Pt, Cu), as described in Chapter 2. A brief derivation of Mie's theory is given in Appendix 2A.

When the size of the scattering medium is much larger than the wavelength of the incident light, then the absorption and the intensity of the scattered light are essentially wavelength independent and produce a white color. This explains the color of gray and white hair, which is due to scattering of light from microscopic (micron-sized) gas bubbles in the hair.

3.4.6.3 Brillouin scattering and the Landau–Placzek ratio

The Landau–Placzek ratio in fluids, which is the ratio of the intensities of the Rayleigh line to the sum of the Brillouin lines, reveals fluctuations in the dielectric constant of the fluid. In single-component fluids, the Landau–Placzek ratio equals $C_P/C_V - 1 = \chi_T/\chi_S - 1$, where the C's are the specific heats at constant pressure and constant volume, respectively, and the χ's are the isothermal and adiabatic compressibilities, respectively, and they reflect the presence of fluctuations in the local density. In multicomponent mixtures, the Rayleigh line is affected by local fluctuations in the composition. Thus Landau–Placzek measurements reveal variations in either local density or composition fluctuations. By studying the temperature dependence of composition fluctuations through the Landau–Placzek ratio, Schroeder (1970) predicted the presence of a second-order phase transition in the potassium–silicate glass binary, although the actual transition occurs below the glass transition temperature and was never achieved.

3.4.7   Phonons, Raman scattering, and infrared absorption

Molecular groups, coordination cells, or unit cells in solids undergo thermal motions that are regular with respect to one an other, so they are characterized by normal modes of vibration. In small molecules, the normal-mode frequencies can be calculated. From simple harmonic motion equations, it is clear that the frequencies will depend on the ratio of the spring constant to the masses. A larger mass will reduce the frequency; a stronger interatomic interaction will increase it.

Measurements that obtain these normal modes of vibration include infrared absorption and Raman scattering spectroscopies, as well as inelastic neutron scattering. In IR spectroscopy, incident electromagnetic radiation interacts with the atomic vibrations of the material through the electric dipole moment. The molecular normal modes of vibration modify the local dipole moments, which induce transitions between the different vibration levels in the presence of an oscillating electromagnetic field. The incident wave, consequently, is absorbed at energies matching these phonon transitions. Infrared spectroscopy can be very useful in determining the presence and concentration of various molecular species in materials. For example, both the water molecule and the hydroxyl radical have well-defined IR absorption lines (the H–O–H bend is at 1640 cm^{-1}; the O–H stretch is at 3450 cm^{-1}), and their presence can often be measured quantitatively by measuring the amount of absorption associated with each line. Organic compounds have many well-defined IR absorption lines.

The transition probability for a photon absorption between two vibration levels that follow the condition $h\upsilon_0 = \xi_2 - \xi_1$ can be written as in Eq. (3.14) with the following dipole expectation value:

$$\mathbf{P}_{nn'} = q_e \langle \mathbf{r}_{nn'} \rangle = \int \Psi_n q_e \mathbf{r} \Psi_{n'} dV \tag{3.58}$$

Infrared absorption lines are generally seen when the phonon vibration is antisymmetric, thus producing a nonzero transition probability (e.g., selection rules discussed earlier). The infrared absorption spectrum of SiO_2 is given in Fig. 3.10.

In Raman scattering, the incident radiation causes an oscillatory electric field in the medium. The local charges oscillate at the same frequency, causing a diffuse emission of light (Rayleigh scattering). However, since the atoms are undergoing thermal vibrations, these acoustic vibrations (phonons) will couple with the polarizability tensor set up by the Rayleigh oscillations, and energy can be exchanged between the Rayleigh oscillations and the phonon density of states. This results in the emission of frequency-shifted light either up or down (anti-Stokes or Stokes) from the incident radiation frequency. In general, Raman scattering peaks are 10^{-3} to 10^{-4} less intense than the Rayleigh peak.

If the polarizability of the material is written as α and the incident light is $E = E_0 \cos 2\pi\nu t$, and if the molecule has a normal mode of vibration ν_1 so that the polarizability may be written as $\alpha = \alpha_0 + (\partial\alpha/\partial x)x$, then the polarization causing the scattered light may be written as follows:

$$P = \alpha_0 E_0 \cos 2\pi\nu t + \frac{1}{2}\left(\frac{\partial\alpha}{\partial x}\right)x_0 E_0 \{\cos[2\pi(\nu + \nu_1)t] + \cos[2\pi(\nu - \nu_1)t]\}$$

$$\tag{3.59}$$

The first term causes the Rayleigh scattering, the first term in curly brackets the anti-Stokes Raman, and the second term in curly brackets the Stokes Raman. In order for a vibration to be Raman active, the polarizability must change during the vibration. This usually is associated with a symmetric vibration.

3.5   Measurement techniques

3.5.1   Measurements of the refractive index

In order to examine the many methods used to measure the refractive index of transparent materials, we will first refer the reader back to Section 3.4.2 on reflection coefficient and particularly to Snell's Law and the definition of total internal reflection.

3.5.1.1   Abbe refractometer

The Abbe refractometer is easy to use and comes at a modest cost of about $3,000–5,000. In principle, it measures the angle of incidence required to just begin total internal reflection (critical angle) for light propagating from a standard glass hemisphere of high index to a sample of lower index. In practice, due to easier alignment of the beams, the light is incident in the reverse direction into the interface between the sample and the glass hemisphere (Fig. 3.26). Accurate measurement of refractive index using this instrument, however, requires a good contact between the sample surface and the top surface of the prism of the instrument. Often, if the sample surface is not perfectly smooth and flat, a small amount of index-matching liquid will be used to improve the quality of the refracted signal. The index of the oil is chosen to be near that of the sample (or of the prism if the sample index is totally unknown). However, as the sample index is measured, the index liquid should be changed to match the unknown more closely in order to obtain a good measurement. The only time we have found this method to be lacking is in the measurement of porous glass samples, due to an infiltration of the index matching oil into the sample pores.

   The measurement procedure involves sending a beam of light, either white light or monochromatic light, at a grazing angle with the interface between the sample and the prism. Then, by adjusting the arm of a telescope to various angles, the investigator looks for the angle that delineates the light field from the dark field. That is the critical angle of

Abbe Refractometer

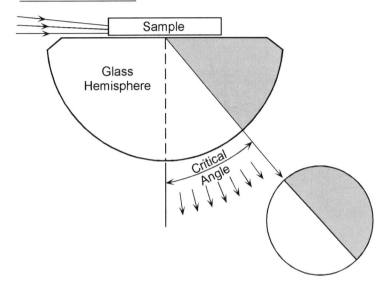

Figure 3.26: Principle of operation of the Abbe refractometer.

the system and it yields the refractive index of the sample according to Eqs. (3.48).

The refractive index of the prism is generally between 1.6 and 1.8, so the sample index must be lower. The accuracy of the measurement is $\pm 2 \times 10^{-3}$. Calibrated prisms with different indices may be used to broaden the range of measurement.

3.5.1.2 *Minimum-deviation prism goniometer*

This is the most accurate method for measuring refractive index. However, it is the most painstaking and has a relatively high cost for the goniometer stage (about $30,000 - 40,000$). The method requires a prism of the unknown material with an accurately measured prism angle w. The method involves the following steps (Fig. 3.27):

(1) Position the source to one side of the unknown prism and measure the minimum angle of deviation.
(2) Rotate the prism $180°$ and repeat.

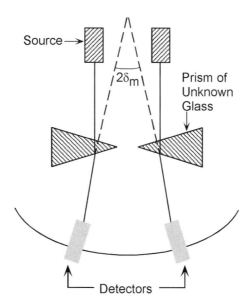

Figure 3.27: Setup for a minimum-deviation prism measurement of refractive index (after Fanderlik 1983).

(3) This gives a measure of $2\delta_m$ from which the refractive index can be calculated:

$$n_\lambda = \frac{\sin\frac{1}{2}(w + \delta_m)}{\sin\frac{1}{2}w} \qquad (3.60)$$

The accuracy of the method is $\pm 5 \times 10^{-6}$ if the angles can be measured with an accuracy of ± 0.5 angular seconds.

3.5.1.3   Refractometers

The Hilger–Chance (60° V-block) and the Grauer (90° V-block) refractometers are simple and can be constructed with little need for precision machining. However, they need a high-quality V-block prism, often made from two pieces. The V-block prism requires high-quality optical glass (no variations in refractive index, no bubbles) and polished, flat surfaces on the internal faces of the V-block. The method consists of the measurement of the angle of a mirror required to send a collimated optical beam back along its incoming track, as shown in Fig. 3.28. The refractive index of the V-block must be known to the same accuracy as that desired for the

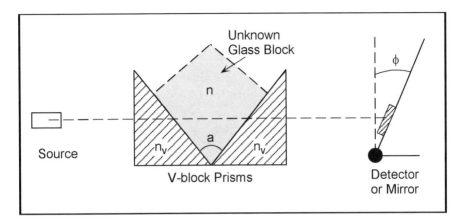

Figure 3.28: V-block refractometer with 90° prisms. If a mirror is used after the V-block, then the light is returned toward the source. A viewer monochromator can be used to detect the returned light. This method has great precision.

unknown sample. However, if that is not possible, the instrument can still be used for relative measurements between samples, since it has good precision. The index of the unknown, n, is found in relation to that of the prism, n_v:

$$\text{For } a = 90° : \quad \frac{n}{n_v} = 1 \mp \left(\frac{1}{n_v}\right)\left[\sin^2 \phi - \frac{\sin^4 \phi}{n_v^2}\right]^{\frac{1}{2}}$$

For $a = 60°$:

$$n = \left\{\frac{1}{3}\left[2\sin^2 \phi + \sqrt{3}n_v \sin \phi + 2n_v^2 + (2\sqrt{3}\sin \phi + n_v)\sqrt{n_v^2 - \sin^2 \phi}\right]\right\}^{\frac{1}{2}}$$

For any angle a :

$$n = \left[\frac{1}{2}\left(\frac{\sin^2 \phi}{\sin^2 a} + \frac{\sin \phi}{\sin a}\right) + \frac{2}{3}n_v^2 + \left(\frac{\sin \phi}{\sin a} + \frac{n_v}{\cos a}\right)\sqrt{n_v^2 - \sin^2 \phi}\right]^{\frac{1}{2}}$$

$$(3.61)$$

3.5.1.4 Ellipsometry

An ellipsometer is usually used to measure the thickness and refractive index of films deposited on a substrate. This application will be discussed in the next chapter in section 4.1 on under thin films. However, with an appropriate standard, the same method can be applied to index measurements on bulk samples.

In the diagram of the ellipsometer in Fig. 3.29, monochromatic light is passed through a polarizer (Glan–Thomson calcite prism) and a quarter-wave compensator (mica plate with 45° retardation) to give elliptically polarized light. This light is incident upon a sample surface at a specified angle. The reflected light is then detected through an analyzer (Glan–Thomson calcite prism), and both the polarizer and the analyzer angles are varied to find the maximum extinction of the reflected light. The values obtained consist of the polarizer angle (P), the analyzer angle (A), and the angle of incidence. The last is critical to the accuracy of the measurement and must be calculated from data taken on a standard sample of the same thickness whose refractive index is well known. Once the angle of incidence is known accurately, the sample refractive index for the wavelength of the incident light can be calculated using the following equation:

$$n = (\sin \phi_0)[1 + \left(\frac{1-\rho}{1+\rho}\right)^2 \tan^2 \phi_0]^{\frac{1}{2}} \qquad (3.62)$$

where:

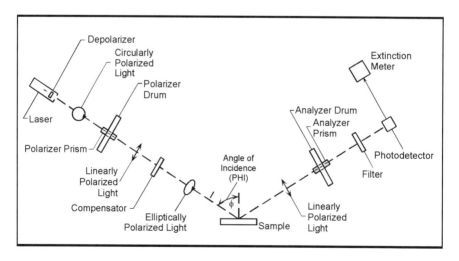

Figure 3.29: Ellipsometer (from Gaertner Instruments).

$\phi_0 =$ angle of incidence of ellipsometer

$P_i =$ polarizer settings for different measurements

$A_i =$ analyzer settings for different measurements

$$\rho = \tan \chi e^{i\Delta}$$

$$\chi = \frac{1}{2}[180° - (A_2 - A_1)] \quad [0 < \chi < 90°]$$

$$\Delta = 360° - (P_1 - P_2) \quad [0 < \Delta < 360°]$$

3.5.1.5 Becke line method

The Becke line method is used with small glass fragments. Here, the index of refraction is compared to that of an immersion liquid using a microscope. The sample is placed in the immersion liquid of known refractive index. The sample must be illuminated by a beam of rays converging at an extremely acute angle. This can be achieved either by reducing the aperture of the condenser on the microscope or by raising the condenser. As the microscope stage is moved away from focus, there will be a bright line contouring the perimeter of the glass specimen. This is the Becke line. If the refractive index of the immersion liquid is lower than that of the glass, the Becke line will move inside the object when the distance between the objective lens and the specimen is increased. Decreasing the distance will cause the Becke line to move outside the specimen. If the immersion liquid index is higher than that of the sample, the Becke line will move in the opposite direction.

Index relationship	Objective lens to sample distance	Becke line
$n_l < n_g$	Increase	Moves inside
	Decrease	Moves outside
$n_l > n_g$	Increase	Moves outside
	Decrease	Moves inside

Once the relationship between the unknown index and the immersion liquid is found, another immersion liquid is chosen so as to bracket the unknown index. Then the bracket is narrowed until the desired degree of accuracy is reached. Finely graduated liquids can yield $\Delta n = 0.02$.

Monochromatic light must be used, and the temperature of the liquids must be kept at the calibration temperature.

Modification: The temperature must be kept relatively constant in the preceding test because the liquid has a strong temperature dependence in its refractive index. This can be used to advantage. Choose a liquid with a higher refractive index than the sample, and then raise the temperature of the liquid until a match is observed. The sample will disappear when the index match is achieved. Since the glass sample will undergo a negligible change in index in a small temperature range, only the liquid index will be assumed to vary with temperature. If the temperature dependence of the refractive index of the liquid is known, then the unknown index can be calculated:

$$n_g = n_\ell(T_c) + (T_m - T_c)\frac{dn_\ell}{dT} \qquad (3.63)$$

where T_m is the measured temperature where the indices match and T_c is the calibration temperature (usually 20 °C).

3.5.1.6   Femtosecond transit time method

A new method, recently developed by one of our former students, provides an excellent means of measuring the entire dispersion curve for any shape of sample, bulk, or thin film. The method uses the white light pulse from an optical parametric amplifier working off a titanium-sapphire laser. The pulse has a duration of about 100 fs. The pulse is passed through the sample and then matched with a second timed pulse that has traveled through a physical delay line. A diagram of the apparatus is shown in Fig. 3.30. By decomposing the transmitted pulse into its component colors via reflection from a grating, the time taken for each wavelength to transmit through the sample can be calculated. This gives the refractive index at each wavelength directly! The range of wavelengths available is between 300 nm and 1.3 μm. The accuracy is the pulse width divided into the travel time through the thickness of the sample, which turns out to be good to the second decimal place in refractive index for a sample 1 mm thick. The method gives the index dispersion over a broad range of wavelengths, but the accuracy is low and the equipment can cost upward of $150,000. Of course, the Ti-sapphire laser has more critical uses in other optical measurements (see Chapter 7, "Nonlinear Optical Behavior of Materials").

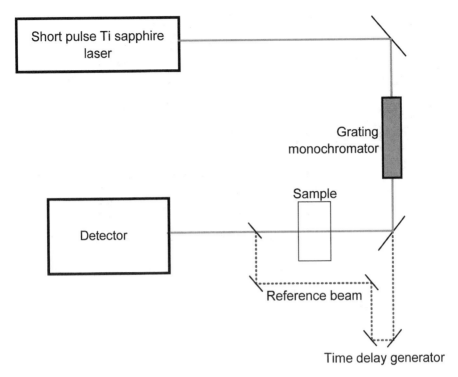

Figure 3.30: Time delay measurement for calculation of index dispersion in a material.

3.5.2 Infrared absorption and Raman scattering measurements

Instruments for IR and Raman spectroscopies are available commercially. Infrared spectroscopy is conducted using either a dispersive spectrometer, which operates much like the UV-visible spectrometer described in the last chapter, or a fast Fourier transform spectrometer. The Fourier transform infrared (FTIR) measurement is conducted using a Michelson interferometer, as illustrated in Fig. 1.25. One of the mirrors is scanned rapidly to give a light pulse in the interferometer. The intensity of the interference pattern is analyzed to give $I(t)$ and Fourier-transformed to yield $I(\omega)$, from which absorbance may be deduced. Raman spectroscopy is conducted by illuminating a sample with monochromatic light from either a laser or a high-intensity lamp. Because of the 10^4–10^5 loss of intensity between the incident light and the Raman-scattered light, the spectrometer must be designed to reject strongly the incident wavelength and remain very sensitive to any other wavelength. Usually, a double monochromator is used, whereby the received light is dispersed through a grating, and passed through a narrow slit dispersed again through a second grating, and then passed through a slit before a photomultiplier (PMT) tube. A typical setup is shown in Fig. 3.31. Modifications include using a linear diode array or a charge-coupled detector (CCD) for broadband detection. The advantage of this approach is in collecting a portion of the spectrum all at once, thus allowing for longer integration times than in the scanned photomultiplier tube (PMT). However, since broad-band detection is used, the spatial extent of the detected dispersed light must be large and slits must be open wide. Consequently, supernotch filters that remove a very narrow slice of the spectrum must be used to remove the incident light wavelength. Other modifications include FT Raman for materials that transmit in the near-IR (1 μm). Fourier transform Raman

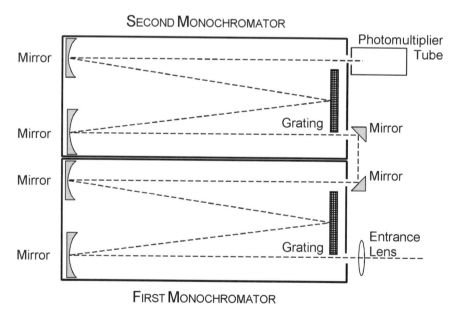

Figure 3.31: Sketch of a typical double-monochromator Raman spectrometer.

has a lower resolution than the PMT method, so it is particularly useful only with systems that have broad Raman lines, since the technique benefits from a rapid scan of the spectrum. Linear diode arrays and CCD detectors can yield high resolution, but they cannot approach the laser line as well as the conventional method, so low-energy vibrations ($<75\,\mathrm{cm}^{-1}$) are generally omitted.

Appendix 3A

Quantum Mechanical Treatment of the Simple Harmonic Oscillator

This problem is treated with no loss of generality in one dimension. The classical approach uses the following equations. The equation

$$m\frac{d^2x}{dt^2} + kx = 0 \qquad (3.64)$$

has the solution

$$x(t) = x_0 e^{\pm i\omega_0 t}$$

with $\omega_0^2 = k/m$.

The quantum mechanical treatment, using the Dirac method of raising and lowering operators, is given here. We begin with the time-dependent Schroedinger wave equation:

$$-\frac{\hbar^2}{2m}\frac{\partial^2\Psi(x,t)}{\partial x^2} + V(x)\Psi(x,t) = -\frac{\hbar}{i}\frac{\partial\Psi(x,t)}{\partial t} \qquad (3.65)$$

Assuming that the wave function can be separated into a time-dependent and a position-dependent factor leads to a time-independent Schroedinger equation:

$$\Psi(x,t) = \Phi(x)e^{\frac{-i\xi t}{\hbar}} \qquad -\frac{\hbar^2}{2m}\frac{d^2\Phi(x)}{dx^2} + V(x)\Phi(x) = \xi\Phi(x) \qquad (3.66)$$

where ξ is the total energy of the system and the potential energy $V(x) = \frac{1}{2}kx^2 = \frac{1}{2}m\omega_0^2 x^2$ for a simple harmonic oscillator.

The following substitutions lead to a simplified operator form of the wave equation:

Substitutions:

$$\beta \equiv \frac{2m\xi}{\hbar^2}; \qquad \alpha^2 \equiv \frac{m^2\omega_0^2}{\hbar^2}; \qquad \zeta \equiv \frac{\beta}{\alpha} = \frac{2\xi}{\hbar\omega_0} \tag{3.67a}$$

Operator definitions:

$$q \equiv \sqrt{\alpha}x; \qquad p \equiv -i\frac{\partial}{\partial q}; \qquad p^2 = -\frac{\partial^2}{\partial q^2} \tag{3.67b}$$

Schroedinger equation:

$$(q^2 + p^2)\Phi(q) = \xi\Phi(q) \tag{3.67c}$$

Since the operators p and q do not commute ($[p,q] = pq - qp \neq 0$), it is not possible to factor the left side of the equation as follows: $(q^2 + p^2) \neq (q + ip)(q - ip)$. This is shown here:

$$[q,p]f(q) = -iq\frac{\partial f(q)}{\partial q} + i\frac{\partial(qf(q))}{\partial q}$$

$$= -iq\frac{\partial f(q)}{\partial q} - iq\frac{\partial f(q)}{\partial q} + if(q) = if(q) \tag{3.68a}$$

This suggests:

$$(q^2 + p^2) = \frac{1}{2}[(q + ip)(q - ip) + (q - ip)(q + ip)]$$

$$= \frac{1}{2}[(q^2 + ipq - iqp + p^2 + q^2 - ipq + iqp + p^2] = (q^2 + p^2) \tag{3.68b}$$

The following definition of two new operators leads to an elegant solution of the wave equation:

$$a \equiv \frac{1}{\sqrt{2}}(q + ip) = \frac{1}{\sqrt{2}}(q + \frac{\partial}{\partial q})$$

$$a^+ \equiv \frac{1}{\sqrt{2}}(q - ip) = \frac{1}{\sqrt{2}}(q - \frac{\partial}{\partial q})$$

resulting in

$$(q^2 + p^2)\Phi(q) = (aa^+ + a^+a)\Phi(q) = \zeta\Phi(q) \tag{3.69}$$

As earlier, it can be shown that the commutator for the new operators is as follows:

$$[a, a^+] = aa^+ - a^+a = \tfrac{1}{2}(q + ip)(q - ip) - \tfrac{1}{2}(q - ip)(q + ip) = 1$$

giving:

$$a^+a = aa^+ - 1 \qquad \text{and} \qquad aa^+ = a^+a + 1$$

$$[aa^+ + a^+a]\Phi(q) = \zeta\Phi(q) = [aa^+ + aa^+ - 1]\Phi(q) = [2aa^+ - 1]\Phi(q)$$

$$[aa^+ + a^+a]\Phi(q) = \zeta\Phi(q) = [a^+a + a^+a + 1]\Phi(q) = [2a^+a + 1]\Phi(q)$$

producing two wave equations:

$$aa^+\Phi(q) = [\tfrac{1}{2}\zeta + \tfrac{1}{2}]\Phi(q)$$
$$a^+a\Phi(q) = [\tfrac{1}{2}\zeta - \tfrac{1}{2}]\Phi(q) \tag{3.70}$$

If we now apply a^+ to the first wave equation, we find that it simply raises the energy eigenvalue by one unit. Application of a to the second equation lowers the energy eigenvalue by one unit:

$$aa^+\Phi_n(q) = \tfrac{1}{2}(\zeta + 1)\Phi_n(q)$$

$$a^+(aa^+)\Phi_n(q) = (a^+a)a^+\Phi_n(q) = (aa^+ - 1)a^+\Phi_n(q) = \tfrac{1}{2}(\zeta + 1)a^+\Phi_n(q)$$

$$aa^+(a^+\Phi_n(q)) = \tfrac{1}{2}(\zeta + 3)(a^+\Phi_n(q)) = \tfrac{1}{2}(\zeta + 3)\Phi_{n+1}(q)$$

$$\tag{3.71a}$$

Similarly:

$$a^+a\Phi_n(q) = \tfrac{1}{2}(\zeta - 1)\Phi_n(q)$$

$$a(a^+a)\Phi_n(q) = (aa^+)a\Phi_n(q) = (a^+a + 1)a\Phi_n(q) = \tfrac{1}{2}(\zeta - 1)a\Phi_n(q)$$

$$aa^+(a\Phi_n(q)) = \tfrac{1}{2}(\zeta - 3)(a\Phi_n(q)) = \tfrac{1}{2}(\zeta - 3)\Phi_{n-1}(q)$$

$$\tag{3.71b}$$

Successive applications of these operators continue to respectively raise and lower the eigenvalues of the energy. Eventually, the lowering operator lowers the system to its ground state, $\Phi_0(q)$. Further application of the lowering operator, a, is not possible, and it is this clue that allows us to calculate the solutions of the wave function:

$$a\Phi_0(q) = 0$$

$$a^+a\Phi_0(q) = a^+(0) = 0 = \frac{1}{2}(\zeta - 1)\Phi_0$$

which is possible only if $\zeta = 1 = 2\xi_0/\hbar\omega_0$, giving

$$\xi_0 = \frac{1}{2}\hbar\omega_0 \tag{3.72}$$

This gives us the ground-state energy of the simple harmonic oscillator. Returning to the operation of the lowering operator on the ground-state wave function allows solution of the wave function:

$$a\Phi_0(q) = 0$$

$$a\Phi_0(q) = \frac{1}{\sqrt{2}}(q + \frac{\partial}{\partial q})\Phi_0(q) = 0$$

which has a solution of:

$$\Phi_0(q) = C_0 e^{-q^2/2} \tag{3.73}$$

Higher energy levels and their associated wave functions are calculated by applying the raising operator, a^+, to $\Phi_0(q)$. The solutions are Hermite polynomials that obey a generating function as follows:

$$\Phi_n(q) = \left(\frac{1}{2^n n!\sqrt{\pi}}\right)^{\frac{1}{2}} e^{-q^2/2} H_n(q)$$

where

$$H_n(q) = e^{+q^2/2}\left(q - \frac{\partial}{\partial q}\right)^n e^{-q^2/2}$$

The Hermite Polynomials are then calculated to be:

$$H_0(q) = 1 \quad H_1(q) = 2q$$

$$H_2(q) = 4q^2 - 2 \quad H_3(q) = 8q^3 - 12q \tag{3.74}$$

$$H_4(q) = 16q^4 - 48q^2 + 12$$

The Hermite Polynomials and full solutions are given in Table 3.6 and shown in Fig. 3.32.

The successive energy levels associated with each wave function are given by incrementing ζ in units of 2 to obtain the following equation:

$$\xi_n = \hbar\omega_0(n + {}^1\!/_2) \tag{3.75}$$

As shown in the text, since the x value of the simple harmonic oscillator corresponds to the displacement of the electron cloud from its neutral position over the ionic core, the induced dipole moment is $p = e\langle x \rangle$. Thus the oscillator strength of a transition between two states of the simple harmonic oscillator is:

$$f_{nm} = \frac{2m\omega_{nm} <x_{nm}>^2}{\hbar}$$

$$<x_{nm}> = \frac{1}{\sqrt{\alpha}} \int_{-\infty}^{\infty} \Phi_n^*(q)q\Phi_m(q)dq = \left(\frac{1}{2^n n!\sqrt{\pi}}\right)\frac{1}{\sqrt{\alpha}} \int_{-\infty}^{\infty} H_n^*(q)qH_m e^{-q^2}dq$$

$$\tag{3.76}$$

However, the Hermite polynomials obey orthogonality requirements, so one sees the following effect:

$$\text{Orthogonality:} \quad \int_{-\infty}^{\infty} H_n^*(q)H_m(q)e^{-q^2}dq = \sqrt{\pi}2^n n!\delta_{nm} \tag{3.77a}$$

The following relation holds between different orders of the Hermite Polynomials:

$$qH_m(q) = {}^1\!/_2[H_{m+1}(q) + 2mH_{m-1}(q)] \tag{3.77b}$$

n	$H_n(q)$	$\Phi_n(q)$	ξ_n
0	1	$(1/\sqrt{\pi})^{\frac{1}{2}}e^{-q^2/2}$	$\frac{1}{2}\hbar\omega_0$
1	$2q$	$(2/\sqrt{\pi})^{\frac{1}{2}}qe^{-q^2/2}$	$\frac{3}{2}\hbar\omega_0$
2	$4q^2 - 2$	$(1/2\sqrt{\pi})^{\frac{1}{2}}(2q^2 - 1)e^{-q^2/2}$	$\frac{5}{2}\hbar\omega_0$
3	$8q^3 - 12q$	$(1/3\sqrt{\pi})^{\frac{1}{2}}(2q^3 - 3q)e^{-q^2/2}$	$\frac{7}{2}\hbar\omega_0$
4	$16q^4 - 32q^2 + 12$	${}^1\!/_2(1/6\sqrt{\pi})^{\frac{1}{2}}(4q^4 - 12q^2 + 3)e^{-q^2/2}$	$\frac{9}{2}\hbar\omega_0$

Table 3.6: Wave functions and energies for the simple harmonic oscillator.

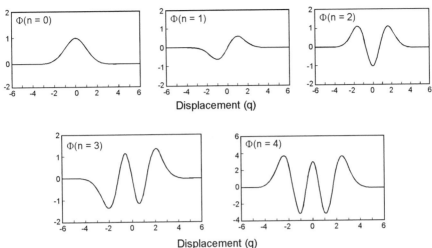

Displacement (q)

Figure 3.32: Simple harmonic oscillator stationary wave functions for $n = 0 - 4$.

When applied to the expectation value of x for transitions between states of index m and index n, the result gives a numerical condition known as the *selection rules for optical transitions between states of the simple harmonic oscillator:*

$$\int_{-\infty}^{\infty} H_n{}^*(q) q H_m e^{-q^2} dq$$

$$= \frac{1}{2} \int_{-\infty}^{\infty} H_n{}^*(q) H_{m+1} e^{-q^2} dq + m \int_{-\infty}^{\infty} H_n{}^*(q) H_{m-1} e^{-q^2} dq$$

(3.78)

Since the orthogonality condition requires that the two indices of the Hermite polynomials be equal in order to obtain a nonzero value, we see that the condition for a nonzero transition oscillator strength becomes

$$n = m \pm 1 \qquad (3.79)$$

Thus Eq. (3.79) is a selection rule of optical transitions between states of the simple harmonic oscillator. It states that the only possible transitions are between a state of index m and state of either index $m + 1$ by the absorption of a photon or of index $m - 1$ by the emission of a photon. That is, only adjacent states in energy can participate in optical transitions. Transitions are allowed only for states with

$$\Delta n = \pm 1 \tag{3.80}$$

The polarizability becomes:

$$\alpha_{\text{pol}} = \sum_{nm} \frac{2q_e^2 <x_{nm}>^2}{\hbar^2} \frac{\hbar \omega_{nm}}{\omega_{nm}^2 - \omega^2} = \frac{8n^2 q_e^2}{m(\omega_0^2 - \omega^2)}) \tag{3.81}$$

If a large change in energy is required, the simple harmonic oscillator can make it only in multiple emissions of the $\hbar \omega_0$ photon. (The simple harmonic oscillator emits a sharp line only at $\hbar \omega_0$.) Materials obey this law strictly, therefore, if the thermal energy is decreased so that $k_B T < \hbar \omega_0$, no photon can be emitted by that oscillator. The low-temperature limit on the emission of radiation is the basis for Planck's radiation law, which marked the birth of quantum mechanics.

Appendix 3B

Calculation of the Refractive Index of Glasses

3B.1 Method of molar refraction

From the Clausius–Mosotti equation it can be shown that the molar refraction of a glass can be related to its refractive index by the following equation:

$$R = \frac{M}{\rho} \frac{n^2 - 1}{n^2 + 2} \tag{3.82}$$

where M is the molecular weight and ρ is the density. The molar refraction is also related to the structure of the glass through the sum of polarizabilities of its components and its Loschmidt number:

$$R = \frac{4\pi N_L}{3} \sum_i A_i \alpha_{poli} = \sum_i A_i R_i \tag{3.83}$$

where A_i is the atomic fraction of each component element and R_i is the ionic refraction. A table of ionic refractions for various elements is required, as in Table 3.7.

For example, consider a glass of composition: 60% SiO_2, 20% B_2O_3, 5% Al_2O_3, 5% MgO, 5% Na_2O and 5% K_2O by mole. Calculate the following components below table 3.7.

Ion	Ionic refraction	Ion	Ionic refraction	Ion	Ionic refraction
Li^+	0.08	Ni^{+2}	1.0	Ti^{+4}	0.6
Na^+	0.47	Cu^{+2}	1.1	Zr^{+4}	1.1
K^+	2.24	B^{+3}	0.006	Hf^{+4}	1.1
Rb^+	3.75	Al^{+3}	0.14	Th^{+4}	3.9
Cs^+	6.42	Ga^{+3}	0.6	P^{+5}	0.05
Ti^+	10.0	In^{+3}	2.0	V^{+5}	0.4
Be^{+2}	0.03	Sc^{+3}	0.9	Nb^{+5}	0.9
Mg^{+2}	0.26	Y^{+3}	1.7	Ta^{+5}	0.8
Ca^{+2}	1.39	La^{+3}	3.5	Cr^{+6}	0.3
Sr^{+2}	2.56	Fe^{+3}	1.1	Mo^{+6}	0.7
Ba^{+2}	4.67	As^{+3}	1.7	W^{+6}	0.6
Zn^{+2}	0.7	Sb^{+3}	2.8	F^-	2.44
Cd^{+2}	2.8	Bi^{+3}	3.8	O^{-2}	6.95
Pb^{+2}	7.1	Si^{+4}	0.08	S^{-2}	22.7
Mn^{+2}	1.7	Ge^{+4}	0.4	Se^{-2}	28.8
Fe^{+2}	1.2	Sn^{+4}	1.5	OH^-	4.85
Co^{-+2}	1.1				

Table 3.7: Ionic refractions for various ions.

	Atomic concentrations of all ions per mole		Atomic fractions	$R_{\{i\}}$	$A_i R_i$
Si	60		0.17	0.08	0.014
B	40		0.12	0.006	0.001
Al	10		0.029	0.14	0.004
Mg	5		0.014	0.26	0.004
Na	10		0.029	0.47	0.014
K	10		0.029	2.24	0.065
O	210=120 + 60 + 15 + 5 + 5 + 5		0.61	6.95	4.240
Sum	345				4.340

Note that in oxide glasses, the oxygen dominates the molar refraction, not just by atomic concentration but also by its atomic refraction.

Component molar masses:

$SiO_2 = 60.1\,g$ $B_2O_3 = 69.6\,g$ $Al_2O_3 = 102\,g$
$MgO = 40.3\,g$ $Na_2O = 62\,g$ $K_2O = 94.2\,g$

The molar mass is

$$M = (60.1g)(0.60) + (69.6g)(0.20) + (102g)(0.05) + (40.3g)(0.05)$$
$$+ (62g)(0.05) + (94.2g)(0.05) = 64.91\,g/mole$$

The density of the glass must be either looked up or estimated: $4.8\,g/cm^3$. Substituting into Eq. (3.82) gives $n^2 = 2.418$ or $n = 1.56$.

Several other methods are available for calculating both the refractive index and the dispersion. One is briefly described next.

3B.2   Method of Demkina (From Fanderlik 1983)

Here coefficients are listed for various molecular components of glasses and summed as follows (k can be either the index, the dispersion, or the Abbe number Table 3.8):

$$K = \frac{\frac{a_1}{s_1}k_1 + \frac{a_2}{s_2}k_2 + \frac{a_3}{s_3}k_3 + \cdots + \frac{a_i}{s_i}k_i}{\frac{a_1}{s_1} + \frac{a_2}{s_2} + \frac{a_3}{s_3} + \cdots + \frac{a_i}{k_i}} \tag{3.84}$$

where

$a_1, a_2, a_3, \ldots, a_i$ = weight percentage of oxides,

$s_1, s_2, s_3, \ldots, s_i$ = structural factors for mass to volume

$k_1, k_2, k_3, \ldots, k_i$ = optical factors

Example:

Component	a_i (wt%)	a_i/s_i	$(a_i/s_i)k(n)$	$(a_i/s_i)k(n_F - n_c)$
SiO_2	45.5%	0.75833	1.11854	527.04
PbO	48.1%	0.14021	0.34497	1079.84
K_2O	6%	0.06383	0.10085	76.59
As_2O_3	0.3%	0.00280	0.00440	4.48
Sb_2O_3	0.1%	0.00065	0.00128	2.46
Sums	100%	0.96582	1.57004	1690.4

Oxides	s_i	$k(n_0)$	$k(n_F - n_c) \times 10^5$	$k(v_D)$
SiO_2 I	60	1.475	695	68
SiO_2 II (>80%)	60	1.458	678	68
$B_2O_3I(BO_4)$	43	1.610	750	81
$B_2O_3II(BO_3)$	70	1.464	670	69
Al_2O_3	59	1.490	850	58
As_2O_3	107	1.570	1600	36
Sb_2O_3	154	1.980	3800	26
MgO	140	1.640	1300	49
CaO	86	1.830	1750	47
BaO	213	2.030	2280	45
ZnO	223	1.960	2850	34
PbO I (<50%)	343	2.460	7700	19
PbO II (5–11%)	223	2.460	7700	19
PbO III(1–3%)	223	2.500	11,600	13
Na_2O	62	1.590	1400	42
K_2O	94	1.580	1200	48

Table 3.8: Optical factors for various glass components.

In contrast to the molar refraction technique, which uses atomic components, this approach allows us to determine which molecular components contribute most to the index and the dispersion. For example, SiO_2 contributes most strongly to the index, while PbO contributes most strongly to the dispersion. The results are actually very close to experimental measurements:

n(calculated) $= 1.6256$ n(measured) $= 1.6259$

$n_F - n_c$(calculated) $= 0.1750$ $n_F - n_c$(measured) $= 0.1756$.

This method is essentially the basis for computer-based look-up tables of optical properties. Additional methods include the Lorentz–Lorenz method and the Sun-Huggins method. Both are similar to the Demkina method. We refer the reader to the following: Fanderlik, (1983) and Sun and Huggins (1945).

Appendix 3C

Ligand Field Theory Concepts

Ligand field theory has been used to calculate the influence of local environment on the energy levels of transition-metal and rare-earth ions. In Ligand field theory, the ligand atoms are represented as point charges, and the electrostatic field of the environment is calculated at the location of the ion of interest. The techniques make rigorous use of the symmetries of the local neighborhood (Bendale, 1996).

The Hamiltonian for the motion of the N electrons of the central ion is written as

$$H = \sum_{i=1}^{N}\left[-\frac{\nabla_i^2}{2} - \frac{Z^*}{r_i}\right] + \sum_{i<j}^{N}\frac{1}{r_{ij}} + \sum_{i=1}^{N}V(r_i) + \sum_{i=1}^{N}H_{s.o.}(i) \qquad (3.85)$$

The first term in the equation covers the kinetic and potential energies of electrons and the central core. The second term covers the interaction between the electrons. The third term covers the interaction between each electron of the central ion and the ligands in the near environment. The fourth term covers the spin-orbit splitting of the central ion electrons. In the calculation, the last three terms can be regarded as perturbations over the degenerate set of wave functions obtained by solving the Schroedinger equation using only the first term.

Comparison of the calculated results with experiment shows three distinct cases due to the relative magnitude of the last three terms.

(1) weak crystalline field case (1st transition metal series):

155

$$\sum_{i<j} \frac{1}{r_{ij}} > \sum_i V(r_{ij}) > \sum_i H_{s.o.}(i) \qquad (3.86)$$

(2) strong crystalline field case (first transition-metal series):

$$\sum_i V(r_i) > \sum_{i<j} \frac{1}{r_{ij}} > \sum_i H_{s.o.}(i) \qquad (3.87)$$

(3) strong spin-orbit coupling case (second and third row transition metals):

$$\sum_i H_{s.o.}(i) > \sum_i V(r_i), \sum_{i<j} \frac{1}{r_{ij}} \qquad (3.88)$$

The ligand field is treated by summing all the neighboring ions and calculating a multipole expansion of the electric potential at the location of the central ion by using the well known expansion in spherical harmonics, as shown in Chapter 1. The calculation is simplified by letting the charge density of the ligands have the same symmetry as the coordination shells. This leads to a simplified expression for the potential. For example, for d electrons, the angular momentum quantum number is $\ell = 2$, so only the 2ℓ terms up to 4 are considered in the expansion:

$$V(r_i) = A_{0,0}r^0 Y_{0,0} + \sum_{m=-2}^{2} A_{2,m} r^2 Y_{2,m} + \sum_{m=-4}^{4} A_{4,m} r^4 Y_{4,m} \qquad (3.89)$$

For d complexes with octahedral symmetry, the five-fold degenerate d orbitals (D) split into a doubly degenerate E_g term and a threefold T_{2g} term, purely from symmetry considerations, with the difference in energy being the splitting energy, Δ. In tetrahedral coordination, the energies are reversed from octahedral symmetry, with a reduced splitting energy. Table 3.9 gives the mean electron-pairing energy, ligand field, splitting energy, and spin state of some transition-metal ions.

Configuration	Ion	Mean electron-pairing energy	Ligand field	$\Delta_0(cm^{-1})$	Spin state
d^4	Cr^{+2}	$23,500\,cm^{-1}$	$6H_2O$	13,900	High
	Mn^{+3}	2800	$6H_2O$	21,000	High
d^5	Mn^{+2}	25,500	$6H_2O$	7800	High
	Fe^{+3}	30,000	$6H_2O$	13,700	High
d^6	Fe^{+2}	17,600	$6H_2O$	10,400	High
			$6CN^-$	33,000	Low
	Co^{+3}	21,000	$6F^-$	13,000	High
			$6NH_3$	23,000	Low
d^7	Co^{+2}	22,500	$6H_2O$	9,300	High

Table 3.9: Ligand field splitting energies.

References

Agrawal, G. P. 1989. *Nonlinear Fiber Optics*. New York: Academic Press.

Anderson, E. E. 1971. *Modern Physics and Quantum Mechanics*. Philadelphia: W. B. Saunders.

Azzam, R. M. A., and N. M. Bashara. 1989. *Ellipsometry and Polarized Light*. Amsterdam: North Holland.

Bendale, R. 1996. Private Discussion.

Boas, M. L. 1983. *Mathematical Methods in the Physical Sciences*. New York: Wiley.

Chen, D. G., B. G. Potter, and J. H. Simmons. 1994. GeO_2-SiO_2 thin films for planar waveguide applications. *J. Non-Cryst. Solids* 178:135–147.

Fanderlik, I. 1983. *Optical Properties of Glass*. Amsterdam: Elsevier.

Fleming, J. W. 1984. Dispersion in GeO_2-SiO_2 glasses. *Appl. Optics* 23:4486.

Gaertner Scientific Corp., 8228 McCormick Blvd., Skokie, IL 60076.

Hecht, E. 1998. *Optics*. Reading, MA: Addison-Wesley.

Hopf, F. A., and G. I. Stegeman. 1985. *Applied Classical Electrodynamics*. Vol. 1, *Linear Optics*. New York: Wiley.

Jackson, J. D. 1975 and later editions. *Classical Electrodynamics*. New York: Wiley.

Keck, D. B., R. D. Maurer, and P. C. Schultz. 1973. *Appl. Phys. Lett.* 22:307.

Kittel, C. 1996. *Introduction to Solid State Physics*. 7th ed. New York: Wiley.

Krane, K. S. 1983. *Modern Physics*. New York: Wiley.

Maurer, R. D. 1973. *Proc. IEEE* 61:452.

Midwinter, J. E. 1979. *Optical Fibers for Transmission*. New York: Wiley.

Osonai, H., T. Shioda, T. Moriyama, S. Araki, M. Horiguchi, T. Izawa, and H. Takata. 1976. *Electron. Lett.* 12:550.

Pauling, L. 1927. *Proc. Roy. Soc. (London)* A114:181.

Potter, K. 1994. Photosensitive mechanisms in germanosilicate optical fiber and planar waveguides. PhD dissertation, University of Arizona, Tucson.

Reilly, M. H. 1970. *J. Chem. Solids* 31:1041.

Rich, T. C., and D. A. Pinnow. 1976. *Appl. Phys. Lett.* 20:550.

Rowe, J. E. 1974. *Appl. Phys. Lett.* 25:576.

Ruffa, A. R. 1968. *Phys. Stat. Solids* 29:605.

Schroeder, J. 1970. Light scattering in glass. In *Treatise on Materials Science and Technology*. Vol. 12, *Glass 1: Interaction with Electromagnetic Radiation*, edited by M. Tomozawa and R. Doremus. New York: Academic Press.

Sigel, G. 1977. Optical absorption in glasses. In *Treatise on Materials Science and Technology*. Vol. 12, *Glass 1: Interaction with Electromagnetic Radiation*, edited by M. Tomozawa and R. Doremus. New York: Academic Press.

Sun, K. H., and M. L. Huggins. 1945. The effect of chemical composition on the relation between refractive index and abbe value for binary systems. *J. Soc. Glass Technol.* 29:192–196.

Tessman, J. R., A. H. Kahn, and Shockley. 1953. Electric Polarizabilities of Ions in Crystals. *Phys. Rev.* 92:890.

Tropf, W. J., M. Thomas, and T. J. Harris. 1995. Properties of crystals and glasses. in *Handbook of Optics*, Vol. 2, edited by M. Bass. New York: McGraw-Hill.

Wallace, S. 1991. Porous Silica-Gel Monoliths: Structural Evolution and Interactions with Water. Ph.D. dissertation, University of Florida, Gainesville, FL.

Weber, M. J., Milam, and Smith. 1978. *Opt. Eng.* 17:1.

Chapter 4

Optical Properties of Insulators — Some Applications

4.1 Thin films

4.1.1 Mathematical treatment

Consider the optics of light passing through multiple thin-film layers. In our theoretical treatment we will assume that the thin-film surfaces are of infinite extent, or at least "infinite" with respect to the size of the optical beams incident upon them, and that incident beams on the surfaces are described by plane waves. As discussed in Chapter 3, the reflectance R and transmittance T of a single film or of a film stack is defined by the ratio of the intensity of light reflected or transmitted (respectively) to the intensity of incident light. We may assume here that the films comprising a film stack are parallel-sided slabs of homogeneous, isotropic material whose optical properties can be described by the complex refractive index, $\eta = n + i\kappa$.

As we have seen, a wave impinging on an interface will give rise to both a reflected and a refracted wave. The angle of reflection will be equal to the angle of incidence, and the angle of refraction will be described by Snell's law. For a given amplitude and polarization of incident wave, the amplitude and polarization of both the reflected and transmitted waves at each interface can be determined by applying the boundary condition that

159

requires continuity of the tangential components of both the electric and the magnetic fields across an interface. The solution of this boundary condition gives rise to the Fresnel equations for s- and p-polarized light at each interface, as described in Chapter 3.

It can be shown that for a normally incident wave, our stated continuity condition leads to the expressions for the amplitude reflection and transmission coefficients, r and t, at a simple interface:

$$r = \frac{n_1 + i\kappa_1 - n_2 - i\kappa_2}{n_1 + i\kappa_1 + n_2 + i\kappa_2}$$

$$t = \frac{2(n_1 + i\kappa_1)}{n_1 + i\kappa_1 + n_2 + i\kappa_2}$$

(4.1)

If the incident medium is nonabsorbing; as with air, then the equations reduce to their commonly recognized form:

$$r = \frac{n_1 - n_2 - i\kappa_2}{n_1 + n_2 + i\kappa_2}$$

$$t = \frac{2n_1}{n_1 + n_2 + i\kappa_2}$$

$$\phi = \tan^{-1}\frac{2n_1\kappa_2}{n_1{}^2 - n_2{}^2 - \kappa_2{}^2}$$

(4.2)

where ϕ is the phase shift in the electric field upon reflection. The reflectance and transmittance, then, are given by:

$$R = rr^* = \frac{(n_1 - n_2 - i\kappa_2)(n_1 - n_2 + i\kappa_2)}{(n_1 + n_2 + i\kappa_2)(n_1 + n_2 - i\kappa_2)}$$

$$T = \frac{n_2 + i\kappa_2}{n_1}tt^* = \frac{4n_1(n_2 + i\kappa_2)}{(n_1 + n_2 + i\kappa_2)(n_1 + n_2 - i\kappa_2)}$$

(4.3)

For a lossless medium in air, the reflectance at the interface will reduce to $(1 - n)^2/(1 + n)^2$. This method of calculating reflectance and transmittance at each interface over a stack of multiple thin-film layers quickly becomes impractical. Thus a powerful matrix approach can be used to calculate the optical response of the film stack. If we consider the complex amplitude of the forward- and backward-traveling electric field as it propagates through a multilayer stack, at two different interfaces it must be true (due to system linearity) that

$$\begin{bmatrix} E_i^{+} \\ E_i^{-} \end{bmatrix} = \begin{bmatrix} S_{11} & S_{12} \\ S_{21} & S_{22} \end{bmatrix} \begin{bmatrix} E_j^{+} \\ E_j^{-} \end{bmatrix} \tag{4.4}$$

Thus, at each interface the electric field propagation may be described by the simple matrix equation $E_i^{-} = IE_i^{+}$. Similarly, a matrix can be developed for the layer whose thickness is d, such that $E_i^{+} = LE_{i+d}^{-}$. Hence, the overall scattering matrix that describes propagation through an entire film stack will be the product of the interface and layer matrices taken in the proper order:

$$\mathbf{S} = \mathbf{I}_{01}\mathbf{L}_1\mathbf{I}_{12}\mathbf{L}_2 \cdots \mathbf{I}_{(m-1)m}\mathbf{L}_m.$$

Here the 01 interface describes the ambient–film interface and the $(m-1)m$ interface represents the final film–substrate interface. (For a complete mathematical treatment of this matrix approach, the reader is referred to Azzam and Bashara 1989.)

It can be shown (Macleod 1989) that if we represent the complex refractive index of the entire film structure as $N = C/B$, then the following equations can be written for normal incidence:

$$\begin{bmatrix} B \\ C \end{bmatrix} = \prod_{j=1}^{m} \begin{bmatrix} \cos \delta_j & \frac{i}{n_j + i\kappa_j} \sin \delta_j \\ i(n_j + i\kappa_j) \sin \delta_j & \cos \delta_j \end{bmatrix} \begin{bmatrix} 1 \\ n_{\text{substrate}} + i\kappa_{\text{substrate}} \end{bmatrix} \tag{4.5}$$

where $\delta_j = 2\pi(n_j + i\kappa_j)d_j/\lambda$, $d_j =$ layer geometrical thickness, and the subscript j denotes the jth film. If n_o is the refractive index of the incident medium, then the reflectance is given by

$$R = \frac{(n_0 B - C)(n_0 B - C)^{*}}{(n_0 B + C)(n_0 B + C)^{*}} \tag{4.6}$$

the phase shift is given by

$$\phi = \tan^{-1}\left[\frac{in_0(CB^{*} + BC^{*})}{n_0^2 BB^{*} + CC^{*}}\right] \tag{4.7}$$

and the transmittance is given by

$$T = \frac{4n_0 n_{\text{substrate}}}{(n_0 B + C)(n_0 B + C)^{*}} \tag{4.8}$$

These equations may be made fully general for an absorbing incident medium by replacing n_o with the complex refractive index $\eta_0 = n_0 + i\kappa_0$.

For EM waves incident at oblique angles (Fig. 4.1) the theory is slightly complicated by the fact that the electric and magnetic fields have

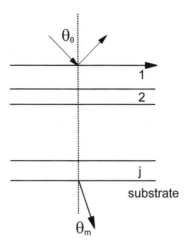

Figure 4.1: Reflection and transmission of a plane wave by a multilayer film stack.

components that are parallel to and normal to the interface. As discussed in earlier chapters, the component of the electric field vector parallel to the plane of incidence (the plane composed of the incident, reflected, and transmitted wave propagation vectors) is known as p-polarized light, and the component normal to the plane of incidence is known as s-polarized light (see Fig. 4.2). In this case the matrix elements and phase shift are modified according to the polarization of the incident beam. For s-

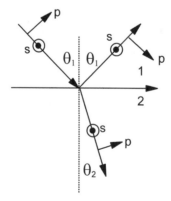

Figure 4.2: Schematic of s and p planes of polarization.

polarized light, $(n_j + i\kappa_j)$ terms are replaced by $\eta_{j,s} = (n_j + i\kappa_j)\cos\theta_j$, for p-polarized light they are replaced by $\eta_{j,p} = (n_j + i\kappa_j)/\cos\theta_j$. The phase thickness of each layer becomes, for each plane of polarization, $\delta = (2\pi/\lambda)(n + i\kappa)d\cos\theta$. Hence, for a multilayer stack the complete matrix expression is given by

$$\begin{bmatrix} B \\ C \end{bmatrix} = \prod_{j=1}^{m} \begin{bmatrix} \cos\delta_j & \frac{i}{\eta_j}\sin\delta_j \\ i(\eta_j)\sin\delta_j & \cos\delta_j \end{bmatrix} \begin{bmatrix} 1 \\ \eta_{\text{substrate}} \end{bmatrix} \tag{4.9}$$

where $\delta_j = 2\pi(n_j + i\kappa_j)d_j\cos\theta_j/\lambda$ and d_j = layer geometrical thickness.

Finally, the reflectance is given by

$$R = \frac{(\eta_0 B - C)(\eta_0 B - C)^*}{(\eta_0 B + C)(\eta_0 B + C)^*} \tag{4.10}$$

the phase shift is given by

$$\phi = \tan^{-1}\left[\frac{in_0(CB^* + BC^*)}{n_0^2 BB^* + CC^*}\right] \tag{4.11}$$

and the transmittance is given by

$$T = \frac{4\eta_0 \eta_{\text{substrate}}}{(\eta_0 B + C)(\eta_0 B + C)^*} \tag{4.12}$$

A more detailed treatment of the calculations for R and T is given in Appendix 4A.

If the optical thickness of the single layer is designed to be any integral multiple of one-quarter of the wavelength of the incident light $(d = m(\lambda_0/4)n_1)$, then the layer matrix for an odd number of quarter waves is easily reduced to

$$\begin{bmatrix} 0 & \frac{i}{\eta} \\ i\eta & 0 \end{bmatrix} \tag{4.13}$$

and for an even number of quarter-waves to

$$\begin{bmatrix} 1 & 0 \\ 0 & 1 \end{bmatrix} \tag{4.14}$$

By reducing the overall system matrix expression, it is easy to show that an odd number of quarter-wave layers of index η_j on a substrate η_{sub} acts as a single surface of index $\eta_j^2/\eta_{\text{sub}}$. By contrast, the half-wave layers act as though they do not exist, so the overall complex refractive index of the structure is simply that of the substrate. As an example, a quarter-wave layer of index n_f in air on a substrate of index n_{sub} is equivalent to a single interface of index n_f^2/n_{sub} in air. The resulting reflectance of the layer is then

$$R = \left(\frac{n_0 - n_f^2/n_{\text{sub}}}{n_0 + n_f^2/n_{\text{sub}}}\right)^2 \qquad (4.15)$$

where, in this case, $n_0 = 1$. Conversely, a stack assembly comprised of two quarter-wave layers on a substrate will have an equivalent index of n_{sub} and the reflectance of the structure will be

$$R = \left(\frac{n_0 - n_{\text{sub}}}{n_0 + n_{\text{sub}}}\right)^2 \qquad (4.16)$$

It is possible to apply these rules to conduct stepwise reduction of multiple quarter-wave layers in a film stack to develop quickly a simple expression for the refractive index and reflectance of the entire structure. For example, a stack made up of five quarter-wave layers of index $n_1, n_2, n_3, n_4,$ and n_5 on a substrate n_{sub} will have an equivalent index of $N = (n_1^2 n_3^2 n_5^2/n_2^2 n_4^2 n_{\text{sub}})$ and a reflectance of $R = (n_0 - N)^2/(n_0 + N)^2$. This example is shown pictorially in Fig. 4.3.

One particular application of this quarter-wave condition is seen in antireflective coatings. As we have just observed, the reflectivity of a single quarter wave on a substrate is given by

$$R_1 = \frac{(n_0 n_s - n_1^2)^2}{(n_0 n_s + n_1^2)^2} \qquad (4.17)$$

which gives a vanishing reflection when $n_1^2 = n_0 n_s$. This structure clearly only provides antireflection (AR) properties at those wavelengths for which the phase thickness of the film layer is exactly an odd number of quarter waves. For a soda-lime silica glass with a refractive index of 1.5, the AR coating material must have an index of 1.22. However, the lowest index obtainable with an inorganic and nonporous medium is 1.35 for Cryolite (sodium aluminum fluoride) and 1.38 for MgF_2, consequently, the minimum reflectivity is not achieved. The coating is usually designed to have a minimum reflectivity in the yellow-green portion of the visible spectrum, so AR-coated lenses appear purple (blue-red) in reflected light. The minimum achieved with a single coating is about 1%. Lower reflectivity is achievable with either a porous coating that can reach the

ambient

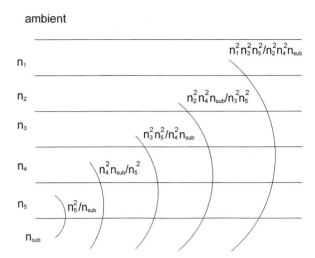

n_1

n_2

n_3

n_4

n_5

n_{sub}

Figure 4.3: Reduction of a quarter-wave stack by determination of the equivalent refractive index for successive layers.

index target of 1.22 or with multiple coatings. Multiple coatings alternate high-index (H) and low-index (L) layers. Minimum reflection is obtained with periodic quarter-wavelength stacks of $g(LH)^n La$, while maximum reflectance is obtained with $g(HL)^n Ha$, where g is the matrix, H is a high-index layer, L is a low-index layer, and a is air. High-index layers are formed using ZrO_2 (2.1), TiO_2 (2.4) and ZnS (2.32).

4.1.2 Fabry–Perot oscillations

In all multilayer media, the potential exists for incident light to interfere constructively with itself over particular wavelength regions. Thus, the measurement of light absorbed in a thin film is often complicated by oscillations in transmission intensity as a function of wavelength in regions where the film is transparent. These so-called Fabry–Perot oscillations correspond to interferences of the transmitted light with light that is multiply internally reflected from the 2 (or more) surfaces of the film (see Figs. 4.4 and 4.5).

Letting the amplitude reflection and transmission coefficients at the first and second surfaces be denoted, respectively, by r_1 and t_1 and by r_2 and t_2, the first transmitted beam amplitude, A_1, can be written as

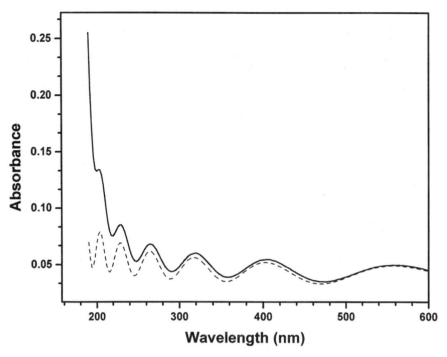

Figure 4.4: Comparison of ultraviolet absorption data taken from a 45 mol% GeO_2–55 mol% SiO_2 thin film (solid line) and a theoretically generated Fabry–Perot fringe fit (dashed line).

$$A_1 = A_0 t_1 t_2 e^{i\delta} \tag{4.18}$$

where A_0 is the incident amplitude and the phase angle $\delta = 2\pi(d/\lambda)(n_1 + i\kappa_1)\cos\phi_1$, where the film index is n_1 and the angle of light propagation in the film is ϕ_1. The transmitted nth reflected beam has an amplitude

$$A_{n+1} = A_1(r_1 r_2 e^{i2\delta})^n \tag{4.19}$$

The total transmitted amplitude (see Fig. 4.6) is therefore

$$A_T = \sum_{n=0}^{\infty} A_0 t_1 t_2 e^{i\delta}(r_1 r_2 e^{i2\delta})^n = \frac{A_0 t_1 t_2 e^{i\delta}}{1 - r_1 r_2 e^{i2\delta}} \tag{4.20}$$

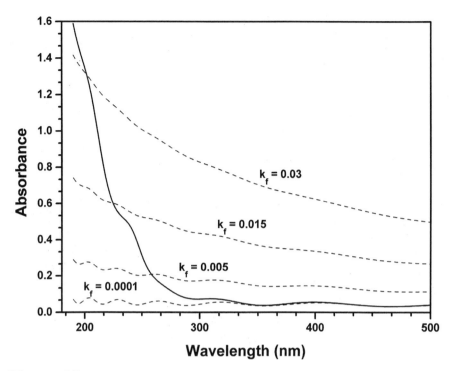

Figure 4.5: A series of curves (dashed lines) generated from a theoretical Fabry–Perot(FP) analysis of a material which exhibits absorption at ultraviolet wavelengths, plotted along with the optical loss spectrum of a photosensitive 45 mol% GeO_2–55 mol% SiO_2 thin film (solid line). The optical losses in the generated spectra have been modeled by constant k (film) values of 0.0001, 0.005, 0.015, and 0.03. This method yields an accurate fit of the FP fringes in the transparent region and gives a good measure of the loss in the transparency region (from Simmons-Potter and Simmons 1996).

For the case of light incidence at θ_0 from a medium of index n_o and transmission at θ_s to a medium of index n_s, the transmission coefficient T is written as

$$T = \left(\frac{n_s \cos\theta_s}{n_0 \cos\theta_0}\right)[A_T^* A_T] = \left(\frac{n_s \cos\theta_s}{n_0 \cos\theta_0}\right)\left[\frac{(A_0 t_1 t_2)^2}{1 + r_1^2 r_2^2 - 2r_1 r_2 \cos 2\delta}\right] \quad (4.21)$$

The angles θ_s and θ_o are related through Snell's law. The amplitude reflection and transmission coefficients are found in Eqs. (3.45).

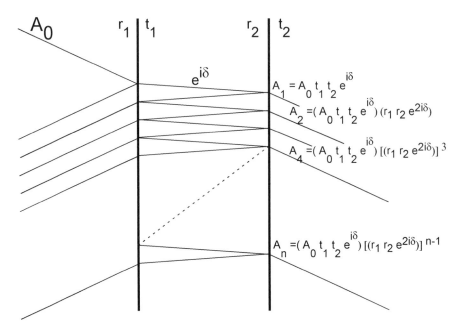

Figure 4.6: Multiple Fabry–Perot reflections in a thin film.

4.1.3   Ellipsometry measurements

Ellipsometry (Azzam and Bashara 1989), covered in the previous chapter, can be used to measure the thickness and refractive index of the coating on a substrate. In principle, this coating may consist of several layers with different thicknesses and indices; however, accuracy decreases rapidly and the technique becomes exponentially more complex beyond a single layer of coating. So we cover only the single film here.

A coated substrate is measured by first obtaining the substrate index following the procedure outlined in Chapter 3 (section 3.5.1.4) on an uncoated substrate material. The procedure is repeated on the coated substrate. Assuming that the film index is n_1 and the thickness is d_1 the equations are:

$$Snell's\ law: \quad n_0 \sin \phi_0 = n_1 \sin \phi_1 = n_s \sin \phi_s$$

$$\beta = 2\pi \left(\frac{d_1}{\lambda} \right) n_1 \cos \phi_1$$

$$r_{01,s} = \frac{n_0 \cos \phi_0 - n_1 \cos \phi_1}{n_0 \cos \phi_0 + n_1 \cos \phi_1}; \quad r_{01,p} = \frac{n_1 \cos \phi_0 - n_0 \cos \phi_1}{n_1 \cos \phi_0 + n_0 \cos \phi_1}$$

$$r_{1s,s} = \frac{n_1 \cos \phi_1 - n_s \cos \phi_s}{n_1 \cos \phi_1 + n_s \cos \phi_s}; \quad r_{1s,p} = \frac{n_s \cos \phi_1 - n_1 \cos \phi_s}{n_s \cos \phi_1 + n_1 \cos \phi_s}$$

(4.22)

$$R_s = \frac{r_{01,s} + r_{1s,s} e^{-2i\beta}}{1 + r_{01,s} r_{1s,s} e^{-2i\beta}} = |R_s| e^{i\delta_s}$$

$$R_p = \frac{r_{01,p} + r_{1s,p} e^{-2i\beta}}{1 + r_{01,p} r_{1s,p} e^{-2i\beta}} = |R_p| e^{i\delta_p}$$

$$\tan \psi = \frac{|R_p|}{|R_s|}; \quad \Delta = \delta_p - \delta_s$$

The values of ψ and Δ can be obtained from the measurements of analyzer and polarizer settings, as in Section 3.5.1.4. Then the index and thickness of the film can be calculated using one of several canned programs available from the instrument providers.

4.2 Glasses, crystals, and birefringence

In some materials it is seen that electromagnetic waves propagate at different velocities along different directions, or axes, in the material or that they experience splitting or rotation of the polarization during propagation. This effect is called *birefringence*. Crystals other than cubic have orientational birefringence. Often this birefringence is used to detect whether a powder is crystalline or glassy. By propagating light at different velocities for different polarization directions, a crystal subjected to polarized light will tend to rotate the polarization direction as the light propagates through the crystal. Thus, if examined between two cross-Polaroids, some crystal grains will rotate the polarization by 90° and will appear bright, whereas others will not and will appear dark. Rotating the stage carrying the crystals without changing the orientation of the polarized sheets (polarizer and analyzer) will cause the bright crystals to extinguish and the dark ones to brighten.

Glasses have isotropic optical properties as long as their structure remains essentially isotropic on the average (macroscopically). Several conditions, however, can alter this behavior and cause a variation in the speed of propagation of electromagnetic waves with direction of polarization, hence birefringence. Anisotropy in the glass is most often caused by nonhydrostatic stresses. This produces stress birefringence through the stress-optic coefficients. This phenomenon is extremely useful in determining the presence of residual stresses in glass optical elements. Polarographs use a white light source with a linearly polarized polarizer and a linearly polarized analyzer. If the glass inserted between the polarizer and the analyzer is stress free, then it appears essentially transparent and colorless. If, however, the glass has residual stresses, the linear polarization will rotate its orientation as it propagates through the glass. The rate of rotation varies with wavelength, so the transmitted light will exhibit a variety of colors, depending on the stress anisotropy in the object, indicating that an additional annealing treatment is required to relax the induced stresses.

Another source of birefringence is the formation of oriented phases in the glass (e.g., long cylinders aligned in the same direction). When the source is structural, the phenomenon is called *anomalous* or *geometric birefringence*. Porous glasses often show birefringence, some geometric and some stress induced.

Let us examine stress birefringence in detail. We do this by considering a block of material such as shown in Fig. 4.7, with an applied stress along the z axis sufficient to change the refractive index in that direction. Light incident along the stress axis will not show any effect, since its polarization directions are along axes that are unstressed. Light incident from any other direction will have a component that is normal to the applied stress direction. The component along the normal direction (say, along the x axis) will have different propagation velocities for its different polarization components. The polarization parallel to the other normal direction (the y axis) will propagate at the normal velocity of the material. It is called the *ordinary* ray. The polarization parallel to the direction of applied stress will propagate at a different velocity. This is called the *extraordinary* ray. The difference in refractive index between the two directions is called the *birefringence* ($\Delta n = n_e - n_o$), while the difference in optical path length between the two polarization directions is called the *retardation* $[r = w(n_e - n_o)]$. The *optical retardation*, often defined as the retardation per unit length (r/w), is thus equal to the birefringence. The optical retardation is used as a measure of the induced stress, σ, through the elasto-optic or stress-optic coefficient:

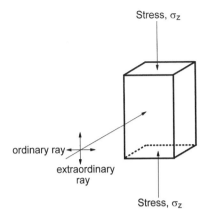

Stress Birefringence in Glass

Figure 4.7: Formation of stress birefringence in a glass block. The ordinary ray has a polarization normal to the direction of application of the stress. The extraordinary ray polarization is parallel to the stress.

$$B\sigma = \frac{\Delta n}{n} = \frac{r}{nw} \qquad (4.23)$$

where B is the stress-optic coefficient in Brewsters and n is the refractive index of the unstrained material. When B is in brewsters, and $\Delta n/n$ is in nm/cm, the stress is in dynes/cm^2. Values for B in glasses vary around $2 - 4 \times 10^{-6}$ Brewsters. Glasses can have either positive or negative birefringence. Most display a positive birefringence $(\Delta n > 0)$ in the direction of an applied compressive stress and a negative change in refractive index in a direction perpendicular to the applied compression.

In crystals, the absorption coefficient of the propagating light often will vary along with the refractive index. Under this condition, the materials will be called *dichroic* and will exhibit a variety of colors, depending upon the direction or orientation of the reflected light. Dichroic jewelry is interesting since it changes color with the angle of light incidence and so seems to sparkle.

A more general expression of the anisotropic propagation of light in materials is found through the Pockels coefficients. In this case, the dielectric constants and the electric susceptibility are expressed as second rank tensors, related to the refractive indices as follows for a Cartesian geometry:

$$n_1 = \sqrt{\epsilon_{11}}; \quad n_2 = \sqrt{\epsilon_{22}}; \quad n_3 = \sqrt{\epsilon_{33}} \qquad (4.24)$$

with $c_i = 1/n_i^2$.

If the c_i are plotted to form an indicatrix ellipsoid in Cartesian space with principal axes m_1, m_2 and m_3, the change in refractive index can be written as:

$$\Delta c_{ij} = z_{ijk}E_k + q_{ijkl}\sigma_{kl}$$
$$\Delta c_{ij} = z_{ijk}E_k + p_{ijkl}e_{kl} \qquad (4.25)$$

where the E_k are the applied electric fields, the σ_{kl} are the applied stress, and e_{kl} are the applied strains. The coefficients z_{ijk} are the electro-optical coefficients; q_{ijkl} are the stress-optic coefficients; p_{ijkl} are the strain-optic coefficients or elasto-optic coefficients. Together the p_{ijkl} and the q_{ijkl} are called Pockels coefficients. There are 81 components to Pockels coefficients, obviously reduced by symmetry. The stress-optic constant $B = -[n^3/2](q_{11} - q_{22})$.

4.3   Photochromic and electrochromic behavior

Under prolonged UV exposure, some glasses will darken. This effect is called *solarization*. One example can be seen in glasses containing Mn and Fe impurities. In such materials, a photochemical reaction occurs that liberates an electron that forms a color center:

$$Mn^{2+} + h\nu \rightarrow Mn^{3+} + e^-$$
$$Fe^{3+} + e^- \rightarrow Fe^{2+} \qquad (4.26)$$

The electron, released by the absorbed photon, is trapped by the iron ion. This leaves behind an Mn^{3+} color center, stabilized by the fact that the electron is trapped to form Fe^{2+}. The trivalent Mn imparts a violet color to the glass.

A reversible solarization or photochromic behavior occurs when Eu^{2+} and Ti^{4+} are present in a highly reduced silicate glass. Under solar exposure (UV light), the following reaction occurs:

$$Eu^{2+} + Ti^{4+} \rightleftharpoons Eu^{3+} + Ti^{3+} \qquad (4.27)$$

As the light is removed, the reaction reverses and the electron is thermally excited away from the Ti ion and returns to the Eu ion. These glasses, however, show fatigue effects and acquire a permanent coloration with time. Alkali-boro-alumino-silicate glasses containing silver are much more popular for eyeglass use (Araujo 1977 and Araujo and Borrelli 1991). These glasses contain ≤ 0.7 wt% AgCl. Other components, such as As, Sb, Sn, Pb, and Cu, increase the sensitivity and photochromic absorbance of the glass. In these glasses, the induced absorption results from the formation of colloidal silver during irradiation. In effect, the glass acts as a reversible photographic plate. The metallic silver appears gray and absorbs light. The absorption bands observed are similar to those produced by elliptical but not spherical colloids of silver randomly distributed on the larger silver halide crystals. Small amounts of copper can increase the photosensitivity, so it is assumed that an electron transfer from the silver colloids to cupric ions is the main mechanism of the reverse process. Here, thermal excitation of electrons from the silver colloids and trapping by the cupric ion removes the color after cessation of the irradiation by inducing the recombination of Ag^+ and Cl^- ions. This process is called *thermal bleaching*. The dominant process is as follows:

$$Cl^- + h\nu \Rightarrow Cl^0 + e^-$$
$$Ag^+ + e^- \Rightarrow Ag^0$$

$$(4.28a)$$

After irradiation is removed:

$$Ag^0 \Rightarrow Ag^+ + e^-$$
$$Cu^2 + e^- \Rightarrow Cu^+$$
$$Cl^0 + Cu^+ \Rightarrow Cl^- + Cu^{2+}$$
$$Ag^+ + Cl^- \Rightarrow AgCl$$

$$(4.28b)$$

Boron trioxide seems to play a major role in controlling the size of colloids produced. The major effect that must be controlled is the balance between thermal bleaching and the solarization of the glass. A rapid thermal bleaching process is desired in spectacles in order to regain transparency after sunlight is removed. However, if the thermal bleaching process is rapid, it is also active during solarization, so it reduces the total amount of color achieved during exposure. These processes are controlled by the size of the silver halide crystals, the size and shape of the precipitated silver colloids, and the diffusion distance between the reacting ions. Optimizing these processes to shorten bleaching time and still obtain sufficient coloration is still the subject of research today.

4.4   Oxides, chalcogenides, and halides

Transparent crystals are an important class of insulating optical materials; however, we have delayed their treatment to later chapters because their main purpose in optical applications are in nonlinear optics. Consequently, we will cover their properties in detail in those chapters. In this chapter, we will examine glasses as optical materials.

Glasses make up an important component of insulating optical materials. The majority of their applications are in linear optics, where they serve an important role. Glasses are desirable for optical elements because of their very high homogeneity, very high transparency, high hardness to allow good polishing and resist scratching, and an excellent optical window over the visible. Their cost is reasonable, but polishing to special shapes can be a labor-intensive process.

Glasses in general are metastable frozen liquids that have a noncrystalline structure. Glasses produced from the melt are cooled through a transformation region in which their fluid flow and diffusional characteristics are slowed down until the structure can no longer rearrange itself to maintain the thermodynamic equilibrium of a liquid during cooling. At that point, the glass-transition temperature, the liquid essentially freezes out of equilibrium and continues to cool as a solid. This behavior is commonly seen in the change in volume with temperature (Fig 4.8). If a glass is cooled rapidly through the transformation region, then the frozen-in structure will form at a higher fictive temperature. So while the glass-transition temperature is defined as the temperature at which the glass has a viscosity of 10^{12} Pa·s, the fictive temperature can vary drastically depending on quenching rate. Often the glass-transition temperature is defined as the fictive temperature corresponding to a quench rate of $1°/s$.

Oxide glasses generally have high bandgap energies and multiphonon absorption peaks in the near-IR. Consequently, they transmit between 200 nm and 2.5 µm. Fig. 3.7 shows their relationship between index and

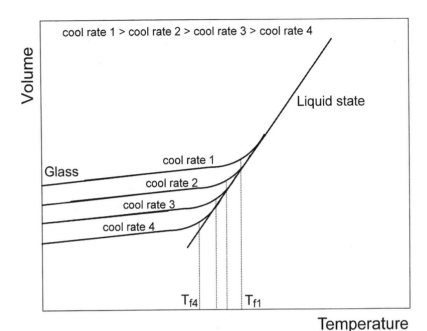

Figure 4.8: Illustration of the variation in the volume and, consequently, the density and refractive index of a glass with cooling rate.

Abbe number. Oxide glasses have indices between 1.4 for the silicates and 2.4 for the titanates.

Much work has been conducted on IR-transmitting glasses for a number of applications. The chalcogenide glasses consist mainly of materials that replace oxygen by heavier group 6 elements (S, Se, Te) and replace silicon by heavier metals (Ge, As, etc.). The chalcogenides have generally low bandgap energies and/or numerous absorption bands in the visible, so they usually appear black. However, they have good transparency windows in the IR between 1 and 12 microns. Because of their heavy constituents, these glasses have indices well above 2.

Halide glasses replace the oxygen anion by fluorine, chlorine, or some of the heavier halides. The more stable halide glasses contain fluorine and heavier metals (heavy-metal fluorides). These materials are characterized by a high bandgap energy and low-frequency multiphonon vibrations. Figure 4.9 gives some representative absorption curves. Consequently, heavy-metal fluorides transmit between 200 nm and

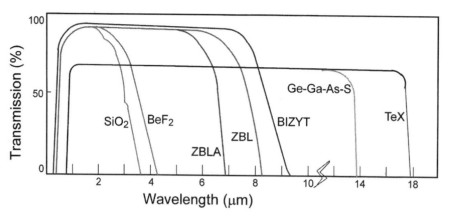

Figure 4.9: Variation in the IR absorption region of different families of non-oxide glasses. The heavy-metal–chalcogenide glasses have a lower transmission in the transparency region because their index is much higher than the halide glasses and they experience a larger reflection during the measurement (after Lucas and Adam 1991).

6–7 μm. Further extension in the IR are possible by using rare-earth fluoride or by replacing some of the fluorine by chlorine. The latter, however, reduces stability.

New glasses recently developed at the University of Rennes extend the transmission window past 10 μm by using compositions based on TeX compositions, where the X stands for halides. This happens with a loss of visible transparency as the band edge moves to 0.7–1.5 μm. Figure 4.9 shows a comparative plot of the transmission of the different classes of glasses. The low transmission in the transparency region of the chalcogenide and heavy-metal glasses results essentially from an enhanced reflection coefficient due to their very high index by comparison to silicate glasses.

4.5   Optical plastics

Under conditions of mass production, plastics are more economical than glass. Compared to glass, plastic optics weigh less and can be made in very complex shapes. Prototypes are expensive, so glass is often used for prototypes. However, new computer designs are taking the guesswork out of plastic lens design, and prototypes are made less often. Plastic optics do have their problems. They can never achieve the optical quality of glass; they have low scratch resistance and lower thermal stability. They are more difficult to AR-coat than glass. They exhibit greater stress birefringence and higher temperature dependence in their refractive index. The latter is generally negative, so plastics can be used to counteract the glasses with positive temperature dependence. Polishing cannot achieve the surface finish of glass, so surface light scattering or haze is often present. Generally plastics cannot achieve the high refractive indices of glasses, but they do achieve lower indices than glass.

In general, plastics do not have as wide a transmission window as glass. Their UV edge is somewhat below 380 nm. If an intrinsic yellow color is present, the UV edge may degrade down to 450 nm. The IR absorption generally cuts the transparency window near 1,000 nm, due to overtones of the C–H vibration. Exposure to UV radiation breaks the $C = C$ double bond and cross-polymerizes the structure. There are transmission windows at 1,300 and 1,500 nm, so plastic fibers have been considered for optical communications; however, transmission in that range is highly susceptible to moisture.

Dyes and pigments can be added to plastic optics to make reasonably sharp filters. Often, suppliers use a "blue" toner in producing clear plastics. This is usually applied as a brightener to overcome the yellowing tendency of a poor manufacturing process. Optical plastics should be obtained without this "blue" toner.

The greatest potential for plastic optics, besides inexpensive or throwaway applications, is in digital and Fresnel optics. Here, flat disks of transparent material are formed as gratings to act as lenses. Glass is very hard to machine into the complex shapes needed; plastic optics are far superior. Fresnel lenses can be used as a thin coating on a lens to correct for chromatic or spherical aberration. Such applications are bringing great promise to plastic optics.

Major uses for plastic optics include lightweight lenses for inexpensive cameras, optical covers on various lighting systems (automotive), space applications, Fresnel lenses, and lenticular arrays.

Transmission curves for several optical plastics are shown in Fig. 4.10. Unlike the chalcogenide glasses, the reduced transmission in the transparent region of plastics is due to surface scattering rather than reflection, since plastics have low refractive indices. Table 4.1 lists some properties of the more commonly used plastics.

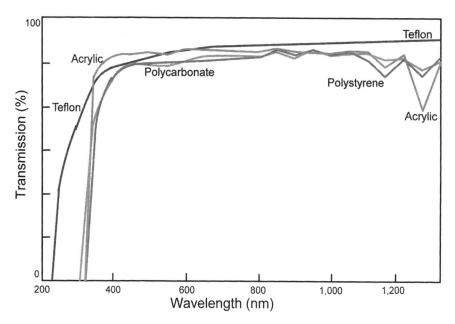

Figure 4.10: Transmission curves for a representative group of polymers. Loss in the IR comes from multiphonon absorption of light carbon-bonded species (C–H) (after Keyes 1995).

4.6   Sources of color

Color in materials comes from many mechanisms, and differences in color may result from subtle changes in the material's structure. This section will review the many processes that impart color to the appearance of materials. These processes have already been covered in detail in the previous chapters, so in this section we will simply elucidate a few of the more interesting issues pertaining to coloration of materials by some fundamental optical processes. The reader who desires a more in-depth coverage of sources of color is referred to the superb book on the subject by K Nassau (1983).

The many sources of color can be grouped into at least 15 categories. These are included in the general topics of: emissive radiation, absorption, reflection, dispersion, scattering, and interference.

Polymer	%T	Index	Abbe No.	Maximum temperature	Characteristics
Allyl diglycol carbonate:				212°F	Castable, thermoset,
CR 39 (PPG)					abrasion-resistant (good for spectacle lenses)
ADC				302°F	
Methyl methacrylate:	92%	1.491	57	200°F	Moldable, impact-resistant, highly stable
Acrylic					
Lucite					
Plexiglass					
Polystyrene:	88%	1.590	31	180°F	Moldable, less UV resistant, water absorbtive
Lustrex, Dylene					
Styron					
Styrene acrylonitrile:	88%	1.571	35	180°F	Yellowish, low expansion coefficient
Lustran, Tyril, SAN					
Copolymer (30/70) styrene/acrylic:	90%	1.562	35	190°F	Similar to SAN but clear
NAS					
Polycarbonate:	89%	1.586	35	255°F	Highly temperature stable, strong, easily scratched,
Lexan, Merlon					injection moldable
Polysulfone		1.633			Difficult to mold, has high shrinkage
Methylpentene polymer		1.466	56		Close to acrylic, more resistant to temperature, tough, hard to mold
Terpolymers of acrylonitrile, butadiene, & styrene, ABS	80%				Very tough
BK-7 glass	>99%	1.517	64	>1,000°F	Excellent as optical material

Table 4.1: Properties of common plastics.

4.6.1   Emission

Emission of light is achieved via a variety of processes. The most fundamental is black-body radiation, which results simply from the thermal agitation of the component atoms of a material. As discussed in Chapter 1, emissive black-body radiation is the source of solar light and light from incandescent bulbs and from fires and hot objects that glow from red to white. Modifications occur when the thermal energy or electric currents are high enough to cause electronic excitations. This is the source of light in sodium and mercury vapor lights, and fluorescent light bulbs.

Fluorescence and photoluminescence derive from another source of colored light: the electronic excitation of atoms and molecules followed by a radiative de-excitation process to lower energy. Fluorescence is usually associated with UV excitation, luminescence is associated with excitation by visible light. These processes occur by means of the electronic excitation of atoms followed by radiative de-excitation between lower energy levels. Colors made vibrant by black light come from fluorescence, such as in special chalks. Dye lasers operate by photoluminescence. The emission from solid-state lasers is stimulated by flash lamps or semiconductor lasers. The colors of the light are determined by the energy difference between the electronic states. The same emission processes can be stimulated by electronic currents through a gas, and this provides the excitation for gas lasers. Carbon dioxide laser emission is determined by the difference in energy between pairs of phonon vibration levels of the CO_2 molecules; that for He–Ne and Ar lasers results from differences in electronic energy levels.

Another source of emissive colors is the recombination of electrons and holes, which is essentially an electronic transition between an excited state and a ground state. Semiconductor lasers owe their emission to this process. All emissive colors are additive, and because of the eye's sensitivity to only red, blue, and green light, all other colors are perceived as the addition of these primary colors (red, green, and blue, RGB).

4.6.2   Absorption

Absorption is a subtractive process in which the color perceived results from the removal of certain frequencies from mixtures of other frequencies or from white light, which is the sum of all colors

$(W = R + G + B)$. Consequently, the primary subtractive colors are made from the removal of the primary additive colors from white light. Thus, magenta is $W - G$, cyan is $W - R$, and yellow is $W - B$. Absorption occurs via a variety of processes. The most common is the absorption of light by electronic excitations between two energy levels. This is the opposite of the emissive process. Thus, transition metals and rare earths have a variety of colors (see Chapter 3) that result from absorption due to electronic excitations of their $3d$ or $4f$ electrons. These states can change with the oxidation level of the atoms. The brown color of beer bottles comes from oxidized iron, while the green color results from the reduced iron. Table 3.5 lists various transition-metal colors. In another instance, the red color of ruby results from absorption of Cr^{+3} ions in corundum (Al_2O_3). The green color of emerald results from absorption of the same Cr^{+3} ion, except that the host is beryl $(3BeO \cdot Al_2O_3 \cdot 6SiO_2)$. In both hosts, the Cr^{+3} ion (with three $3d$ electrons) is octahedrally coordinated by oxygen ions. In ruby, the next-nearest neighbors are Al^{+3} ions. The Cr^{+3} absorption bands are in the violet and the green, thus giving a strong red color with a weak blue shade. In beryl, the next nearest neighbors are Si^{+4} ions, which bond more tightly to the oxygen ions. Consequently, the cage around Cr^{+3} ions is slightly larger, and the absorption bands shift down to the blue and red frequencies, giving the bright green color of emerald.

Some absorption processes result from a continuum set of electronic transitions above a given energy level. These materials act as low-pass filters. Semiconductors achieve their color via this process, for the bandgap energy determines the absorption edge. Very low bandgap energies produce black semiconductors, since all visible frequencies are absorbed. As the bandgap energy increases, the absorption edge moves up in frequency and the semiconductors acquire a red, then an orange, and then a yellow color. Eventually, the semiconductors pass all visible frequencies without absorption, and they appear colorless. Oxide glasses are transparent for the same reason.

Another mechanism of absorptive color is associated with color center defects in crystals. These result from cation or anion vacancies, interstitials, substitutional, or Frenkel defects (paired vacancies). Amethyst and fumed quartz acquire their color from the formation of color centers in SiO_2. In amethyst, the substitution of Al^{+3} for Si^{+4} traps a hole at the substitutional site. The resulting hole center gives amethyst its purple color. Heating the stone will allow the trapped charge to relax and will remove the color.

Another source of absorptive coloration comes from charge transfer. The transfer of electrons between Ti^{+4} and Fe^{+2} to form Ti^{+3} and Fe^{+3} in sapphire induces light absorption due to the difference in energy between the two states. The absorption process that corresponds to a transition of $2.2\,eV$ is broad and covers from green to red wavelengths, giving a strong transmission in the blue.

4.6.3   Reflection

Reflective coloring is found in metals, as described in Chapter 2. Here, the metal reflects low frequencies and absorbs frequencies higher than its plasma resonance. Copper is red because its plasma resonance is near $2.0\,eV$ (just above the red color, $620\,nm$). Brass and gold are yellow because they have a plasma resonance at higher energy. Finally, silver reflects across the visible because its plasma frequency is in the near UV ($4.0\,eV$).

4.6.4   Dispersion

Dispersive coloration is found in prisms and in water droplets. Here, the difference in refractive index between different frequencies, or colors, leads to different reflection angles. Dispersive separation of light in glass forms color bands from prisms. In water, the process leads to the formation of rainbows even though the difference in refractive index between the blue and red colors is only between 1.3435 ($404\,nm$) and 1.3307 ($710\,nm$), or only 0.0128.

The green flash is a change in color of the sun as it sets. Because of dispersion of the air, there are several images of the sun that are displaced from each other at sunset or sunrise, as would happen with a prism. The blue image is not present due to Rayleigh scattering. The green image is weak and sits above the others. The images then go down in order of yellow, orange, red. Thus the bottom of the sun is red while the top is green. The size of the sun is 30 arc minutes, and the amount of refraction from dispersion is about an arc minute in the vertical direction (e.g., 1/30 the diameter of the sun). The green image of the sun is visible when only the top edge of the rim is visible. The duration of the green image is dependent upon the rate of descent of the sun, so it is very short at the equator and in spring and fall. Consequently, the green flash is more visible at higher latitudes and in the summer. It is best observed over water. And, since the image is weak due to atmospheric Rayleigh scattering (below), the absence of moisture helps, since water absorbs weakly in the green. Lynch and Livingston (1995) report that members of the Byrd Antarctic Expedition to Little America saw the "green flash" for a period of 35 minutes (Haines 1931).

4.6.5 Scattering

Rayleigh scattering is the source of blue sky, because the strength of the scattering coefficient is inversely proportional to the fourth power of the wavelength. The ratio of the Rayleigh scattering power for a visible wavelength to the scattering power of the red wavelength changes as detailed in Table 4.2.

The table shows that the violet wavelength has scattering 12 times greater than the red wavelength. The blue wavelength has 6 times. The result is that the strength of the violet and blue scatterings make the sky blue in color. Our eyes are not as sensitive to violet, and the solar intensity for blue is much higher than violet, so blue dominates. As the optical path length of light in the atmosphere increases, and as the light intensity decreases, Rayleigh scattering is more effective at removing the progressively longer wavelengths from the transmitted light. Thus at zenith, solar light loses only its blue component, but during sunset or sunrise the optical path of sunlight is much longer through the atmosphere and components of longer wavelength are removed. So if the sun is low on the horizon, the rays that graze the Earth have the longest path through the atmosphere and are essentially red, having lost all other wavelengths to Rayleigh scattering. Higher in the sky, the rays traverse a shorter path through the atmosphere and the color of solar light shifts to orange and then yellow. This is the reason why clouds high in the sky reflect yellow while lower clouds will reflect more orange and reddish colors. As the sun sets, the intensity of the light decreases and

Color	violet	blue	green	yellow	orange	red
Wavelength	400 nm	475 nm	535 nm	590 nm	610 nm	750 nm
Relative Rayleigh scattering	12.4	6.20	3.90	2.60	2.30	1

Table 4.2: Relative Rayleigh scattering for different colors.

more of the rays become red. So what makes lavender clouds? These occur at low light levels after the sun has set. Here, weak red light from the sun mixes with blue scattered light from the stronger illumination higher in the atmosphere to produce the lavender color on the clouds.

Mie scattering from larger particles (greater than the wavelength of light) causes essentially wavelength-independent scattering. Mie scattering from micron-sized air bubbles produces grey and white hair color. Mie scattering from bubbles in ice is the source of its blue or white color. The same kind of scattering from grain boundaries in polycrystalline materials forms their white color, or from water droplets forms clouds.

4.6.6   Interference colors

Multiple layers with different or alternating refractive index will cause patterns of interference across the visible spectrum if the layer thicknesses are less than a few multiples of the optical wavelengths. For a single layer, interference patterns are formed by adding the light reflected from the air–layer interface to the light reflected from the layer–bulk interface. A calculation of the path difference yields:

$$\textit{Constructive interference}: \quad m\lambda \cos\theta = 2d(n_2 - n_1^2/n_2)$$

$$\textit{Destructive interference}: \quad (m + 1/2)\lambda \cos\theta = 2d(n_2 - n_1^2/n_2)$$

(4.29)

where the index of the medium is n_1, the index of the film is n_2, m is any integer, d is the film thickness, and $\cos\theta$ is related to the angle of incidence and reflection of the light, ϕ, as: $\cos\theta = \sqrt{1 - (n_1/n_2)^2 \sin^2\phi}$.

Interference from a thin layer of oil produces the colors in oils slicks or the colors in soap bubbles from a thin layer of soap. Interference processes are a major source of biological coloration. Many biological colors are formed by interference due to regularly spaced reflective surfaces. The hair on butterfly wings forms biological gratings that produce color by interference of light to reinforce various wavelengths either constructively or destructively. The *Serica-sericae* beetle and the indigo or gopher snake are other well-known examples of iridescent interference coloration. Since interference is strongly dependent on the presence of directional light, this coloration method has distinct advantages for biological systems, because the amount of color diminishes rapidly as direct sunlight decreases. Consequently, objects that are brightly colored by interference mechanisms during the day become colorless in low light conditions (possibly a survival advantage).

Interference is also responsible for the colors on ancient glass. Here, multiple layers of different index are formed by wet–dry cycling of the glass surface, which is etched by water during the wet periods to form a dehydroxylated surface layer. This layer then cracks away from the glass during the dry periods. The fresh glass is etched again as the cycle repeats. The difference in refractive index between the dehydroxylated layers and the air gap between them causes the periodic variations necessary for interference coloring.

Figure 4.11: Rows of precipitated silica colloids in the structure of opals. These produce strong scattering and interference colors.

The color of opal gemstones arises from interference produced by precipitated amorphous silica colloids arranged in regular layers in the structure of the material. The variation in color arises because the order occurs in limited regions and the orientation of the layers varies from region to region. A micrograph of the structure of an opal is shown in Fig. 4.11.

Appendix 4A

Alternate Calculation of Multiple Film Stacks

In the calculation of the optics of multiple thin films, an alternate matrix approach may be used. It is similar to that presented in the body of the text in that it calculates the electric and magnetic fields at each interface using the boundary condition that requires continuity of the parallel components of the electric field and magnetic displacement vectors at each boundary. The final results are the same, with a difference only in the expressions used.

The electric fields and magnetic displacement vectors can be expressed as follows:

$$
\begin{bmatrix} E_i \\ H_i \end{bmatrix} = \begin{bmatrix} \cos k_0 h_i & \frac{i}{\gamma_i} \sin k_0 h_i \\ i\gamma_i \sin k_0 h_i & \cos k_0 h_i \end{bmatrix} \begin{bmatrix} E_{i+1} \\ H_{i+1} \end{bmatrix} = M_i \begin{bmatrix} E_{i+1} \\ H_{i+1} \end{bmatrix} \qquad (4.30)
$$

where the subscript i denotes the ith film. The subscript O denotes the incident medium (normally air), and the subscript s denotes the last medium or substrate.

$$
k_0 h_i = \frac{2\pi}{\lambda_0}[(n_i + i\kappa_i)d_i \cos \theta_i]
$$

$$\gamma_{s,i} = \sqrt{\frac{\epsilon_0}{\mu_0}} n_i \cos\theta_i \qquad\qquad \gamma_{p,i} = \sqrt{\frac{\epsilon_0}{\mu_0}} \frac{n_i}{\cos\theta_i}$$

$$\gamma_{s,0} = \sqrt{\frac{\epsilon_0}{\mu_0}} n_0 \cos\theta_0 \qquad\qquad \gamma_{p,0} = \sqrt{\frac{\epsilon_0}{\mu_0}} \frac{n_i}{\cos\theta_i} \qquad (4.31)$$

$$\gamma_{s,s} = \sqrt{\frac{\epsilon_0}{\mu_0}} n_s \cos\theta_s \qquad\qquad \gamma_{p,s} = \sqrt{\frac{\epsilon_0}{\mu_0}} \frac{n_s}{\cos\theta_s}$$

Snell's law : $n_0 \sin\theta_0 = n_1 \sin\theta_1 = n_2 \sin\theta_2 = \cdots = n_i \sin\theta_i = \cdots = n_s \sin\theta_s$

Denoting the layers by Roman numerals, the matrices for successive interfaces are written as

$$\begin{bmatrix} E_{\mathrm{I}} \\ H_{\mathrm{I}} \end{bmatrix} = M_{\mathrm{I}} M_{\mathrm{II}} M_{\mathrm{III}} \cdots M_p \begin{bmatrix} E_{p+1} \\ H_{p+1} \end{bmatrix} \qquad (4.32)$$

with the product matrix written as:

$$M = \prod M_i = M_{\mathrm{I}} M_{\mathrm{II}} M_{\mathrm{III}} \cdots M_p = \begin{bmatrix} m_{11} & m_{12} \\ m_{21} & m_{22} \end{bmatrix} \qquad (4.33)$$

The reflection and transmission coefficients are written as:

$$r = \frac{\gamma_0 m_{11} + \gamma_0 \gamma_s m_{12} - m_{21} - \gamma_s m_{22}}{\gamma_0 m_{11} + \gamma_0 \gamma_s m_{12} + m_{21} + \gamma_s m_{22}}$$

$$t = \frac{2\gamma_0}{\gamma_0 m_{11} + \gamma_0 \gamma_s m_{12} + m_{21} + \gamma_s m_{22}} \qquad (4.34)$$

$$R = r^* r$$

$$T = \left(\frac{n_s \cos\theta_s}{n_0 \cos\theta_0}\right) t^* t$$

This approach allows calculation of the reflection and transmission coefficients for any number of layers of known index, thickness, and loss coefficient.

At normal incidence for a single film these equations reduce to

$$R_1 = \frac{n_1^2 (n_0 - n_s)^2 \cos^2(k_0 h_1) + (n_0 n_s - n_1^2)^2 \sin^2(k_0 h_1)}{n_1^2 (n_0 + n_s)^2 \cos^2(k_0 h_1) + (n_0 n_s + n_1^2)^2 \sin^2(k_0 h_1)} \qquad (4.35)$$

As with Eq. (4.17), if the optical thickness of the single layer is designed to be any odd multiple of one one quarter of the wavelength of the incident light $(d = \lambda_0/4n_1)$, the condition derived for an antireflective coating is the same:

$$R_1 = \frac{(n_0 n_s - n_1^2)^2}{(n_0 n_s + n_1^2)^2} \tag{4.36}$$

which gives a vanishing reflection when $n_1^2 = n_0 n_s$.

References

Araujo, R. J. 1977. Photochromic Glass. In *Treatise on Materials Science and Technology—Glass 1: Interaction with Electromagnetic Radiation*, edited by M. Tomozawa and R. H. Doremus. New York: Academic Press.

Araujo, R. J., and N. F. Borrelli. 1991. Photochromic glasses. In *Optical Properties of Glass*, edited by D. R. Uhlmann and N. J. Kreidl. Westerville, OH: American Ceramic Society.

Azzam, R. M. A., and N. M. Bashara. 1989. *Ellipsometry and Polarized Light*. Amsterdam: North Holland.

Haines, W. C. 1931. The green flash observed October 16, 1929, at Little America by members of the Byrd Antarctic Expedition. *Mon. Weather Reviews* 59:117.

Keyes, D. 1995. Optical Plastics. In *Handbook of Laser Science and Technology*. Suppl. 2, *Optical Materials*, edited by M. J. Weber. Boca Raton, FL: CRC Press.

Lucas, J., and J. L. Adam. 1991. Optical properties of halide glasses. In *Optical Properties of Glass*, edited by D. R. Uhlmann and N. J. Kreidl. Westerville, OH: Americam Ceramic Society.

Lynch, D. K., and W. Livingston. 1995. *Color and Light in Nature*. Cambridge: Cambridge University Press.

Macleod, H. A. 1989. *Thin Film Optical Filters*. 2nd Ed. New York: McGraw-Hill.

Nassau, K. 1983. *The Physics of Color*. New York: Wiley.

Simmons-Potter, K., and J. H. Simmons. 1996. Modeling of absorption data complicated by Fabry–Perot interference in germanosilicate thin film waveguides. *J. Opt. Soc. Amer.* B13:268–72.

Chapter 5

Optical Properties of Semiconductors

5.1 Introduction

Semiconductors are characterized by electron mobilities and free carrier densities that are greater than those of insulators but less than those of both good conductors (metals) and semimetals. Typically, semiconductor free charge carrier densities lie between 10^{13} and 10^{21} carriers/cm^3. Unlike metals, semiconductors have an electronic energy gap (0.1–5.0 eV). This energy gap is a nonzero difference in energy between the highest occupied states and the lowest unfilled states in the electronic band structure. This means that, in principle, undoped semiconductors have no free electrons at absolute zero. All electrons are in valence states. As with insulators, at absolute zero the bandgap energy value defines the energy required to raise valence electrons to free-electron states.

There are two types of charge carriers that can be present in semiconducting materials: intrinsic and extrinsic. These types are differentiated by the source of the free charges that dominate transport in the semiconductor and indirectly by the band structure of the material. Intrinsic semiconductors have a bandgap energy sufficiently low that the thermal energy of room temperature ($k_B T = 26$ meV) can excite valence band electrons into the conduction band. There they become "intrinsic" charge carriers that can diffuse freely in the material. Intrinsic charge carriers are also those that come from the principal components of the material in pure crystalline form. In contrast, extrinsic charge carriers come from dopants, impurities, or structural defects in the semiconductor. Extrinsic charges come from defect levels that may be located either in the bandgap or in one of the bands (conduction or valence) of the material. Since extrinsic carriers arise from local charge imbalances, they do not require thermal activation and so are the only means to provide free charge carriers in semiconductors with larger bandgaps. The same material may be produced in an intrinsic (pure) form or an extrinsic (doped) form.

When designing a semiconductor for a given optoelectronic device (i.e., a Si logic chip in a computer), it is common to consider whether or not its operation is one that requires large free charge carrier densities to produce efficient conduction through the device. If the answer is yes, dopant impurities must be introduced into the crystal to produce "extrinsic" charge carriers. Clearly, since intrinsic charge carriers are thermally excited, their concentration is strongly temperature dependent. The concentration of extrinsic charge carriers has a weaker temperature dependence. Extrinsic charge is added by having either excess electrons or holes; thus, different dopants are used to increase the concentration of specific defects. If the dopants are donors, they have excess electrons and increase the electron concentration. These make the semiconductor *n type*. If the dopants are acceptors, they have too few electrons; thus they provide excess positive charge, or holes. These make the semiconductor *p type*.

Both intrinsic and extrinsic charge carriers in semiconductors can be thought to exist in "free" states. As free carriers, they travel through the material with essentially little interaction. However, it is the nature of the "essentially little interaction" that influences the behavior of semiconductors. Clearly, the charge carriers are not really free, since they must interact with the ionic cores forming the matrix. However, within the confines of the material, they act as nearly free carriers (Fig. 5.1), and any interaction is taken as the source of perturbations in their behavior. Consequently one turns to a "nearly free" electron model to describe the basics of semiconductor behavior. The fundamental principle here is that the free-electron behavior is dominated by wave-like characteristics. So understanding the behavior of electrons in a semiconductor requires an examination of both real and wave-vector spaces. To understand the behavior of electrons in real space we consider a reasonably homogeneous material at large sizes. Generally, one need only consider the graininess of the matrix when sizes on the order of the

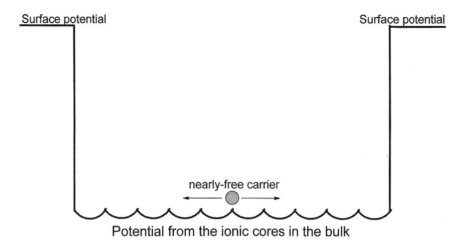

Figure 5.1: Illustration of nearly free charge carrier in a semiconductor. Away from the surface of the material, it is affected by the periodic potential of the ion cores.

unit cell are important. In reciprocal space the opposite is true: Long-wavelength or small wave-vector behavior can be treated as a continuum in bulk materials, and the crystal structure of the matrix comes into play only at short wavelengths or large wave-vector values. Recent studies of quantum confinement behavior have examined semiconductor samples whose size is of the order of several unit cells of the crystal. At these small sizes, both real and reciprocal spaces are affected, even at small k-vector values. The next sections examine the free-electron and nearly free-electron models of semiconductor behavior.

5.2 Free-electron gas (Sommerfeld theory)

The free-electron approach solves the Schroedinger wave equation with no potential energy term (the electrons are totally free). This is clearly useful only in the continuum approximation and must fail at large wave vectors (k) and short wavelengths, when structural, periodic graininess becomes important. The nearly free-electron model (section 5.3) applies modifications through the Bloch theorem, which requires that the wave function be periodic in the same way as the ion cores.

The Schroedinger wave equation without potential energy simplifies to

$$-\frac{\hbar^2}{2m}\nabla^2\psi = \frac{\hbar}{i}\frac{\partial\psi}{\partial t} = \xi\psi \tag{5.1}$$

where $\xi = \hbar\omega$ is the free-electron energy, a constant independent of time and position. The wave function solution, ψ, without boundary conditions is

$$\psi_k = \frac{1}{\sqrt{V}}e^{i\vec{k}\cdot\vec{r}-i\omega t} \tag{5.2}$$

The free-electron energy is calculated from the solution of the differential equation with the following result:

$$\xi(k) = \frac{p^2}{2m} = \frac{\hbar^2 k^2}{2m} \tag{5.3}$$

The parabolic dependence of the free-electron energy, ξ, on momentum or k-vector is shown in Fig. 5.2. The addition of boundary conditions forms the 'Nearly-Free Electron Model'.

5.3   Nearly free-electron model

The Schroedinger wave equation for electrons in a crystal involves terms for the kinetic energy of the electrons, terms for the kinetic energy of the ionic cores, and terms for the potential energies of the core–core Coulomb interaction, core–electron Coulomb attraction, and electron–electron Coulomb repulsion. These terms must be summed over all ionic cores and all electrons. It is possible to simplify the problem first with the Born–Oppenheimer approximation. Here, we note that the ionic cores are much more massive than the electrons, so the electronic motion is much faster than that of the ions. Thus, the electrons are moving in an essentially static field of ions. The ions see only the time-averaged motion of electrons (adiabatic approximation). The Schroedinger wave equation is then separable into three terms: the ion interaction, the electron interaction, and the ion–electron interaction. The ion interaction is important in calculating materials structure and makes up an extensive field of quantum chemistry, beyond the scope of this book. We will ignore this part and assume that when the materials are crystalline they are perfect crystals, and deviations from perfections are discussed in terms of defects.

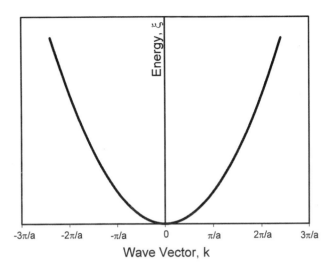

Figure 5.2: Parabolic dependence of the energy on wave vector for a free carrier wave function.

The electron–electron interaction is generally handled as a perturbation of the electron–ion interaction. So we are left with considering the ion–electron interaction, which is divided into two parts: the *ion–electron Coulomb interaction* with frozen ions, which we analyze shortly, and the interaction between electrons and moving ions (thermal vibrations of the lattice), which is treated through *electron–phonon interactions*.

5.3.1   Bloch theory

In treating the *ion–electron Coulomb interaction*, the effect of the periodic lattice on the nearly free electron is treated by considering a periodic potential function in the Schroedinger equation:

$$-\frac{\hbar^2}{2m}\nabla^2\psi + \sum_{i,j} U_{i,j}\psi = \xi\psi \qquad (5.4)$$

where the potential energy is summed over all ions and electrons. For the nearly free-electron approximation, we replace the sums with a periodic function: $U(\mathbf{r}) = U(\mathbf{r}+\mathbf{R}_q) = U(\mathbf{r}+n\mathbf{R}_q)$, with $\mathbf{R}_q$ representing the lattice vectors in the direction of the periodicity.

The Bloch theorem suggests a solution that requires the wave to exhibit the periodicity of the lattice through the following solution to the spatial part of the Schroedinger equation:

$$\psi_{nk}(r) = e^{ik \cdot r} U_{nk}(r) \qquad (5.5a)$$

where $U_{nk}(r + R) = U_{nk}(r)$. Thus,

$$\psi_{nk}(r + R) = e^{ik \cdot (R+r)} U_{nk}(r) = e^{ik \cdot R} \psi_{nk}(r) \qquad (5.5b)$$

The wave function component, $U_{nk}(r)$ reflects the periodicity of the lattice with the spatial period R. The condition expressed in Eq. (5.4) is true for all integers (n_i) that are multiples of the unit cell dimension (a) as follows: $R = n_i a$. One can calculate the allowed values of k as follows (with n = integer), by taking the highest periodicity with $R = a$:

$$\psi_{nk}(r + a) = e^{ika} \psi_{nk}(r) = \psi_{nk}(r)$$
$$e^{ika} = 1 \qquad (5.6a)$$

Thus,

$$k_n = \frac{2\pi n}{a} \qquad (5.6b)$$

A sketch of the effect of a periodic line of atoms on a series of 1D wave functions is shown in Fig. 5.3. The total solution sums over the various k values and multiplies with the temporal term to produce the following:

$$\Psi_n(r,t) = \sum_k g(k)\psi_{nk}(r)e^{-\frac{i}{\hbar}\xi_n(k)t} \qquad (5.7)$$

where the $g(k)$ are the weighing functions for the various k components.

The condition just set up for the carrier wave function reflects the periodicity of the lattice of ionic cores. This means that when the k-vector is small (e.g., the wave functions have long wavelengths), the periodicity spans many unit cells of the lattice. As k grows, the wavelength shortens and spans fewer lattice sites. At some point the graininess (atomic nature) of the lattice begins to influence the dispersion relation (energy vs wave vector: ω vs k) behavior of the material. At the limit, it is clear that when the wavelength of the carrier wave function is twice the atomic separation, a, in the lattice, this forms the minimum wavelength that the wave may sustain, since there are no atoms in between to define a higher frequency or shorter wavelength. This limit occurs when the k-vector equals π/a. This point is defined as the edge of the Brillouin zone of the crystal *for the particular lattice direction in which a is the lattice*

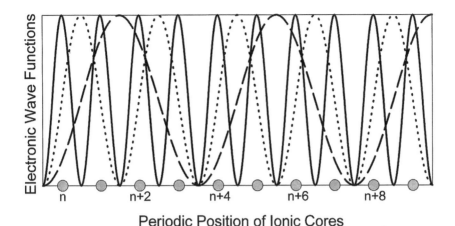

Periodic Position of Ionic Cores

Figure 5.3: Allowed wave functions that satisfy the Bloch condition at high frequencies. The solid line represents the highest wave vector allowed at a value of π/a. The dotted and dashed lines correspond to smaller k values (e.g., longer wavelengths).

spacing. It is clear that this atomic separation is different for different crystallographic directions. The Brillouin zone changes with crystalline orientation. Thus, it reflects the structure and symmetry of the reciprocal of the crystal lattice. As the k-vector is increased beyond the zone boundary, the wavelength of the wave function cannot get any shorter, so the energy dispersion folds upon itself. Thus, only the first Brillouin zone $(-\pi/a \leq k \leq \pi/a)$ is necessary to describe the dispersion behavior of the material. If one looks at the momentum of the particle associated with the wave vector, it is clear that the momentum reverses at the zone boundary; thus the dispersion curve approaches the boundary with a flat slope. This is often referred to as Bragg reflection of the quantum mechanical wave. Higher energy is possible with another branch of the dispersion curve, and since it must have close-to-parabolic behavior far from the zone boundary and exhibit a flat slope at the boundary, the result is a gap of forbidden energies, as shown in Fig. 5.4. Essentially, the maximum energy for the lower branch is ξ_A. The minimum energy for the upper branch is ξ_B. This leaves an energy gap of $\Delta\xi = \xi_B - \xi_A$.

The region between $-\pi/a$ and $+\pi/a$ is called the first Brillouin zone. It contains the valence (occupied-electron) bands and the free-electron or conduction bands. Additional bands appear at higher energies, corre-

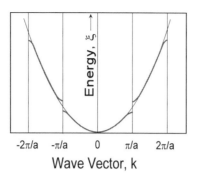

Wave Vector, k

Figure 5.4: Effect of the Bloch condition on the energy dispersion. Because the shortest wavelength allowed is 2a, there must be a break in the energy dispersion at that value as shown in the diagram. The region between $-\pi/a$ and $+\pi/a$ is called the 1st Brillouin zone. The higher k-values only repeat the same wavelengths in the 1st Brillouin zone.

sponding to electronic states from the second, third, fourth, etc. Brillouin zones.

At small values of k, the effect of the underlying ionic core crystal structure is negligible, and the energy still follows a parabolic dependence on k. At large values, the effect of a grainy matrix becomes important and one must consider the effect of the Brillouin zone boundary.

An important readjustment is necessary in the definition of both velocity and mass in order to treat the nearly free electron within the confines of energy and momentum conservation. The electron's velocity is written as the group velocity of the wave:

$$v_g = \frac{d\omega}{dk} = \frac{1}{\hbar}\frac{\partial\xi}{\partial k} = \frac{1}{\hbar}\nabla_k \xi \qquad (5.8)$$

An effective mass is defined for the free electron and the free hole by the rate of curvature of the band-dispersion (ξ vs k) curve:

$$m_{\mathrm{eff}} = \hbar^2 \left[\frac{d^2\xi}{dk^2}\right]^{-1} \qquad (5.9)$$

For electrons in a band, there can be regions of high or low curvature (different from the parabolic curvature). This reduces or enhances the effective mass. The foregoing definition allows us to apply the same dispersion relations to the Bloch waves as are used for the free electrons. Consequently, they maintain the same energy dispersion equation for the entire band, whether near or far from the Brillouin zones. Thus, we may write in general:

$$\xi = \frac{\hbar^2 k^2}{2m_{\text{eff}}} \tag{5.10}$$

The crystal does not weigh less if $m_{\text{eff}} < m_{\text{electron}}$, but Newton's second law is obeyed throughout the crystal by use of this definition. Clearly, to a first approximation (parabolic approximation), the effective mass is approximately the electronic mass at $k = 0$, but it goes to infinity as k approaches $\pm$ π/a at the Brillouin zone (BZ) boundaries. In the region near $k = 0$, the dispersion is parabolic; thus the momentum is represented by the Bloch wave vector ($p = \hbar k$). Nearer the BZ boundary, there is an increase in the reflected component of the wave function due to Bragg reflection from the periodic lattice. This corresponds to a negative momentum transfer to the electron from the lattice, and the effective mass grows. At the BZ boundary, the reflected wave equals the incident component and the wave function becomes a standing wave. The velocity of the electron goes to zero as the effective mass goes to infinity. Effectively, what happens is that as the electron approaches the zone boundary its exchange of energy with the lattice increases, since the periodicity of its wave function begins to match the lattice spacing. When a perfect match is achieved ($k = \pm\pi/a$), a standing wave is formed by which all the electron's momentum is transferred to the lattice. A negative effective mass corresponds to the case where the lattice is transferring net energy to the electron and the force acting on it is in the opposite direction to its momentum (e.g., a negative effective mass means that in going from k to $k + \Delta k$, the momentum transfer to the lattice from the electron is larger than the momentum transfer from the applied force to the electron).

5.3.2   Density of states

A calculation of the density of states available in reciprocal space that still reflects the periodicity of the Bragg lattice can be obtained by using some sort of boundary condition. A simple way to produce a periodic wave function is to choose an arbitrary cube to represent a unit volume and to apply periodic boundary conditions on the surfaces of this cube. These surfaces are not necessarily the outer surfaces of the sample, so the boundary conditions do not require reflection of the wave functions at the cube faces but rather that the wave passes into the adjacent cube with the same value and phase with which it entered the opposite side of the cube. These are called *periodic* boundary conditions. They force the wave to have periodicity without reflection. They are written as follows:

$$\psi(x + L_x, y, z) = \psi(x, y, z)$$
$$\psi(x, y + L_y, z) = \psi(x, y, z) \qquad (5.11)$$
$$\psi(x, y, z + L_z) = \psi(x, y, z)$$

This modifies the wave functions by forcing the wave-vector components, k_x, k_y, and k_z, to be quantized as follows:

$$k_x = \frac{2n_x \pi}{L_x} \qquad n_x = 0, \pm 1, \pm 2, \pm 3, \ldots$$

$$k_y = \frac{2n_y \pi}{L_y} \qquad n_y = 0, \pm 1, \pm 2, \pm 3, \ldots \qquad (5.12)$$

$$k_z = \frac{2n_z \pi}{L_z} \qquad n_z = 0, \pm 1, \pm 2, \pm 3, \ldots$$

Since the wave functions are now periodic with quantized k-vectors, it is instructive to examine the effect of this quantization on the density of states of the semiconductor. The quantization in k-vector limits the number of energy states that are available to the nearly free electrons. Since the electrons are fermions (fractional spin particles), and the Pauli exclusion principle requires that only one fermion with each of two spins occupy each available state, only two electrons can occupy each k-space state. Thus the energy levels occupied must increase in energy for each pair of electrons added. From quantization, a region of k-space will contain a fixed number of allowed k-values (k-space density of states). Therefore, it is possible to calculate the maximum occupied region of k-space for a given number of nearly free electrons. Each change in integer in Eq. (5.12) occupies a fixed volume of k-space calculated to be:

$$\Delta V_k = \left(\frac{2\pi}{L_x}\right)\left(\frac{2\pi}{L_y}\right)\left(\frac{2\pi}{L_z}\right) \qquad (5.13)$$

The reciprocal (k-space) volume for a 2-spin state is written as n_k:

$$n_k = \frac{2}{\Delta V_k / V} = \frac{2}{(2\pi)^3} \qquad (5.14)$$

For a given value of k, there are $n(k)$ possible states:

$$n(k) = \frac{2(4/3\pi k^3)}{(2\pi)^3} = \frac{k^3}{3\pi^2} \tag{5.15}$$

If there must be N_F orbitals to account for all the electrons in a material of volume V, the number n of available states per unit volume may be written in terms of the maximum occupied k-value at absolute zero, denoted by k_F, as follows:

$$n = \frac{N_F}{n_k} = \frac{(4/3)\pi k_F^3}{n_k} = \frac{2(4/3)\pi k_F^3}{(2\pi)^3} = \frac{k_F^3}{3\pi^2} \tag{5.16}$$

and

$$k_F = (3\pi^2 n)^{1/3} \tag{5.17}$$

where k_F is the Fermi wave vector that represents the radius of a sphere containing the occupied one-electron levels. The maximum (Fermi) energy, ξ_F, reached by the uppermost filled state is written as follows:

$$\xi_F = \frac{\hbar^2 k_F^2}{2m} = \frac{\hbar^2}{2m}(3\pi^2 n)^{2/3} \tag{5.18}$$

The density of orbitals $N(\xi)$ per unit energy range $d\xi$ can be calculated as follows:

$$N(\xi)d\xi = \frac{dn}{d\xi}d\xi = \frac{dn}{dk}\frac{dk}{d\xi}d\xi = \frac{(2m)^{3/2}}{2\pi^2\hbar^3}\xi^{1/2}d\xi \tag{5.19}$$

This gives the density of states in the nearly free-electron model. Equation (5.19) holds for the valence band, but the conduction band must be corrected by the bandgap offset if the zero energy point is taken to be the top of the valence band. Under this condition, the valence and conduction band densities of states become:

$$N(\xi)_v d\xi = \frac{(2m_h)^{3/2}}{2\pi^2\hbar^3}(-\xi)^{1/2}d\xi$$

$$N(\xi)_c d\xi = \frac{(2m_e)^{3/2}}{2\pi^2\hbar^3}(\xi - \xi_G)^{1/2}d\xi \tag{5.20}$$

The radius of real space per valence electron is related to the number of valence electrons per unit volume as $n = 1/((4/3)\pi r_s^3)$, and the Fermi wave vector and Fermi energy may be written as:

$$k_F = \frac{(9\pi/4)^{1/3}}{r_s}$$

$$\xi_F = \frac{\hbar^2 k_F^2}{2m} = \left(\frac{e^2}{2a_B}\right)(k_F a_B)^2 = R_y(k_F a_B)^2 = \frac{50.1\,\mathrm{eV}}{(r_s/a_B)^2} \tag{5.21}$$

where a_B is the Bohr radius of the one-electron hydrogen atom ($a_B = \hbar^2/me^2 = 0.529\mathrm{\AA}$) and R_y is the Rydberg unit, which is the ground-state energy of the one electron hydrogen atom ($R_y = e^2/2a_B = 13.6\,\mathrm{eV}$).

Above absolute zero, the electron occupancy is no longer 100% for all energy levels up to the Fermi energy; instead it becomes a probability function dependent on temperature given by the Fermi–Dirac distribution:

$$f(T) = \frac{1}{1 + e^{(\xi - \xi_F)/k_B T}} \tag{5.22}$$

By combining the density of states with the electron occupancy probability, it is possible to calculate the concentration of intrinsic electrons or holes in various bands of the semiconductor as a function of temperature. This calculation is conducted later in the chapter, after we have examined the band structure and defect structure of semiconductors.

5.4   Band structure

Most semiconductors of interest have a face-centered cubic (FCC) lattice with tetrahedral bonding. When the lattice contains only a single element, the structure is diamond cubic, which consists of an FCC structure. In this case, all atoms are tetrahedrally coordinated. If we define the position of atoms in the primitive cell, the smallest possible unit cell of the structure, as a set of basis positions, then a two-atom basis at (0, 0, 0) and (1/4, 1/4, 1/4) also describes an FCC structure for the two-atom unit cell (Yu and Cardona 1996). If the lattice contains two elements, then the structure is zinc blende, which is identical to the diamond cubic structure, except that the basis consists of two kinds of atoms. The FCC structure consists of atoms at all eight cube corners ($8 \times 1/8$) and atoms on all six faces ($6 \times 1/2$) to have an occupancy of four atoms per cell. Then there are eight tetrahedral (four-coordinated) holes inside each corner, 12 octahedral (six-coordinated) holes in the middle of the cube edges, and one octahedral hole at the cube body center. Both the diamond cubic and the zinc blende structures occupy the FCC sites and half the tetrahedral holes, as shown in Fig. 5.5. In the FCC *diamond cubic* structure, the occupied sites contain the same atoms (diamond, Si, and Ge). In the *zinc blende* structure, the anion (which is larger) occupies the FCC sites and the small cation occupies half the tetrahedral holes (ZnS, GaAs, CdTe, SiC(3C), InSb). If all the tetrahedral holes are filled by cations, the structure formed is called *antifluorite* (Li_2O, Na_2O). If the cations are bigger, they go to the octahedral holes instead of the tetrahedral holes; the structure formed with all octahedral holes filled is called *rock salt* (NaCl, MgO, NiO, TiN, ZrN).

The reciprocal lattice of an FCC unit cell is a body-centered cubic (BCC) cell. The Brillouin zone is the Wiegner–Seitz cell of the reciprocal lattice, as shown in Fig. 5.6. It forms a truncated octahedron. It can be shown that all the properties of the reciprocal crystal can be represented inside the first Brillouin zone, so we need only examine a single reciprocal cell. Another crystal structure that can be found in semiconductors is the Wurtzite structure, in which the anions occupy the hexagonal close-packed (HCP) structure and the cations occupy two-thirds of the tetrahedral holes. (More discussion is found in Appendix 5C.)

One interesting difference between various crystal compositions in either the zinc blende or the diamond cubic structure arises from the ratio of the ion size to the size of the tetrahedral hole available for it. In general, the tetrahedral hole in the close-packed FCC structure made up of atoms

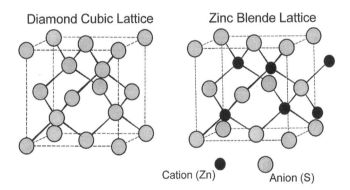

Figure 5.5: Atomic positions for the diamond cubic (DC) and the zinc blende (ZB) lattices. Note that the DC lattice has only one kind of atom. The ZB lattice has two kinds of atoms, with the smaller in the tetrahedral holes.

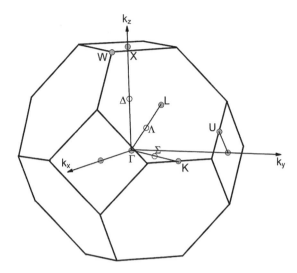

Figure 5.6: Reciprocal lattice for the DC and ZB lattices. This has the BCC symmetry and represents the first Brillouin zone of the crystal. The points marked on the reciprocal lattice correspond to the crystallographic directions in the real-space lattice.

with radius a_0 has a radius of $0.225a_0$. A diatomic structure will assume the zinc blende form if the cation size is sufficiently smaller than the anion size $(r_c/r_a \leq 0.225)$ to fit in the hole. If not, the FCC lattice must be stretched to fit around the larger cation. This happens until the cation size grows to a radius ratio of $r_c/r_a = 0.414$. The latter corresponds to the size of the octahedral hole in the FCC lattice. If the cation size is larger than $0.414a_0$, the cation of an ionic crystal will occupy the larger octahedral hole up to a size of $0.732a_0$. In the diamond cubic cell, the strong *covalent* bonding structure is 4-coordinated; consequently, the "interstitial" ion must fit into the tetrahedral hole even though its size is the same as that of the ion making up the FCC cell. This results in considerable distortion of the FCC lattice, and the cell cannot remain close packed. The consequences of these distortions are that the reciprocal cells are not symmetric. For example, the sides of the four-sided face are not equal, so it is not a square. The same holds for the tetragonal face. The amount of distortion in the ZB cell depends on the radius ratios (r_c/r_a) of the semiconductor components.

The reciprocal lattice is determined by the periodicity of the real lattice.

Since the real lattice is not isotropic, neither is the reciprocal lattice. Therefore, the band structure will vary with position and orientation. Points on the Brillouin zone unit cell have letters that differentiate between various points and directions of symmetry in reciprocal space (Fig. 5.6). The Γ point represents the center of the unit cell and corresponds to the $< 000 >$ position in the reciprocal crystal. The X point is at the center of the four-sided face and corresponds to the $< 100 >$ direction. The L point is the center of the six-sided face and corresponds to the $< 111 >$ direction. These three point are the most important in the description of the band structure in reciprocal space. Other points are as follows: the K point bisects the line joining two tetragonal faces; the U point bisects the line joining a tetragonal face to a four-sided face; the W point is at the corner of a four-sided face; the Δ-point bisects the Γ-X line and corresponds to the $< 101 >$ direction; the Σ point bisects the Γ-K line and corresponds to the $< 110 >$ direction; and the Λ point bisects the Γ-L line. Typical band-structure diagrams give values of the different bands at all these point.

Cubic semiconductors have three hole bands. Two are degenerate at the Γ ($k = 0$) point. These are the heavy-hole (HH) and light-hole (LH) bands. The heavy-hole band rests above the light-hole band and displays a lower curvature (larger effective mass), hence its name. The third band, the split-off (SO) band, is displaced in energy from the others by the spin-orbit splitting energy of the crystal. These band structures can be obtained to a first approximation for k-values near zero by using:

$$\xi_c \approx \xi_G + \frac{\hbar^2 k^2}{2m_e^*}$$

$$\xi_v(HH) \approx -\frac{\hbar^2 k^2}{2m_{hh}^*}$$

$$\xi_v(LH) \approx -\frac{\hbar^2 k^2}{2m_{lh}^*} \tag{5.23}$$

$$\xi_v(SO) \approx -\Delta -\frac{\hbar^2 k^2}{2m_{so}^*}$$

where ξ_c is the conduction band energy; $\xi_v(HH)$, $\xi_v(LH)$, and $\xi_v(SO)$ are the valence-band energies for the heavy-hole, the light-hole, and the split-off bands, respectively; ξ_G is the bandgap energy; Δ is the spin-orbit energy; and m^* are the corresponding effective masses. Cubic semiconductors exhibit a degeneracy in the hole bands as the heavy-hole and the light hole bands coincide at the Γ point $(k = 0)$. Hexagonal semiconductors have separate light-hole and heavy-hole bands, even at the Γ point.

The parabolic or effective mass approximation for describing the energy dispersion in semiconductors is reasonably good up to about 15% of the distance to the zone boundary. Beyond that, deviations become noticeable. The empirical 20-band or $30 \times 30\, k \cdot p$ calculation (Pollack, Higginbotham, and Cardona 1966; Bailey, Stanton, and Hess 1990) does a good job of approximating the electronic-band structure of semiconductors, and it is reasonably easy to use if sufficient data is available. Using only four bands (Kane 1966), it can produce a close estimate of the energy dispersion for the conduction and the split-off hole bands in the Brillouin zone. However, the four-band $k \cdot p$ approximation yields energy dispersion functions for the heavy- and light-hole bands that turn upward. Consequently, it is necessary to use a higher order approximation that takes into account higher-energy conduction-band states. Finally, a reasonable estimate of the band structure is obtainable over the entire k-range of the Brillouin zone using 20-band $k \cdot p$ theory, with the following results:

$$\xi_c(k) \approx \xi_g + \frac{\hbar^2 k^2}{2m_e^*} + \frac{\hbar^2 P_{cv}^2 k^2}{3m_e^*}\left(\frac{2}{\xi_G} + \frac{1}{\xi_G + \Delta}\right)$$

$$\xi_{LH,HH}(k) \approx -\frac{\hbar^2}{2m_{lh,hh}^*}\left[Ak^2 \pm (B^2 k^4 + C^2(k_x^2 k_y^2 + k_y^2 k_z^2 + k_z^2 k_x^2))^{1/2}\right] \quad (5.24)$$

$$\xi_{SO}(k) \approx -\Delta + \frac{\hbar^2 k^2}{2m_{so}^*} - \frac{P_{cv}^2 k^2}{3(\xi_G + \Delta)m_{so}^*}$$

where A, B, and C are constants to be determined empirically and P_{cv} is the optical matrix element for a transition between the valence and conduction bands. Valence and conduction bands also show warping that is direction dependent (anisotropy).

Another zeroth-order band structure can be obtained by just folding the empty real-space lattice into the first Brillouin zone. This approach gives all the band features if you replace all the level crossings by level repulsion (e.g., round off sharp corners of the Brillouin zone surface), as done schematically in Fig. 5.4.

When the lowest valley of the conduction band is aligned with the highest peak of the valence band, the minimum bandgap occurs at the Γ point, and the material is considered to be a direct-gap semiconductor

(Fig. 5.7). This is because electron excitations can occur with no change in momentum. When the lowest valley of the conduction band does not align with the top of the valence band (which occurs at the Γ point), then electron excitations by photons require a change in momentum (Fig. 5.7). Since photons carry negligible momentum ($p = \xi/c$), then any change in momentum during the excitation requires an interaction with the lattice (e.g., phonon interaction). These materials are called indirect-gap semiconductors since carrier excitations must be coupled with phonon interactions. In metals, electron–phonon coupling is very low, so indirect transitions are improbable. In semiconductors, a higher electron–phonon coupling increases the probability of indirect transitions.

Illustrations are presented of the results of calculations of the band structures of two direct-gap semiconductors (CdTe and GaAs) in Fig. 5.8, and two indirect-gap semiconductors (Si and Ge) in Fig. 5.9. CdTe has the simplest band structure. It has a zinc blende structure and shows a clear minimum in conduction-band energy at the Γ point that matches the maximum in the three distinct hole bands. Its bandgap energy is 1.6 eV. GaAs shows a similar structure and a bandgap at 1.52 eV. However,

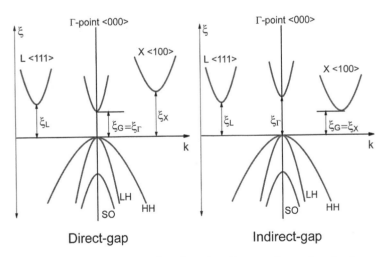

Direct-gap   Indirect-gap

Figure 5.7: Energy diagram showing the three valence bands: heavy-hole (HH), light-hole (LH) and split-off (SO) bands, and three conduction-band valleys (Γ, L, and X points) for a direct-gap and an indirect-gap semiconductor. The direct-gap semiconductor exhibits the narrowest energy gap ($\xi_c - \xi_v$) at the Γ point. The indirect-gap semiconductor exhibits the narrowest gap ($\xi_c - \xi_v$) at some other point (shown here as the X point).

Energy (eV)

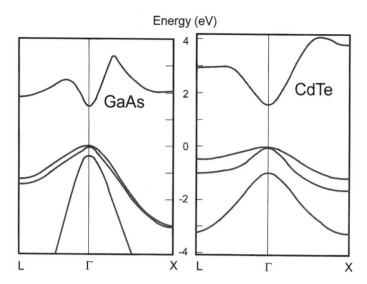

Figure 5.8: Sketch of the band structures of GaAs and CdTe. Both are direct-gap semiconductors.

Energy (eV)

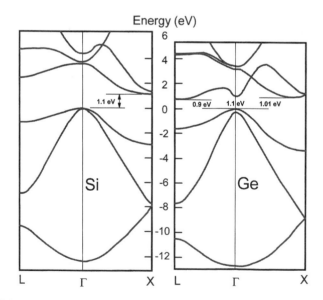

Figure 5.9: Sketch of the band structures of Si and Ge at room temperature. Both are indirect-gap semiconductors.

unlike CdTe, the satellite conduction-band valleys at the L and X points are not much higher than the Γ point. The L point is only 0.3 eV above the Γ point, while the X point has two valleys 0.5 and 0.9 eV above the Γ point. These valleys are sufficiently close to the bandgap energy that energetic electrons can enter the L valleys from the Γ point. The density of states is much higher in the L valley and the mobility is lower.

Both semiconductors support the formation of excitons at the Γ point. Excitons are correlated electrons and holes that remain bound to each other. The curvature of the conduction and valence bands can be opposite at the Γ point, since the momentum (and consequently the velocity) is zero for both particles. Away from the Γ point, the band curvature must be opposite so that the electron and hole can remain correlated at the velocity depicted. Both CdTe and GaAs show the same kind of inverse curvature at the L point. This indicates an ability to sustain excitons. The surface curvature in both semiconductors is flat in one direction, giving the L point excitons a two-dimensional structure. The L point band structure shows two parallel heavy- and light-hole valence bands separated by the spin-orbit splitting energy. The two-dimensional excitons in CdTe have been measured (Potter and Simmons 1991), and their behavior follows theory closely, showing the presence of two bands separated by the predicted spin-orbit splitting at the L point.

Silicon and Germanium both have indirect gaps (Fig. 5.9). The conduction-band structure of Si is somewhat anomalous, since it does not display a valley at the Γ point. The lowest conduction-band valley in Si occurs along the Δ direction about 85% to the zone boundary at the X point. There are six Δ directions, each giving rise to a conduction-band valley. The indirect bandgap of Si is 1.1 eV. What makes Si an ideal material for electronics is the large heavy-hole mass and the very small spin-orbit splitting (0.044 eV). Ge is different from Si in that it has a distinct valley at the Γ point; however, the energy minimum is the L point (0.67 eV). The direct gap is at 0.8 eV. The region near the X point is next, with a valley at 0.83 eV. The spin-orbit splitting value for Ge at the Γ point of 0.29 eV is much larger than that for Si.

States in the band structure that contain "free" electrons are often called *n-type* states; those that contain holes (e.g., missing valence electrons) are often called *p-type* states.

5.5   Impurity states and lattice imperfections

5.5.1   Donor and acceptor bands

Numerous defects can be formed in the crystal structure. Their behavior and their effect on the band structure of the semiconductor vary, depending on the type of defect and its charge.

Substitutional impurities:
If substitutional impurities are present, there are three possibilities, depending on their charge or valence: (1) same valence substitutions cause little change in the band structure; (2) substitutional defects with more electrons become donors; (3) substitutional impurities with fewer electrons become acceptors. If the material contains a relatively small concentration of impurities, the defects form isolated defect states. At higher concentrations, they form defect bands.
Interstitial impurities:
Regardless of the valence of the interstitial impurity, the outer shell electrons are available for excitation to "free" states, so they form donor defects.
Vacancy defects:
The absence of constituents corresponds to missing electrons and forms acceptor defects.
Frenkel defects:
The defects are formed by paired vacancies and interstitials; consequently, the charges are added up to determine whether the defects are donors or acceptors. Sometimes, however, one or the other kind of defect dominates, regardless of the number of electrons. In that case, the behavior will be controlled by the dominating defect.

Acceptor defects form *p-type states* and generate excess holes. Donor defects form *n-type states* and generate excess electrons. Thus, *p*-type semiconductors have excess holes and *n*-type semiconductors have excess electrons. The donor states are one binding energy below the conduction band. The acceptor states are one binding energy above the hole band. Since the effective masses of the electron, m_e^*, and hole, m_h^*, are not equal, donor and acceptor binding energies are also different.

Generally, electrons from donor states are most strongly attracted to the positive charge of the impurity nucleus. This gives hydrogenic-type states with the dielectric constant of the matrix:

$$\xi_d = -\frac{m_e^* q_e^4}{2h^2 \epsilon^2 n^2} = \frac{\xi_H m_e^*}{\epsilon^2 n^2} = -13.6\,\text{eV}\,\frac{m_e^*}{\epsilon^2 n^2} \qquad (5.25)$$

where m_e^* is the electron effective mass and ξ_H is the hydrogen atom ground-state energy (-13.6 eV), also known as the Rydberg ($-$R). For a dielectric constant of about 10, which is typical of semiconductors, and an electronic effective mass of 0.1, the ground-state binding energy of donor defects is about -15 meV. The electron associated with the defect orbits the nucleus at a radius, a_d, of

$$a_d = \frac{\hbar^2 \epsilon}{q_e^2 m_e^*} = \frac{\epsilon a_B}{m_e^*} = 0.53\text{Å}\,\frac{\epsilon}{m_e^*} \qquad (5.26)$$

where a_B is the hydrogen Bohr radius. The orbit radius of the electron about the donor impurity is about 50 Å. Therefore, it is *delocalized* As the impurity concentration increases, the donor electron wavefunctions overlap, and, at about $1/a_d^3$, the associated impurity carriers (donor electrons) are free. (This corresponds to a concentration of about $8 \times 10^{18}\,\text{cm}^{-3}$.) In practice, overlap begins at about $10^{16}\,\text{cm}^{-3}$, and this causes the formation of donor impurity bands in the region of overlap. A similar calculation can be conducted for acceptor impurities in which the hole orbits the impurity nucleus. The same equations apply with the replacement of the hole effective mass, m_h^*, for the electron effective mass, m_e^*, in Eqs. (5.25 and 5.26), except that the hole effective mass is larger than the electron by about a factor of 5–10. Thus, the acceptor state binding energy is of the order of $+100$ meV, while the orbit radius of the hole about the acceptor impurity is of the order of 8–10 Å. Thus it takes a larger concentration of acceptor impurities to form acceptor bands ($10^{18}\,\text{cm}^{-3}$).

5.5.2   Band tails

Another form of distortion of the band structure can come as band tailing. This effect generally results from three causes: impurities, strain, and/or structural defects. In each case, an imperfection in the crystal lattice can perturb the conduction or valence bands. For example, an ionized donor attracts conduction-band electrons and repulses valence-band holes, decreasing the coulomb interaction energy between them. This causes a local variation in the bandgap energy. Local inhomogeneous strain shifts the bandgap by altering the unit cell dimension (uniform pressure reduces the unit cell dimension and thus increases the Coulomb

interaction and the bandgap energy). Structural defects cause disturbances in the lattice crystal symmetry and likewise induce local
variations in the bandgap. When summed over the entire material,
these local bandgap distortions appear as band tails in the density of
states. These tails are intrinsic and therefore, do not change with
temperature. An exception occurs if the semiconductor is in the form of a
film deposited on another film or substrate with vastly different
coefficient of expansion. In this case, strain-induced tailing may exhibit
some temperature dependence.

5.5.3 Excitons

If an incident photon does not have enough energy to create a "free"
electron and a "free" hole, it can sometimes form a stable structure called
an *exciton* by exciting a valence or bonding electron out of its equilibrium
state and into an orbit about its potential minimum. The electron is thus
stably bound to a hole. As the electron $(-q_e)$ orbits the hole $(+q_e)$, a set of
hydrogen-like states are formed. However, unlike the donor and acceptor
defects that have a massive nucleus about which the much lighter
electron or hole orbits, the two components of the exciton have reasonably
close masses. For analysis, the orbiting masses are replaced by the
reduced effective mass of the pair: $\mu^{-1} = (m_e^*)^{-1} + (m_h^*)^{-1}$, and the binding
energy of excitons is considerably less than the binding energy of donor or
acceptor defects, as follows:

$$\xi_{nx} = -\frac{\mu^* q_e^4}{2\hbar^2 \epsilon^2 n^2} = \frac{\xi_H \mu^*}{\epsilon^2 n^2} \quad \text{and} \quad a_x = \frac{\epsilon \hbar^2}{\mu^* q_e^2} = \frac{\epsilon a_B}{\mu^*}$$

$$\xi_{nx} = -\frac{\hbar^2}{2\mu^* a_x^2}$$

(5.27)

where the exciton binding energy is ξ_{nx} and the exciton Bohr radius is a_x.
The ground-state binding energy of the exciton and the exciton Bohr
radius depend upon the masses of the electron and hole, and they vary
with semiconductor composition, as shown in Table 5.1. [So far, we have
discussed Wannier–Mott excitons. These are weakly bound and have a
large orbit radius compared to the lattice constant (unit cell dimension).
These are the most common excitons and are the subject of most studies in
semiconductors. However, there is another type of exciton, the Frenkel
exciton, that occurs in alkali halides and inert gases. It is localized to a

Semiconductor	m_e^*	m_h^*	Exciton binding energy (meV)	Exciton Bohr radius (Å)
Copper chloride (CuCl)			190	6.5
Gallium nitride (GaN)	0.2	0.8		
Zinc selenide (ZnSe)	0.17			21
Cadmium sulfide (CdS)	0.2	0.9	29	28
Cadmium selenide (CdSe)	0.13	0.8		53
Cadmium telluride (CdTe)	0.11	0.35		75
Gallium arsenide (GaAs)	0.07	0.5	4.2	140
Silicon (Si)	0.98ℓ, $0.19t$	0.52		20ℓ, $45t$
Germanium (Ge)	1.58ℓ, $0.08t$	0.3	4	25ℓ, $200t$
Lead sulfide (PbS)	0.1	0.1		200

Table 5.1: Table of Wannier exciton binding energies and Bohr radii.

single atom and propagates through the lattice by hopping between adjacent atoms.]

Although the hole state consists of an empty lattice bonding structure, because it is neutralized by the orbiting electron it can hop freely across the lattice. This forms a *free exciton*. Free excitons are delocalized spatially and have three translational degrees of freedom for motion through the lattice. However, it is possible to encounter a variety of bound excitons that are either bound to a donor impurity (donor-bound exciton) or bound to an acceptor impurity (acceptor-bound exciton). The diagram of Fig. 5.10 illustrates exciton complexes that can be formed.

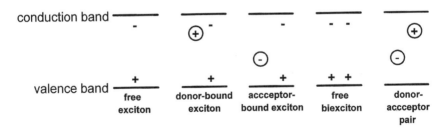

Figure 5.10: Schematic of the simpler exciton-associated defects.

As with the donor defect states, excitons have a negative binding energy, and their states lie below the conduction band. Thus the energy of free excitons with respect to the valence band can be written as

$$\xi_{x} = \xi_{G} + \frac{\hbar^2 k^2}{2M} + \frac{\xi_H \mu^*}{n^2 \epsilon^2} \qquad (5.28)$$

where $M^* = m_e^* + m_h^*$ is the total mass of the exciton. The second term of the equation represents the kinetic energy of the exciton. The third term is the exciton binding energy, which is negative since $\xi_H = -13.6\,\text{eV}$. If the exciton is a bound exciton, then another binding-energy term is subtracted from the free exciton energy. This lowers the total energy of the bound excitons but increases their binding energy (since the binding energy is negative and measured relative to the bottom of the conduction band).

As shown earlier, we can evaluate the velocity of the electron and hole as follows:

$$v_{e} = \frac{1}{\hbar}\left(\frac{d\xi}{dk}\right)_n \quad \text{and} \quad v_{h} = -\frac{1}{\hbar}\left(\frac{d\xi}{dk}\right)_p \qquad (5.29)$$

where the n subscript indicates the conduction band and the p subscript indicates the hole band. It is clear that the two bands change curvature with k in different directions. Consequently, electrons and holes have different velocities in general. The exciton, therefore, can exit only in the few regions of k-space where the two translational velocities are equal. These are defined as critical points. The Γ point in direct-gap semiconductors supports excitons, since the two velocities are zero at the same point. Sometimes the curvature is equal but the sign is opposite. Thus, instead of forming an exciton, the electron and hole drift apart and the coupled pair dissociates.

5.5.4 Donor–acceptor pairs

Since donors and acceptors have opposite charges, they can bind together into pairs and act as stationary molecules embedded in the host crystal. As the two defects approach each other, they become progressively more ionized and their energy increases. The energy of the pair can be expressed as follows:

$$\xi_{\text{pair}} = \xi_D - \xi_A - \frac{q_e^2}{\epsilon r} = \xi_G - \xi_d - \xi_a - \frac{q_e^2}{\epsilon r} \qquad (5.30)$$

where ξ_D and ξ_A are the ionization and affinity energies of the isolated impurities ($\xi_d = \xi_v - \xi_D$ and $\xi_a = \xi_v - \xi_A$ where ξ_v is the energy difference between the vacuum level and the top of the valence band). The impurities are located at precise lattice sites and do not move after the material is formed; consequently, there is a broad distribution of separation distances between donor and acceptor impurities. Since the r-values in Eq. (5.30) span a wide range, the optical absorption peaks associated with donor–acceptor (D–A) transitions are very broad (see Section 5.7.6).

5.5.5 Amorphous semiconductors

The existence of amorphous semiconductors presents an interesting problem, since the Bloch theorem and translational periodicity form the foundation of semiconductor physics. However, the problem is relieved by the realization that amorphous materials do not have a fully random structure and that in general their unit cell structures are reasonably regular and similar to some crystalline solids of the same composition. The nearest-neighbor environments are identical in the structures of amorphous and crystalline semiconductors; in fact, the differences occur in the next-nearest-neighbor region. This does not hold for all amorphous materials, and some have unit cells very different from any crystal. However, they are still regular, and disorder occurs in the packing of the unit cells.

Anderson (1958) and Mott and Davis (1971) realized that the configurational disorder could be viewed as spatial fluctuations in the potential function that would cause the formation of localized states. Thus, the band structure of amorphous semiconductors could be viewed as consisting of tails and midgap or deep defects added to a regular band structure. In some materials, these localized defect states may reach across the entire bandgap, removing semiconductor behavior. But, in general, they form a series of defect bands and band tails only on the conduction and valence bands.

5.6 Carrier densities

It is instructive to look at carrier densities in semiconductors as a function of temperature. Intrinsic carrier concentrations give the densities of carriers that are formed from the major components of the semiconductors and are thermally excited. Extrinsic carrier concentrations relate to carriers that result from dopants and impurities. If $n(\xi)$ is the density of states and $f(\xi,T)$ is the carrier occupancy probability, then the number of electrons and holes per unit volume that will be formed in a semiconductor at temperature T is given by:

$$n = \int_{\xi_c}^{\infty} n_c(\xi) f(\xi, T) d\xi$$

(5.31)

$$p = \int_{-\infty}^{\xi_v} n_v(\xi) \, [1 - f(\xi, T)] d\xi$$

with the parabolic approximation discussed previously, and using the Fermi–Dirac statistics also discussed earlier, it is possible to obtain an integral equation for the carrier densities using the convention that $\xi_v = 0$ and $\xi_c = \xi_G$:

$$n = \frac{(2m_e^*)^{3/2}}{2\pi^2\hbar^3} \int_{\xi_G}^{\infty} \sqrt{\xi - \xi_G} \; \frac{1}{e^{(\xi - \xi_F)/k_B T} + 1} d\xi$$

(5.32)

$$p = \frac{(2m_h^*)^{3/2}}{2\pi^2\hbar^3} \int_{-\infty}^{0} \sqrt{-\xi} \; \frac{1}{e^{-(\xi - \xi_F)/k_B T} + 1} d\xi$$

where ξ_F is the Fermi energy of the semiconductor. These two equations are not integrable in closed form, so we turn to two approximations.

5.6.1   Nondegenerate semiconductors

These are undoped semiconductors (intrinsic carrier concentrations) with the Fermi energy near the middle of the bandgap so that:

$$\xi_c - \xi_F \gg k_B T$$
$$\xi_F - \xi_v \gg k_B T$$

(5.33)

This limit allows approximation of the Fermi–Dirac equation by the Boltzmann distribution, as follows:

$$\frac{1}{e^{(\xi-\xi_F)/k_BT}+1} = e^{-(\xi-\xi_F)/k_BT} \qquad \text{for } \xi > \xi_G$$

$$\frac{1}{e^{-(\xi-\xi_F)/k_BT}+1} = e^{(\xi-\xi_F)/k_BT} \qquad \text{for } \xi < 0 \tag{5.34}$$

Equations (5.32) now become integrable, with the following result:

$$n_i = N^c e^{-(\xi_G-\xi_F)/k_BT} \qquad \text{with} \qquad N^c = 2\left(\frac{m_e^* k_B T}{2\pi\hbar^2}\right)^{3/2}$$

$$p_i = N^v e^{-\xi_F/k_BT} \qquad \text{with} \qquad N^v = 2\left(\frac{m_h^* k_B T}{2\pi\hbar^2}\right)^{3/2} \tag{5.35}$$

This relation gives the value of the Fermi energy in terms of the bandgap energy and the carrier effective masses:

$$\xi_F = {}^1\!/_2\xi_G + {}^1\!/_2 k_B T \, \ln\left(\frac{N^v}{N^c}\right) = {}^1\!/_2\xi_G + {}^3\!/_4 k_B T \, \ln\left(\frac{m_h^*}{m_e^*}\right) \tag{5.36}$$

Thus, the Fermi energy is exactly half the bandgap at absolute zero or if the electron and hole effective masses are equal.

The law of mass action gives the product of the electron and hole densities:

$$np = 4\left(\frac{k_B T}{2\pi\hbar^2}\right)^3 (m_e^* m_h^*)^{3/2} e^{-\xi_G/k_BT} \tag{5.37}$$

and it shows that the product is dependent only on temperature. Thus, increasing the electron density without changing the bandgap requires that the hole density decrease.

In an intrinsic semiconductor, the two carrier species must have the same density:

$$n_i = p_i = 2\left(\frac{k_B T}{2\pi\hbar^2}\right)^{3/2} (m_e^* m_h^*)^{3/4} e^{-\xi_G/2k_BT} \tag{5.38}$$

$$= N^c e^{\xi_G/k_BT} e^{\xi_F/k_BT} = N^v e^{\xi_F/k_BT}$$

For calculations, the constants can be evaluated to give the following simplified equations:

$$n_i = 2.5 \times 10^{19} (m_e/m_0)^{3/2} (T/300\,\text{K})^{3/2} e^{-(\xi_G-\xi_F)/k_BT} / \text{cm}^3$$

$$p_i = 2.5 \times 10^{19} (m_h/m_0)^{3/2} (T/300\,\text{K})^{3/2} e^{-\xi_F/k_BT} / \text{cm}^3 \tag{5.39}$$

5.6.2   Degenerate semiconductors

These are doped semiconductors, so we now have to account for donor and acceptor sites and divide them between neutral and ionized sites. Degenerate semiconductors do not satisfy Eq. (5.33). The law of mass action is still valid, and the carrier densities must satisfy a charge neutrality condition:

$$n_d + n_i + N_A^- = p_a + p_i + N_D^+ \qquad (5.40)$$

with $N_D = N_D^0 + N_D^+$ and $N_A = N_A^0 + N_A^-$, where n_d = electrons from donors; p_a = holes from acceptors; N_D = number of donors; N_A = number of acceptors. In this case, because of the extrinsic doping, there is an imbalance in the carrier density:

n-type semiconductors   excess of ionized donors $n_d \gg p_a$
p-type semiconductors   excess of ionized acceptors $p_a \gg n_d$

In addition, the Maxwell-Boltzmann approximation is no longer valid. In order to calculate the effect of degeneracy, we calculate the occupation probability of each state for both acceptors and donor levels. In general, the occupation probability is written as

$$\langle n \rangle = \frac{\sum N_j e^{-(\xi_j - N_j \xi_F)/k_B T}}{\sum e^{-(\xi_j - N_j \xi_F)/k_B T}} \qquad (5.41)$$

We may now calculate the occupation of the donor and acceptor states as follows:

(a) *Monovalent donor state*:
(i)   no electrons in the state $\xi_j = 0$ $N_j = 0$
(ii) one electron ↑ with ξ_D $\xi_j = \xi_D$ $N_j = 1$
(iii) one electron ↓ with ξ_D $\xi_j = \xi_D$ $N_j = 1$
(iv) two electrons ↑↓ with $2\xi_D + \Delta_e$ $\xi_j = 2\xi_D + \Delta_e$ $N_j = 2$

where Δ_e is the electron–electron repulsion energy $\Delta_e = e^2/(4\pi\epsilon r_{12})$, with r_{12} as the electron–electron distance, which is very large, so Δ_e is small.

$$\begin{aligned}
\langle n \rangle &= \frac{0 + e^{-(\xi_D - \xi_F)/k_B T} + e^{-(\xi_D - \xi_F)/k_B T} + 2e^{-(2\xi_D + \Delta_e - 2\xi_F)/k_B T}}{1 + e^{-(\xi_D - \xi_F)/k_B T} + e^{-(\xi_D - \xi_F)/k_B T} + e^{-(\xi_D + \Delta_e - 2\xi_F)/k_B T}} \\
&= \frac{1}{1 + \frac{1}{2}e^{(\xi_D - \xi_F)/k_B T}}
\end{aligned} \qquad (5.42)$$

The second equation is obtained by neglecting terms containing $\exp(-\Delta_e/k_B T)$. The ionized donor density is then calculated as follows:

$$N_D^+ = N_D[1 - \langle n \rangle]$$

$$N_D^+ = N_D \left[\frac{1}{1 + 2e^{-(\xi_D - \xi_F)/k_B T}} \right] \tag{5.43}$$

If we use the binding energy $\xi_d = \xi_G - \xi_D$ and the difference between the gap energy and Fermi energy $\Delta = \xi_G - \xi_F$, then the ionized donor concentration simplifies to

$$N_D^+ = N_D \left[\frac{1}{1 + 2e^{-(\Delta - \xi_d)/k_B T}} \right] \tag{5.44}$$

In n-type semiconductors, holes are neglected and all acceptor sites are ionized, so the carrier concentration becomes

$$n = N_D^+ - N_A = N_D \left[\frac{1}{1 + 2e^{-(\Delta - \xi_d)/k_B T}} \right] - N_A \tag{5.45}$$

(b) *Monovalent acceptor state*:
(i) no holes in state gives two electrons $\uparrow \downarrow$ $\xi_j = 2\xi_A$ $\quad N_j = 2$
(ii) one hole , one electron $\quad \xi_j = \xi_A$ $\quad N_j = 1$
(iii) one hole $\downarrow$, one electron $\quad \xi_j = \xi_A$ $\quad N_j = 1$
(iv) two holes, no electrons $\quad \xi_j = 0 + \Delta_h$ $N_j = 0$

This gives an electron occupancy probability of:

$$\begin{aligned}
\langle n \rangle &= \frac{0 \cdot e^{-\Delta_h/k_B T} + e^{-(\xi_A - \xi_F)/k_B T} + e^{-(\xi_A - \xi_F)/k_B T} + 2e^{-(2\xi_A - 2\xi_F)/k_B T}}{e^{-\Delta_h/k_B T} + e^{-(\xi_A - \xi_F)/k_B T} + e^{-(\xi_A - \xi_F)/k_B T} + e^{-(2\xi_A - 2\xi_F)/k_B T}} \\
&= \frac{1 + e^{-(\xi_A - \xi_F)/k_B T}}{1 + \frac{1}{2} e^{-(\xi_A - \xi_F)/k_B T}}
\end{aligned} \tag{5.46}$$

The second equation is simplified by neglecting the small exp $(-\Delta_h/k_B T)$ terms. The hole concentration is then calculated to be

$$\langle p \rangle = 2 - \langle n \rangle = \frac{1}{1 + \frac{1}{2} e^{-(\xi_A - \xi_F)/k_B T}} \tag{5.47}$$

The concentration of ionized acceptor states is then calculated to be:

$$N_A^- = N_A(1 - \langle p \rangle)$$

$$N_A^- = N_A \left[\frac{1}{1 + 4e^{-(\xi_F - \xi_A)/k_B T}} \right] \tag{5.48}$$

where we have introduced the two-band degeneracy (light- and heavy-hole) of the zinc blende structure into the factor in front of the exponential in the denominator.
The results obtained are subject to:

(1) Charge neutrality [Eq. (5.40)], which yields the following:
number of negative charge carriers: $n = N_D{}^+ - N_A{}^- + p_i$
number of positive charge carriers: $p = N_A{}^- - N_D{}^+ + n_i$
(2) The Fermi energy level that readjusts to provide charge compensation, which is calculated by applying Eqs. (5.32) in which the carrier
concentration n is solved and related to the Fermi energy level, ξ_F.
The solution without the Boltzmann simplification is not solvable in
closed form. Appendix 5A shows the general solution, which must be
evaluated numerically to give the value of ξ_F that produces the same
electron concentration from both Eqs. (5.32) and Eqs. (5.42). The
latter is given as follows in terms of the Fermi function, $F_{1/2}(\eta)$,
defined in Appendix 5A:

$$n = \frac{(2m_e^* k_B T)^{3/2}}{2\pi^2 \hbar^3} F_{1/2}(\eta) \qquad (5.49)$$

where $\eta = (\xi_F - \xi_G)/k_B T$.

5.7 Absorption and photoluminescence

As shown in Chapter 3, light incident on a material is absorbed if it can
cause an electronic transition. In semiconductors, this process can occur
by means of several mechanisms, including the following.

Direct interband (band-to-band) transitions.
An electron may absorb a photon via excitation from an occupied state
to an empty state with the same wave vector, k. Transitions are allowed
between the various valence bands and the conduction bands, since
valence bands are p states and conduction bands are s states ($\Delta \ell = \pm 1$).
Direct dipole-mediated transitions are not allowed between the heavy-
hole and light-hole valence-band states, since they form states with the
same angular momentum. Transitions are allowed between the split-off
($j = 1/2$) and the heavy-hole or light-hole ($j = 3/2$) bands. However,
since these bands are essentially full at reasonable temperatures, no
transitions take place. At moderate temperatures, therefore, electron
excitations are possible only between any of the hole bands and the
conduction band. Emission of a photon occurs in the reverse process. In

these transitions, since the photon carries essentially no linear momentum $(p = \xi/c)$, the wave vector of the initial and final states must be the same $(\Delta k = 0)$.

Indirect interband transitions.
These are the same as direct interband transitions, except the initial and final states have different wave vector values. Consequently, the transitions require a change in momentum in the electronic states, along with the energy change. Since the photon cannot provide this momentum, the transition is possible only with either the absorption or emission of a phonon (interaction with a lattice vibration). Indirect transitions have lower probability than direct transitions, since one must include the likelihood of an electron–phonon interaction, which contains the probability of creation or annihilation of a phonon (only the creation of phonons is likely at low temperatures).

Impurity-to-band and impurity–impurity transitions.
Doped semiconductors have impurities that act to form either donor or acceptor states or bands. In p-type materials, a low-energy photon can excite an electron from the valence band to the acceptor band, where it is trapped by an acceptor defect (hole transition from the acceptor to the valence band). Thus the valence band acquires a hole and the acceptor atom is ionized. In an n-type material, an electron in the donor band may be excited by a very low-energy photon to the conduction band. This process is not usually observed, because the donor-band ionization (binding) energy is often less than thermal, so most donor impurities are ionized (see Section 5.4.1). Consequently, one also sees absorption processes arising from the excitation of an electron from the valence or the acceptor band to the donor band. In general, photon emission or absorption occurs through electronic transitions between the donor and acceptor bands, with associated nonradiative transitions between conduction and donor bands and between acceptor and valence bands.

Excitonic transitions.
Excitonic states in direct-gap semiconductors can accept excitations of electrons from the valence band. These are direct transitions and do not involve phonons. Excitonic transitions occur if the photon energy is not sufficient to ionize the valence electron but can form a bound-electron–hole complex. Excitonic states have a large cross section in direct-gap semiconductors, so photon emission from excited carriers in the

conduction band occurs by a nonradiative transition to the exciton band followed by a radiative transition (photoluminescence) from the exciton state to the valence band.

Intraband transitions.
If free carriers are present, as in n-type semiconductors or at temperatures with thermal energies of the order of the bandgap energy, photons may give kinetic energy to free electrons. These are intraband transitions (e.g., the initial and final states are in the same band), and they involve a change in momentum and wave vector. This process is generally followed by carrier thermalization as the excited carriers lose kinetic energy through interaction with the lattice (phonons).

Phonon transitions. Photons can be absorbed in materials and re-emitted with either the creation or annihilation of an optical phonon. This is a direct interaction of the photon with the lattice and it occurs in all materials. This has been described in Chapter 3, under Raman scattering. Since the optical phonon band has a weak dispersion, the Raman energy is well defined and narrow. Interaction of photons with acoustic phonons and atomic thermal motions are also possible in semiconductors, as described for insulators, and correspond to Brillouin and Rayleigh scattering, respectively.

All these mechanisms lead to some form of absorptive or emissive behavior. Most processes are sufficiently distinct to be easily identified. However, in studying new materials, it is often not clear whether a transition is direct or indirect. Later, we will discuss the principal processes mathematically so that the reader may use dispersion analysis to distinguish between the various transitions possible, and we will discuss appropriate absorption and emission examples.

In semiconductors at low temperatures, there are few (intrinsic) electrons in the conduction band. Consequently, if the incident light has a frequency lower than the bandgap frequency ($\nu_G = \xi_G/h$), there is no optical absorption. When the optical frequency is above the bandgap frequency, electronic interband transitions are possible and absorption is high. Thus, semiconductors act as low pass filters. They pass light below the bandgap frequency and strongly absorb light above the bandgap frequency. Defect states will alter this behavior in two ways. One is by the formation of defect states or bands within the bandgap. This lowers the effective absorption threshold through impurity-to-band transitions or impurity-to-impurity band transitions. The second is by introducing carriers into either the conduction band (donor impurities) or the valence band (acceptor impurities). These carriers have no threshold energy for absorption. Consequently, if there is a high concentration of defect-induced (extrinsic) carriers, the associated absorption can be considerable and the material may act as a metal, exhibiting a significant reflectivity.

At higher temperatures, intrinsic carriers can be thermally excited across the bandgap and their concentration raised accordingly. At high enough free carrier concentrations, sufficient surface current is formed to induce a large reflection coefficient. In the discussion to follow, the behavior of semiconductors is separated between direct-gap and indirect-gap materials.

5.7.1 Direct-gap semiconductors

Direct band-to-band transitions dominate in direct-gap semiconductors. These transitions do not involve any energy exchange with the phonon field, so there is no change in the wave vector ($k_f = k_i$). The absorption coefficient, α, can be written in terms of the optical transition matrix element squared, P_{if}, and the occupancy of the initial, n_i, and final, n_f, states:

$$\alpha = A \sum P_{if} n_i n_f \tag{5.50}$$

Following Fig. 5.11, there is a relationship between the initial-state and final-state energies and the incident photon energy:

$$\xi_f = h\nu + \xi_i \tag{5.51}$$

Assuming as a first-order approximation that the bands follow parabolic behavior:

$$\xi_f = \xi_G + \frac{\hbar^2 k_f^2}{2m_e^*} \quad \text{and} \quad \xi_i = -\frac{\hbar^2 k_i^2}{2m_h^*} \tag{5.52a}$$

$k_i = k_f = k$, since photons have essentially no momentum,

$$h\nu - \xi_G = \frac{\hbar^2 k^2}{2} \left(\frac{1}{m_e^*} + \frac{1}{m_h^*} \right) = \frac{\hbar^2 k^2}{2\mu^*} \tag{5.52b}$$

$$k^2 = \frac{2\mu^*}{\hbar^2} (h\nu - \xi_G)$$

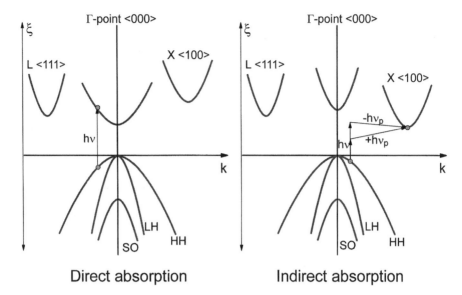

Direct absorption   Indirect absorption

Figure 5.11: Optical absorption transitions that promote a valence electron to a conduction-band state. In *direct-gap* materials, the k-values of the initial and final states are the same, so the electron goes directly to the central Γ point valley. In the drawing, due to the high energy of the absorption process, the free electron in the Γ valley has enough energy to scatter into the lower L point valley. In *indirect-gap* materials, the k-values of the initial and final states for the lowest-energy transition must be different, so the valence electron reaches the X point valley. In this case, a phonon must take part in the excitation, through either absorption $(+h\nu_p)$ or $(-h\nu_p)$.

In semiconductors with dipole-allowed transitions, the density of states of the initial and final states (optical joint density of states) can be calculated from the functions in Eqs. (5.20), with the condition that only those initial and final states with the same k wave vectors can participate in the transition:

$$n_{if}(\nu) = \frac{(2\mu^*)^{3/2}}{\pi^2\hbar^3}(h\nu - \xi_G)^{1/2} \qquad \text{for } h\nu \geq \xi_G \qquad (5.53)$$

Combined with the optical matrix element, this term leads to the following absorption coefficient for a direct interband transition:

$$\alpha \approx \frac{q_e^2}{nch^2}\frac{(2\mu^*)^{3/2}}{m_e^*h\nu}(h\nu - \xi_G)^{1/2} \qquad (5.54)$$

A detailed derivation of this result is found in Appendix 5B.

Very few semiconductors have dipole-forbidden transitions (Cu_2O, SnO_2, TiO_2). Their direct-gap absorption is:

$$\alpha = A(h\nu - \xi_G)^{3/2} \tag{5.55}$$

5.7.2 Indirect-gap semiconductors

Indirect gap semiconductors have a conduction-band minimum away from the Γ point. Therefore the electronic transition requires a change in wave vector and an associated change in momentum (Fig. 5.11). Consequently, optical excitations at the minimum energy require either the absorption or the emission of a phonon in order to change the momentum of the electron sufficiently to reach the conduction-band side valley with the energy minimum. Direct transitions are also possible, but their threshold energy is higher than that for the indirect transitions. Thus, as the photon energy is increased, the indirect transition will take place first, to be followed by a combination of direct and indirect transitions when the direct gap energy threshold is crossed by the incident photon energy. The relative contributions of the direct and indirect transitions is determined by their respective oscillator strengths.

If we examine the indirect transitions, we see that both absorption and emission of phonons is possible; therefore the energy balance is written as:

$$\begin{aligned} h\nu_e &= \xi_f - \xi_i + \xi_p \\ h\nu_a &= \xi_f - \xi_i - \xi_p \end{aligned} \tag{5.56}$$

where ξ_p is the phonon energy. If we define the energy between the indirect conduction-band minimum and the valence-band maximum as $\xi_G{}'$, we may rewrite the absorption energy threshold for the two indirect processes as follows:

$$\begin{aligned} h\nu_e &\geq \xi_G' + \xi_p \quad \text{for phonon emission} \\ h\nu_a &\geq \xi_G' - \xi_p \quad \text{for phonon absorption} \end{aligned} \tag{5.57}$$

In this case, the initial states can come from all states in the valence band, since there is no condition on the k-values, so the initial and final density of states are:

$$n_i(\xi) = \frac{(2m_h^*)^{3/2}}{2\pi^2 \hbar^3} |\xi_i|^{1/2}$$

$$n_f(\xi) = \frac{(2m_e^*)^{3/2}}{2\pi^2 \hbar^3} (\xi_f - \xi_G')^{1/2} = \frac{(2m_e^*)^{3/2}}{2\pi^2 \hbar^3} (h\nu - \xi_G' \mp \xi_p + \xi_i)^{1/2} \qquad (5.58)$$

The product of these equations forms the optical joint density of states for the transition, so the absorption is calculated from the product of the integration of all possible initial and final states separated by $\xi_G' - \xi_p$, the phonon occupation probability, and the optical matrix element. Since phonons are bosons (integer spin values), the phonon occupation probability is written according to Bose–Einstein statistics:

$$n(\xi_p) = \frac{1}{e^{\xi_p/k_B T} - 1} \qquad (5.59)$$

This leads to an absorption coefficient of:

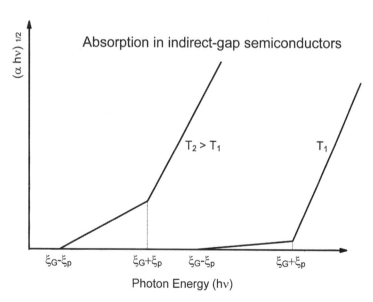

Figure 5.12: Typical absorption behavior for an indirect-gap semiconductor. At low energy, only phonon absorption can mediate the electron excitation. As the temperature is lowered, the smaller phonon population decreases the absorption probability, as shown.

$$\alpha_a = \frac{A}{h\nu} \frac{(h\nu - \xi'_G + \xi_p)^2}{e^{\xi_p/k_B T} - 1} \qquad \text{for phonon absorption}$$

$$\alpha_e = \frac{A}{h\nu} \frac{(h\nu - \xi'_G - \xi_p)^2}{1 - e^{-\xi_p/k_B T}} \qquad \text{for phonon emission}$$

(5.60)

Both terms contribute with their respective energy thresholds. At low temperatures, the phonon occupation is low, so the phonon-absorption contribution to the total photon absorption is low. A schematic of the indirect absorption behavior is shown in Fig. 5.12.

5.7.3 Heavily doped semiconductors

In heavily doped n-type semiconductors, the lowest states of the conduction band are occupied by free electrons, thus the Fermi level is inside the conduction band. Optical excitation of valence electrons to the bottom of the conduction band is no longer possible, since the states are filled. Thus the absorption threshold is shifted to higher energies (blue shift). This shift, due to band-filling, is called the Burstein–Moss effect.

In heavily doped indirect-gap semiconductors, momentum conservation no longer requires only phonons. Electron–electron scattering and impurity scattering can also provide the necessary change in wave vector.

5.7.4 Transitions between band tails

Band tails result from local structural defects, impurities, phonons, and stresses. Impurities can perturb both the conduction and the valence bands. An ionized donor attracts conduction-band electrons and repulses valence-band holes, causing a local variation in band energies. Deep impurities have high binding energies, so they can significantly affect the local bandgap. Local variations in bandgap energy appear as band tails when averaged over the sample or the size of the measuring beam. Impurities can change the density of states significantly, and, when present in high enough concentration, they can form distinct impurity states or impurity bands.

Band tails can also result from phonon vibrations of the lattice. These correspond to temporal variations in the lattice parameter. These tails are temperature dependent.

Band tails are normally due to a random broadening of the lowest-energy transitions. Consequently, the shape of the bandtail is exponen-

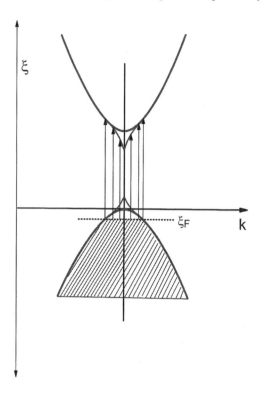

Figure 5.13: Transitions in a heavily doped semiconductor with band tails. All transitions originate from the Fermi energy and terminate on the tail of the conduction band.

tial. This leads to what is called an Urbach tail, or the absorption is said to follow the Urbach rule, written as follows:

$$\frac{d\ (\ln \alpha)}{d\ (h\nu)} = \frac{1}{k_B T} \tag{5.61}$$

Transitions between bandtails appear to follow the Urbach rule, by which the absorption edge of a semiconductor varies exponentially with frequency. This gives it a more gradual increase with frequency than the law of Eq. (5.53). Thus the effect appears as a band tail.

Consider an absorption process in a degenerate p-type material (ξ_F is inside the valence band, as shown in Fig. 5.13). Since the Fermi energy is below the valence-band tail, its effect is neglected in calculating the initial density of states for the transition:

$$n_i = A|\xi_v|^{1/2} \tag{5.62}$$

The final states are inside the conduction-band tail, so the density of states is exponential:

$$n_f = n_0 e^{h\nu/\xi_0} \tag{5.63}$$

where ξ_0 gives the curvature of the band tail and is proportional to $k_B T$ if the Urbach rule holds.

The absorption is then calculated as:

$$\alpha = An_0 \int_{\xi_F}^{h\nu - \xi_F} |\xi_v|^{1/2} \, e^{\xi/\xi_0} d\xi = An_0 \int_{\xi_F}^{h\nu - \xi_F} |\xi - h\nu|^{1/2} e^{\xi/\xi_0} d\xi \tag{5.64}$$

The opposite is also true for n-type semiconductors.

Consequently, the variation of the absorption coefficient with frequency gives the shape of the conduction-band and the valence band tails in p-type and n-type semiconductors, respectively. Strain can also cause shifts in bandgap energy by altering the unit cell dimension. Pressure reduces the unit cell constant; thus it increases bandgap energy. Both strain and impurities are considered to be sources of *intrinsic* impurities. The band tails associated with intrinsic impurities are not temperature dependent.

5.7.5 Exciton Absorption

Exciton effects are observed only if pronounced band tails are not present, and then only at very low temperatures (20 K). Exciton absorption is distinguishable from the band-edge absorption only in high-quality crystals. There, one or several sharp peaks appear in the absorption spectrum. The main characteristic of an exciton peak is the abrupt onset of absorption. Figure 5.14 shows the absorption spectrum of semiconductor quantum dots of CdS_xSe_{1-x} in a glass matrix. The sharp peaks above the absorption edge correspond to exciton transitions. These are clearly evident in the larger samples and disappear as the size of the crystallites is reduced to the size of the exciton Bohr diameter (8 nm). Since the exciton corresponds to a narrow energy level, only the homogeneous line width is present in a perfect crystal. (The next chapter will show that the homogeneous line shape is Lorentzian.) The homogeneous line width is generally less than 10 meV. In a real crystal, the line is broadened by structural defects and acquires an inhomogeneous (Gaussian) line shape with a much larger width (100–200 meV). This is further broadened in the samples studied in Fig. 5.14, since there

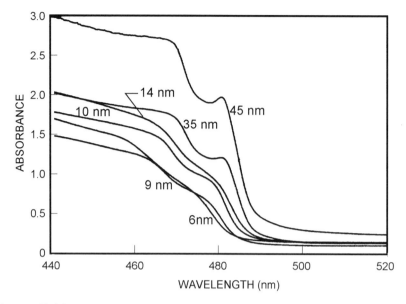

Figure 5.14: Absorption data at 10 K for CdS_xSe_{1-x} quantum dots. The peaks near the absorption edge are due to exciton transitions. These transitions change the nature of the absorption edge and give it a Gaussian shape. The different curves correspond to crystals of different sizes, as marked. Note the gradual loss of distinction of the exciton peaks as the quantum dots shrink to the size of the exciton Bohr diameter (8 nm).

is a dependence of the exciton energy on particle size and the samples measured have a size distribution of about 30% of the average size quoted in the figure. Since the exciton is located just below the conduction-band energy, the Gaussian tail of the exciton transition determines the shape of the absorption edge. Exciton-associated photoluminescence peaks (free excitons, donor-bound or acceptor-bound excitons, biexcitons, etc.) are still sharper than other defects in the semiconductor.

CdTe has exhibited exciton absorption at both the Γ point (Potter and Simmons (1990)) and the L point (Potter and Simmons (1991)). The latter corresponds to a two-dimensional exciton, as noted earlier. Exciton peaks are also apparent in some indirect-bandgap semiconductors, such as Ge, where an exciton absorption peak is present near 0.885 eV.

5.7.6 Defect-associated transitions

If both donor (D) and acceptor (A) defects are present in the same semiconductor, a different degree of compensation is achieved depending on the D/A ratio:

$$D/A = 1 \qquad \text{fully compensated}$$
$$D/A < 1 \qquad \text{partially compensated}$$
$$D/A > 1 \qquad \text{overcompensated}$$

The energy-matching condition for a donor–acceptor transition reduces to

$$h\nu = \xi_G - \xi_d - \xi_a + \frac{q^2}{\epsilon r} \tag{5.65}$$

where r is the distance between the donor and acceptor impurities, ξ_d is the donor binding energy, and ξ_a is the acceptor binding energy, as shown here:

$$\xi_d = \xi_G - \xi_D$$
$$\xi_a = \xi_v + \xi_A = \xi_A \tag{5.66}$$

with ξ_D and ξ_A as the donor defect and the acceptor defect energies, respectively, and with ξ_v as the valence band energy, which is generally defined as zero. The great variability in donor-to-acceptor distance leads to a large variation in the energy-matching condition, which ranges from the value at $r = $ infinity to $r = $ nearest-neighbor distance. Thus, there is a wide range of optical transition energies, giving a broad donor–acceptor absorption peak:

$$\xi_G - \xi_d - \xi_a \leq h\nu \leq \xi_G - \xi_d - \xi_a + \frac{q^2}{\epsilon r_{NN}} \tag{5.67}$$

Since the Coulomb energy of the donor-acceptor pair is still less than the sum of the donor and acceptor binding energies, the entire broad donor–acceptor absorption peak occurs below the band-edge absorption energy.

5.7.7 Luminescence

Luminescence is the emission of light from the radiative de-excitation of carriers. It corresponds to the emission of light as carriers (electrons and holes) recombine. Often the recombination is from the conduction to the valence band. However, if intermediate states are present, they may serve as a source or sink of carriers. For example, electrons may drop from the conduction band to donor states by a nonradiative process and then radiatively drop to the valence or an acceptor band. The photoluminescence probability is high in direct-gap and low in indirect-gap semiconductors. It is also high in materials having large concentrations of defects as these defects may trap carriers and release them through a radiative recombination. Phosphor materials work by this mechanism.

Surface states can play a major role in luminescence, since they have well-defined defect states. Materials with large surface areas show enhanced luminescence. In addition, if impurities or defects concentrate at the surface of an interface, they may act as a source of strong luminescence.

The means of excitation of carriers prior to luminescence de-excitation usually determines the name of the process. For example, electroluminescence is caused by the electrical excitation of a large density of carriers by a current. Cathodoluminescence excites carriers by an electron beam, often in an electron microscope. X-ray luminescence excites carriers with a beam of x-rays, triboluminescence by mechanical excitation, chemiluminescence by chemical reactions. The most widely used is photoluminescence (PL), in which the carriers are photoexcited by UV or visible light.

Thermoluminescence (TL) is not a result of thermal excitation of carriers; rather, it is a thermal trigger used to release trapped carriers in defects due to excitation by other means, such as UV exposure. This is the basis of thermoluminescence dating that is used on artifacts that have had a prior heating during formation. Pottery lends itself well to TL dating, since it is formed by firing at high temperatures. After that, as the pottery is kept outdoors, incident UV causes photoexcited carriers that become trapped in the ceramic. If the pottery is buried underground, residual radioactivity (uranium) will also induce photoexcitation of carriers. In order to determine the age of the object, it is placed in a dark oven and heated. The heating releases the trapped carriers, some of which recombine radiatively and emit a luminescence signal. The age of the artifact is proportional to the amount of luminescence thermally released. Often a more detailed analysis is sought. In this case, the sample is heated very slowly, allowing the lower-energy traps to be excited first. The resulting TL will increase until the traps corresponding to the thermal energy are depleted. As the temperature is raised further, higher-energy traps will become activated and depleted. Such a temperature scan can give the distribution of traps in energy and the associated population densities.

5.7.8 Nonradiative processes

Nonradiative processes are interesting in the studies of optical materials, since they are important competition to the production of light by luminescence. There are three major sources of nonradiative recombination processes: Auger, multiphonon emission, and surface states and defects. Nonradiative processes are important and dominate the recombination processes of many semiconductors.

The Auger process consists of a transfer of energy from the recombination process to a second electron that subsequently dissipates the energy by a variety of processes with no photon emission. The second electron is usually a core electron and loses its energy by creating phonons. This is the most common process for nonradiative recombinations. It is rapid and can easily supplant normal luminescence.

Multiphonon emission is a common process by which excited carriers can lose energy. If photocarriers are excited to states well above the bottom of the conduction band (e.g., incidence of high-energy light), the electrons will cascade down to the bottom of the conduction band by emitting phonons. Generally, however, it is more difficult to satisfy energy and momentum conservation if phonons are used to de-excite carriers across the bandgap. In such cases, Auger processes have a greater cross section.

Surface states and defects often act as intermediate real levels that electrons can seek in place of direct recombination. Thus electrons may jump down in energy to surface states or defect states and then jump down to the valence band. These intermediate transitions are most often nonradiative. Due to the breadth of the defect and surface states, energy and momentum conservation are more easily satisfied, thus increasing the probability of multiphonon emission over that for the direct transition.

As a rule of thumb, nonradiative processes are favored by wide distributions of state energies. Such distributions improve the matching of multiphonon emission energies with the available states (e.g., the sharper the transition, the better is the chance for a radiative transition, since energy and momentum conservation for phonon emission is more difficult to achieve).

5.7.9   Polaritons

Polaritons pertain to the interaction between a photon and a phonon. Their concept is important to the science of optical materials, especially the spectroscopy of materials. Polaritons are important in the region where the optical field interacts with the vibrational motions of the underlying lattice (phonons). Since these processes occur generally far from the bandgap, they are germane to both semiconductors and insulators. The reader is referred to Ibach and Luth (1996) for more detailed coverage.

Figure 5.15 shows what happens in k-space. The photon interacts with the transverse waves of a material through a set of equations that satisfy

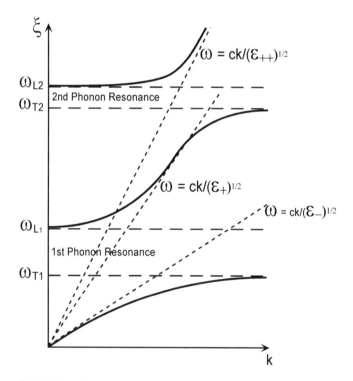

Figure 5.15: Plot of energy vs wave vector for a photon in the frequency region near two phonon resonances. At small k-vector values (typical of the photon dispersion), the transverse and longitudinal modes of the system (dashed lines) appear flat. The dotted lines show the solutions of the polariton wave.

Maxwell's equations. Solutions for coupled electromagnetic and mechanical waves (polaritons) are possible under the following condition:

$$\omega = \frac{ck}{n} = \frac{ck}{\sqrt{\varepsilon(\omega)}} \tag{5.68}$$

As the frequency of the wave is increased, the photon approaches an optical phonon resonance. Let us assume that this resonance is an IR absorption process. An interaction now takes place between the photon and the phonon (absorption of the photon, or Raman scattering of the photon by the phonon). The photon interacts with the transverse optical phonon and this interaction is called a polariton. What happens here is that the real part of the dielectric function of the material goes through an oscillation, as shown in Fig. 3.4 and reproduced here (Fig. 5.16), while the imaginary part shows a loss peak. This is expressed mathematically for a single homogeneous phonon absorption as:

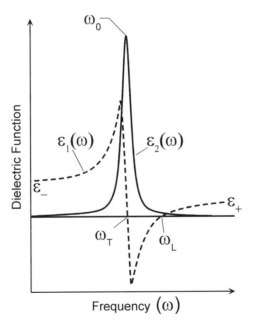

Frequency (ω)

Figure 5.16: Resonance in the dielectric function (real and imaginary parts) showing the values of ω_0, ω_T, and ω_L. Normally, the transition is sharp and $\omega_T = \omega_0$. The difference has been exaggerated in the drawing. ω_L occurs when $\epsilon_1 = 0$. Note that ϵ_- is larger than ϵ_+.

$$\varepsilon_1(\omega) = \varepsilon_+ + [\varepsilon_- - \varepsilon_+]\frac{\omega_{01}^2(\omega_{01}^2 - \omega^2)}{(\omega_{01}^2 - \omega^2)^2 + \gamma^2\omega^2}$$

$$\varepsilon_2(\omega) = [\varepsilon_- - \varepsilon_+]\frac{\omega_{01}^2\gamma\omega}{(\omega_{01}^2 - \omega^2)^2 + \gamma^2\omega^2}$$

$$(5.69)$$

where ω_{01} is the phonon resonance frequency, ε_- is the dielectric constant value before the resonance, and ε_+ is the dielectric constant after the resonance. The $+$ and $-$ symbols are used to denote frequencies that are respectively higher and lower than ω_{01}. The value of ε_+ is lower than ε_- because once the optical frequency has passed the ω_{01} frequency, that (phonon) process can no longer participate in the polarization process so the dielectric constant is lower. As a result, the velocity of the optical wave between resonances is progressively higher (higher slope in Fig. 5.15) with increasing frequency. The transverse resonant frequency $\omega_T \simeq \omega_0$ which correponds to the peak in ε_2. For an ideal crystal, ε_2 is a delta function and ε_1 goes to $\pm \infty$ at ω_T. For a real material, the resonance is broader and corresponds to the interaction linewidth (IR or Raman). The longitudinal frequency is the point where $\varepsilon(\omega)$ crosses 0 on the return to positive values. There is a relationship between the transverse and longitudinal resonances and the dielectric constants derived by Lyddane–Sachs–Teller (LST):

$$\frac{\omega_L^2}{\omega_T^2} = \frac{\varepsilon_-}{\varepsilon_+}$$

$$(5.70)$$

The polariton dispersion behavior shows (Fig. 5.15) an asymptotic approach to the transverse optical phonon near ω_{T1}. A longitudinal wave can exist only when the real part of the dielectric function vanishes. This happens in the dielectric function oscillation after the resonance (Fig. 5.16), and it defines the longitudinal optical phonon and the frequency where the polariton regains normal dispersion.

Between these transverse optical (TO) frequencies and the longitudinal optical (LO) frequencies, the propagation vector is imaginary, and the electromagnetic wave cannot propagate in the medium.

Once the photon returns to normal dispersive behavior, it is likely to encounter another phonon resonance at a higher frequency. The same process takes place again. The lower-frequency dielectric constant is now near ϵ_+ and the upper frequency dielectric constant goes to ϵ_{++}, as shown in Fig. 5.15. The new upper dielectric constant ϵ_{++} is lower than the other two, since the optical electromagnetic field oscillation has become too fast for the second polarization process at ω_{02}. Finally, the combined LST Equations become:

$$\frac{\varepsilon(0)}{\varepsilon_\infty} = \prod_i \frac{\omega_{Li}^2}{\omega_{Ti}^2} \tag{5.71}$$

where $\varepsilon(0)$ is the DC value of the dielectric constant and ϵ_∞ is the dielectric constant at infinite frequency. The polariton resonances are normal modes of the dielectric function of the system and are measurable with Raman and IR spectroscopy if the symmetries permit it.

5.8 Measurements

5.8.1 Polarized light

In order to understand more fully the results of various measurements conducted on semiconductors, we need to revisit the interaction between photons and electronic states. This interaction is mediated by the product of the initial and final densities of states in the electron transition and the optical matrix element, as expressed in Eq. (5.50). It is instructive to examine the wave functions of the conduction-band and valence-band states. To accomplish this, we need to use a quantum mechanical shorthand to express the properties of the wave functions. These follow bracket notation, in which the expression $|s\rangle$ means a wavefunction of the s-state and the expression $\langle p|$ means a wave function of the p-state, with the direction of the bracket indicating that it is the complex conjugate function. The addition of spin information is done the same way, with spin direction indicated by an up or down arrow. The expectation value of a physical quantity such as the x-component of the polarization P_x is expressed as $\langle s|P_x|s\rangle$.

The conduction band has spherically symmetric s-type states, with the following quantum numbers and wavefunction:

$$\textit{Conduction band}: \qquad L = 0 \qquad J = 1/2 \qquad m_j = \pm 1/2 \tag{5.72}$$

$$\textit{Wavefunctions}: \qquad (|s\rangle)|\uparrow\rangle \qquad \text{and} \qquad (|s\rangle)|\downarrow\rangle$$

where the term $|s\rangle$ indicates the s orbital quantum mechanical wave function and the term $|\uparrow\rangle$ indicates the spin state.

The valence bands involve the p-type orbitals and are described by the following quantum numbers and wavefunctions:

Heavy-hole band : $L = 1$ $J = 3/2$ $m_j = \pm 3/2$ (5.73)

Wavefunctions : $(1/\sqrt{2})(|x\rangle + i|y\rangle)|\uparrow\rangle$ and $(1/\sqrt{2})(|x\rangle - i|y\rangle)|\downarrow\rangle$

Light-hole band : $L = 1$ $J = 3/2$ $m_j = \pm 1/2$ (5.74)

Wavefunctions : $(1/\sqrt{6})(|x\rangle - i|y\rangle)|\uparrow\rangle + 2|z\rangle|\downarrow\rangle$ and $(-1/\sqrt{6})(|x\rangle + i|y\rangle)|\downarrow\rangle - 2|z\rangle|\uparrow\rangle$

Split-off band : $L = 1$ $J = 1/2$ $m_j = \pm 1/2$ (5.75)

Wavefunctions : $(-1/\sqrt{3})(|x\rangle - i|y\rangle)|\uparrow\rangle - |z\rangle|\downarrow\rangle$ and $(-1/\sqrt{3})(|x\rangle + i|y\rangle)|\downarrow\rangle + |z\rangle|\uparrow\rangle$

The effect of light of different polarization can be studied by examining the transition matrix element for the various polarizations:

Linear polarization : P_x P_y (5.76)

Circular polarization : $\sigma_+ = (1/\sqrt{2})(P_x + iP_y)$ and $\sigma_- = (1/\sqrt{2})(P_x - iP_y)$

(5.77)

Symmetry attributes of the wave functions give the following results for the various components of the transition matrix elements:

$$\langle s|P_x|x\rangle = \langle s|P_y|y\rangle = \text{nonzero transition matrix elements} \quad (5.78)$$

$$\langle s|P_x|y\rangle = \langle s|P_x|z\rangle = \langle s|P_y|x\rangle = \langle s|P_y|z\rangle = 0 \quad (5.79)$$

Since the dipole moment, P, has no effect on spin, the spin must be unchanged and the following relationships hold:

$$\langle\uparrow|\uparrow\rangle = \langle\downarrow|\downarrow\rangle = 1 \quad \text{while} \quad \langle\uparrow|\downarrow\rangle = \langle\downarrow|\uparrow\rangle = 0 \quad (5.80)$$

Consider the dipole matrix elements for transitions between the valence and conduction bands:

Heavy-hole transitions to the conduction band under linearly polarized light:

$$(1/\sqrt{2})\langle\downarrow|\langle s|P_x|(|x\rangle - i|y\rangle)|\downarrow\rangle = (1/\sqrt{2})\langle\downarrow|\langle s|P_x|x\rangle|\downarrow\rangle = (1/\sqrt{2})P \quad (5.81)$$

and the spin is unchanged in the transition. The same is true for transitions from the other valence-band states under linearly polarized light, so electrons photoexcited by linearly polarized light have both types of spins equally distributed in the conduction band. The following diagram of Fig. 5.17 shows results for transitions between the valence- and conduction bands when spin is taken into account.

Polarized transition probabilities

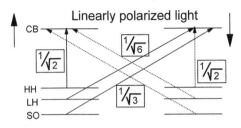

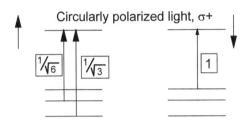

Figure 5.17: Interband transition probabilities when considering spin (shown as an up or down arrow in the sketch) for linearly and circularly polarized light. The states on the right all have ↓ spin; those on the left have ↑ spin.

Heavy-hole transitions to the conduction band under circularly polarized light: Expressed as $\sigma_+ = (1/\sqrt{2})(P_x + iP_y)$, with $P_x = P_y = P_0$, these can be calculated as follows:

$$(1/2)\langle\uparrow|\langle s|(P_x + iP_y)|(|x\rangle + i|y\rangle)|\uparrow\rangle = (1/2)[\langle s|P_x|x\rangle - \langle s|P_y|y\rangle]\langle\uparrow|\uparrow\rangle = 0 \quad (5.82)$$

$$(1/2)\langle\downarrow|\langle s|(P_x + iP_y)|(|x\rangle - i|y\rangle)|\downarrow\rangle = (1/2)[\langle s|P_x|x\rangle + \langle s|P_y|y\rangle]\langle\downarrow|\downarrow\rangle = P_0 \quad (5.83)$$

Thus, if the incident light is circularly polarized in the σ_+ direction, then no spin-up transitions are possible and the photoexcited electrons from the heavy-hole band are spin-down only.

The same treatment applied to the light-hole and the split-off bands gives a zero transition probability for the spin-down electrons and transition probabilities of $(1/3)\,P_0^2$ and $(2/3)\,P_0^2$ for the spin-up electrons. In summing the various contributions from the three valence bands, it is necessary to take into account the density of states for the various bands. Equation (5.53) shows that the density of states varies as $\mu^{3/2}$, where μ is

Valence band	$|\langle p \rangle|^2$	μ/m_e	$|\langle p \rangle|^2 \mu^{3/2}$	Relative strength
Heavy-hole	1	0.9	0.85	8
Light-hole	0.33	0.55	0.13	1.2
Split-off	0.67	0.71	0.40	3.8

Table 5.2: Relative transition strengths for circularly polarized light.

the reduced effective mass for each band involved in the transition as follows:

$$\frac{1}{\mu} = \frac{1}{m_e} + \frac{1}{m_h} \tag{5.84}$$

where m_h must be calculated for each hole band. Taking an example from GaAs, we find the reduced masses and density of states calculated in Table 5.2.

Since the valence bands are degenerate at $k = 0$, both the heavy-hole and light-hole transitions will take place under illumination with light of bandgap energy. Under this condition, the spin-down population is larger by a factor of 8/1.2, and it makes up 87% of the total population. Therefore, the spin-down population dominates the photoexcited electrons (by a factor of 6.5), and a spin-polarized population of electrons is formed in the conduction band by exposure to circularly polarized light. The split-off band contributes only when the illumination has an energy equal to the sum of the bandgap energy and the spin-orbit splitting energy. Even in that case, over 60% of the pins will be down. Electron–electron, impurity, and phonon scattering do not flip the electron spin. Therefore, this spin-polarized population has a long relaxation time and can be measured by polarized photoluminescence. The spin relaxation process in the semiconductor can be measured by photoluminescence depolarization (e.g., measuring the emitted light polarization). Magnetic impurities, if present, can interact with the electronic spin and flip it.

Biexcitons are made up of two spin-paired excitons. Linearly polarized light will form both excitons and biexcitons, since it can excite carriers of both polarization with equal probability. However, circularly polarized light can excite only excitons. Biexcitons are formed only when two beams of opposite circular polarization are superimposed on the sample. This test is the best method for demonstrating that an observed PL emission corresponds to a transition from biexcitons.

$$\sigma_+ \to \text{excitons } (\downarrow)$$
$$\sigma_- \to \text{excitons } (\uparrow) \tag{5.85}$$
$$\sigma_+ + \sigma_- \to \text{excitons } (\uparrow) + \text{excitons } (\downarrow) \to \text{biexcitons}$$

Excitons are formed from two fermion particles, so they should be bosons. But the particles keep their identity in the exciton, and it can be polarized with circularly polarized light. This behavior is of great importance to nonlinear optical behavior of materials, as we shall see in Chapter 7, since exciton transitions can be bleached only if they act as fermions and not as bosons.

5.8.2 Absorption

Semiconductors absorb light with energy larger than their bandgap, so they act as low-pass filters. Absorption measurements can therefore be used to estimate their bandgap energy. Measurements at low temperatures are more meaningful due to the reduced amount of band tailing. Equation (5.54) gives the dependence of the absorption coefficient on frequency for direct-gap semiconductors. Equations (5.60) give the same dependence for indirect-gap semiconductors. In both cases, the material is made thin, and its surfaces are polished to reduce scattering. The absorbance is measured in a UV–visible–near-IR grating spectrometer. For direct-gap materials, one plots the product of absorption coefficient and energy squared against frequency [$(\alpha h \nu)^2$ vs ω]. If a straight line is obtained, it indicates a direct-gap absorption. Its extrapolation to zero absorption gives the bandgap energy. For indirect-gap materials, the analysis is more complex, for the absorbance curve must include the absorption and emission of phonons, as shown in Fig. 5.12. Here, one plots the square root of the same product of absorption coefficient and energy against frequency [$(\alpha h \nu)^{1/2}$ vs ω]. The curve should yield a straight line, with a break at low absorption values to give the effects of phonon absorption and emission.

Band tails are often present, and they add to the absorbance measured. If band tails are present, a thin sample is desired in order to obtain absorbance data up to higher frequencies and to measure the intrinsic absorption edge. In some materials, the absorption edge appears to have an exponential behavior. Exponential behavior is often associated with band tails or highly defected or amorphous semiconductors. The exponential behavior results from subband states buried in the edge.

The result is an Urbach tail, as described in Section 5.7.4. At low temperatures, the exciton absorption will become more distinct, and the band edge should acquire the form of a Gaussian tail, as shown in Fig. 5.14.

In measurements on amorphous semiconductors, authors have often referred to Tauc plots. These are calculations of optical absorption in regions dominated by absorption tails. As already discussed, one assumes Gaussian tails on both sides of the bandgap. Tauc and co-workers (1966) show that the absorption edge of amorphous Ge follows the following relation:

$$(h\nu)^2 \epsilon_2 = A(h\nu - \xi_{bG})^2 \tag{5.86}$$

where ϵ_2 is the imaginary part of the dielectric constant.

5.8.3   Photoluminescence

Many direct-gap semiconductors exhibit strong photoluminescence at low temperatures. Apparatus for measuring photoluminescence consists simply of an intense monochromatic light source (most often a laser) and a grating monochromator and detector. The sample is often held in a cold chamber, which is either a closed cycle He refrigerator or in liquid He. Optical-quality windows allow the light in and out of the chamber. Good contact is required between the sample and the cold heat sink in order to avoid heating the sample under the light exposure. Raman spectrometers can be used for detection, but isolation from the excitation line is not as critical as in Raman scattering.

Photoluminescence is an excellent method for evaluating the purity of semiconductors. At temperatures below about 10 K, intrinsic semiconductors will exhibit sharp PL lines, corresponding to transitions between the free-exciton states and the valence band. If impurities are present in small concentrations, donor-bound excitons will exhibit broader PL peaks at slightly lower energy than the free excitons. The peaks are broadened by the variation in structure around the impurities. Acceptor-bound exciton PL peaks, also broadened, are found at lower energies. If higher concentrations of impurities are present, very broad donor-to-acceptor peaks are found at much lower energies. These are broadened by the wide variation in separation between the donor and acceptor sites in the lattice.

In order of decreasing frequency, the observed processes are:

1. free excitons,
2. free excitons shifted by LO phonons and TO phonons,
3. excitons bound to impurities (donor-bound excitons, acceptor-bound excitons),
4. bound excitons shifted by LO and TO phonons,
5. donor and acceptor peaks (donor to valence band, conduction band to acceptors, donors to acceptors),
6. deep-level defects.

All these peaks, except the very narrow free-exciton peaks, will display single or multiple phonon echoes, due to the emission of phonons concurrent with the optical transition.

At higher temperatures, the PL emission corresponds to interband transitions (electron–hole recombination). Free excitons have low binding energies, and their states become blurred into the conduction band tail. Some bound exciton transitions may remain distinct up to room

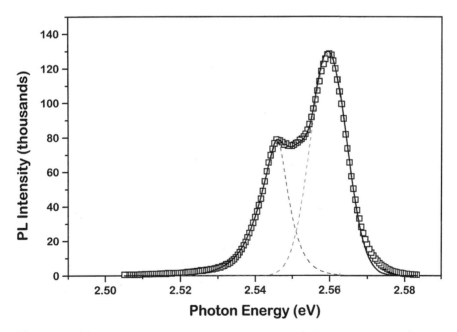

Figure 5.18: Photoluminescence data (Wang and Simmons 1995) from a $Cd_{0.2}Zn_{0.8}Se/ZnSe$ multiple quantum well with 10 periods, made by molecular beam epitaxy (MBE) showing both exciton and biexciton transitions.

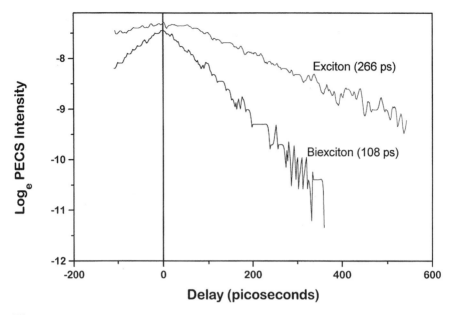

Figure 5.19: Photoluminescence decay times (Wang and Simmons 1995) measured using picosecond photoexcitation correlation spectroscopy (PECS) and a population-mixing technique (Rosen et al. 1984).

temperature. Donor-to-acceptor transitions are observable at room temperature.

Figure 5.18 shows the photoluminescence from a $Cd_{0.2}Zn_{0.8}Se/ZnSe$ multiple quantum well grown by molecular beam epitaxy (see Section 5.9.1). The data, taken at 10 K, shows two distinct peaks, fitted with points representing a theoretical model. The higher-energy peak at 2.562 eV is attributed to the exciton. The peak at 2.548 eV is attributed to the biexciton. The theoretical calculations of line shape are based on a reverse Boltzmann distribution, expected of a biexciton population. Furthermore, these assignments were based on the linear dependence of the 2.548 eV peak intensity on the square of the incident intensity and on the disappearance of that peak under circularly polarized light (σ_+ or σ_-). The low-energy peak reappears under illumination either by 2 cross-circularly polarized (σ_+ and σ_-) beams or by a linearly polarized single beam, as already discussed in Section 5.8.1. Figure 5.19 presents a measurement of photoluminescence decay, showing the exciton with twice the lifetime of the biexciton, as expected from theory.

Stimulated photoluminescence (SPL) is measured at a given wavelength as a function of variable, stimulating light frequency. Thus, as one lowers the incident light energy below the energy of the initial state in the PL transition, the measured intensity drops. By scanning the incident light frequency over a broad range, various PL transition energy levels may be identified.

Often the number of defects is so high that PL is not observed. This results from the presence of so many traps with associated nonradiative recombinations that the emitted radiative light is too low in intensity for detection. However, cathodoluminescence (CL) is often possible. This measurement uses an electron beam intensity high enough to generate light even though many defect traps are present. The high number of excited electrons is often greater than the trap concentration, and a strong radiative optical emission is possible for defect analysis. With CL measurements, it is possible to scan a small electron beam across the sample and observe spatial differences in defect concentration if present.

5.8.4 Heavily doped semiconductors

Doped semiconductors exhibit different behaviors, depending on whether they are n-type or p-type semiconductors. The absorption and PL transitions take place between either the donor levels and the valence band or the conduction band and the acceptor levels. Since the defect levels show essentially no dispersion, the variation of absorption or luminescence with energy reflects only the dispersion of the corresponding intrinsic band.

5.8.4.1 Acceptor-band absorption and luminescence

Semiconductors doped to form p-type carriers have acceptor bands. Photoluminescence transitions go from the conduction band or exciton states to the acceptor band. The measurements generally show free excitons, bound excitons, and D–A transitions.

In hot-acceptor luminescence, the intensity of the PL is measured as a function of the stimulating light frequency, as in SPL. The transition intensity, I, is given in terms of the optical matrix elements and the joint density of states as

$$I = |\langle i|\vec{p}|f\rangle|^2 n_i(\omega)n_f(\omega) \qquad (5.87)$$

Since the acceptor levels have no dispersion, $n_i(\omega)$ is frequency independent. The resulting frequency dispersion of I reflects the shape of the density of states of the conduction band. This measurement is conducted using the frequency dependence of absorption.

In photoluminescence, the incident light excites a delta function of population that relaxes to the bottom of the conduction band by the emission of optical phonons. Luminescence takes place between the bottom of the conduction band and the acceptor states. By monitoring the intensity of the PL at the wavelength corresponding to the conduction-to-acceptor-band transition and by varying the incident light frequency, the electrons are photoexcited from the heavy-hole and light-hole valence bands. The data shows both HH and LH dispersion and phonon replicas (Stanton and Bailey 1991).

Measurements of the band structure of semiconductors are done either with ultraviolet photon spectroscopy (UPS) or by the more accurate method of hot-electron acceptor luminescence.

5.8.4.2   Donor-band absorption

Semiconductors doped with n-type carriers have donor bands. Photoluminescence transitions go from the donor to the valence band. The measurements cannot show any exciton transitions, since these states are overcome when the donor impurities reach high concentrations, as in intentional doping. The donor band has no dispersion, so the shape of the heavy-hole valence band is accessible to absorption spectroscopy. If band tails are present, they can also be measured.

5.8.5   Differential reflection spectroscopy

In many semiconductors, such as GaAs, the conduction band has several distinct valleys. Often the central valley (Γ point) has only a slightly lower energy than the other valleys. Thus photoexcitation of carriers, followed by phonon interactions, can send carriers rapidly into the side valleys. This does not occur in the hole bands, and the holes relax rapidly to the top of the hole band. The electrons, however, take more time to scatter back to the Γ point (times on the order of picoseconds).

Differential reflection spectroscopy has been described in Appendix 2A.

The reflection coefficient is sensitive to the total number of carriers in each band as follows:

$$\frac{\delta R}{R} = -\left(\frac{4\pi e^2}{\varepsilon_0 m_e \omega_L^2}\right) \sum_j \frac{n_j}{m_j} \tag{5.88}$$

where n_j is the number of carriers in each band, m_j is the effective mass of the bottom of that band, and ω_L is the LO phonon frequency. The reflectivity is most sensitive to the population of the band with the lightest effective mass. Since the effective mass of electrons is lowest at the Γ point, differential reflectivity measurements essentially obtain the concentration of electrons in the central valley of the conduction band. As electrons are photoexcited with light of higher energy than the bandgap energy $(h_v > \xi_G)$, they rapidly scatter to the conduction-band side valleys. The time-dependent differential reflectivity measurement thus gives the rate at which the electrons thermalize back to the central valley minus the rate at which they recombine. Associated with time-dependent PL measurements, the differential reflectivity can yield the electron intervalley scattering times and their recombination times.

5.8.6 Summary of optical methods

Photoluminescence, Raman scattering, and infrared absorption spectroscopies combine to reveal much information about semiconductors. Table 5.3 lists the appropriate methods for measuring various properties.

5.9 Materials and properties

5.9.1 Fabrication and growth

Most optical applications of semiconductors involve either photon absorption to form free carriers (photodetectors, solar cells) or the formation of photons by free-carrier recombination (light-emitting diodes, lasers). Both kinds of applications require high-quality single crystals, devoid of defects, grain boundaries, or impurities that act as electron scatterers, traps, and nonradiative recombination centers. As a consequence of the requirement of high structural quality, fabrication and growth conditions for the semiconductors must be carefully controlled for

Property	UV-VIS	Raman	PL	IR	Other
Band structure:					
energy gap	X		X		UPS
excitons, biexcitons	X		X		TRS
effective mass				X	
band offset			X		
band tails	X		X		
Free-carrier properties:					
free carrier density		X		X	Hall
mobility		X		X	Hall
scattering time					DRS, TRS
resistivity					DCR
Lattice properties:					
crystallinity		X	X		
alloy composition		X	X	X	XPS
orientation		X			XRD
stress		X	X		XRD
Impurities and defects:					
type	X	X	X	X	XRF
concentration	X	X	X	X	XRF
trapping efficiency			X		TRS
Microstructure:					
interface traps		X	X		
surface effects		X	X		NSOM
quantum well tunneling			X		

UPS = UV photoelectron spectroscopy; TRS = time-resolved optical spectroscopy; DRS = differential reflection spectroscopy; Hall = Hall effect; DCR = DC resistivity; XPS = x-ray photoelectron spectroscopy; XRD = x-ray diffraction; XRF = x-ray fluorescence; NSOM = near-field scanning optical microscopy or spectroscopy

Table 5.3: Semiconductor property measurements by optical spectroscopy.

most applications. Most fabrication processes fall into two categories: fabrication of substrates and fabrication of films. In both approaches, the key is fabrication of single crystals with few structural defects.

Typical methods for the growth of large boules of semiconductors are the Czachrolsky (CZ), the hydrothermal (HT), and the Bridgman hot-pressing methods. All these methods are complex; the reader is referred to Campbell (1996) and Mahajan and Sree Harsha (1999) for adequate detail. Here, we only sketch out the principal points of each.

The Czachrolsky method essentially draws a seed crystal from the top surface of a homogeneous melt. The crystal is rotated slowly to keep cylindrical symmetry while being gradually pulled from the melt. Growth speeds are a few mm/min. The draw process is conducted under conditions that favor only low-roughness crystal growth conditions (e.g., atoms may deposit on a step in the structure but cannot form a nucleus; thus, only layer growth occurs). A spiral dislocation will thus form a single crystal by layer growth as a single plane is extended continuously to form the crystal. To achieve this condition, the condensation energy and surface entropy are kept very low so that only monolayer growth is possible. The low condensation energy avoids nucleation by reducing the probability of growth sufficiently that only atoms with two solid surfaces can condense. Thus, there is not enough energy to grow a new nucleus on the free surface. The draw speed is very slow to allow the extended atomic diffusion necessary to place the proper atom at the edge of the growth step. These conditions are difficult to maintain, so the CZ method requires strict control over the draw speed and temperature profile in the draw region. In-line sensors and dynamic computer controls have achieved boules of Si large enough to produce 300-mm wafers. Impurities in Si growth are carbon from the crucible and oxygen from the air. GaAs and InP crystals are drawn under a molten boron trioxide layer to avoid volatilization of the As and the P. This is called liquid encapsulated Czachrolsky (LEC). Often, boron is a contaminant.

Hydrothermal methods are used for the fabrication of more complex crystal compositions that are not possible with the CZ method. The HT method yields smaller crystals and is less controllable. An aqueous solution of the desired material is supersaturated at high temperatures, and the crystal grows slowly around a seed. The supersaturation is controlled with temperature and pressure so that the atoms condense only on the seed and cannot nucleate freely or condense on the container walls or any other heterogeneous nucleation site. This is a slow process that can take up to several months per crystal. A variety of complex compositions have been grown by this technique, including the majority of ionic crystals.

Bridgman hot pressing is the most difficult process to control adequately because of the demands of extreme temperature and pressure on the container material and the high restriction on molecular diffusion. Growth which can be either vertical or horizontal, is from a seed crystal. It is accomplished by varying a temperature distribution along the length of the crucible. The distribution slowly lowers the temperature at the growth front below the melting temperature/pressure of the solid.

Among interesting variations on these methods is the CZ skull-melting method, which uses induction heating of a melt in a cooled vessel so that the container walls are never in contact with the liquid. The melt region is restricted to the central part of the material, so contamination and heterogeneous nucleation at the wall are essentially eliminated. High-quality, single-crystal cubic zirconia has been grown by this method. Also notable is float-zone refining, which seeds with a crystal touching the top of the melt region and consists of an RF induction-heating source that is slowly moved downward, forming a molten band in the starting polycrystalline rod and sweeping that band along the length of the rod. Float-zone refining is also used to purify the starting rod, since it allows the impurities to remain in the molten band that passes to the ends of the rod.

Space-borne melters are also being investigated for a true containerless process. The induction-heated melt is kept localized by acoustic standing waves. So while the process must be conducted in space to remove the effect of gravity, the method still needs an ambient gas to transmit the acoustic waves.

Film growth is conducted on a substrate; consequently, layered single-crystal formation (epitaxial growth) is highly dependent on lattice matching of the film to the substrate crystal structure. If the substrate and the film are the same material, growth is called *homoepitaxy*. If the substrate is different, the growth is called *heteroepitaxy*. A small mismatch will cause dislocations to grow from the interface. This lattice-matching requirement is critical in all film growth, including the formation of single and multiple films. In the formation of multiple quantum wells, it is necessary to tailor the bandgap energies of adjacent films. Lattice-matching conditions play a major role in the selection of compound semiconductors that satisfy the lattice-matching require-

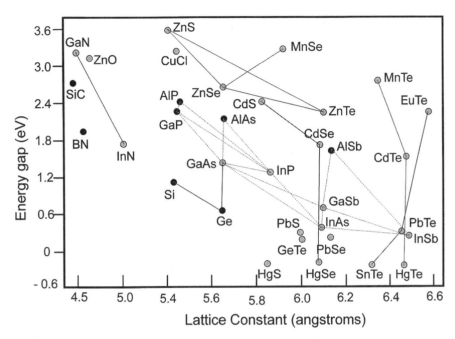

Figure 5.20: Typical energy gap vs lattice constant diagram for various semiconductors. The dark points are indirect-gap semiconductors. The lines joining different materials show alloys that have been formed.

ments. Figure 5.20 shows a plot of bandgap energy vs lattice constant for the major semiconductor compositions, and it is used universally to determine what systems are compatible with one another. Alloy systems that have been investigated are connected by a line in the diagram. In using this graph for compound semiconductors, one assumes the validity of Vagard's law, which predicts a linear relationship between bandgap energy and composition in compound semiconductors. It is customary also to assume that the effective lattice constant varies linearly with composition (x). In the formula $A_x B_{1-x}$ for the compound semiconductor of A and B components, or $AB_x C_{1-x}$ for the compound semiconductor of AB and AC components; the lattice constant and bandgap energy are expressed as:

Lattice constant: $a_{\text{alloy}} = a_1 x + a_2(1 - x)$

Bandgap energy: $\xi_{\text{alloy}} = \xi_1 x + \xi_2(1 - x)$

$$(5.89)$$

Film growth methods must form a high-purity vapor of each component of the semiconductor in a very high vacuum to avoid contamination, followed by slow, low-condensation energy deposition to form single crystals. Again, layer growth is desirable for high-quality crystals. There are many film growth methods, too numerous to cover here. The reader is referred to Campbell (1996) for more detail.

The importance of high-quality, single-crystal structures cannot be overemphasized. For example, ZnSe was a very promising material for the formation of blue-green lasers. However, it was abandoned, even after laser operation was demonstrated, because of the large structural defects that propagated in the films during operation. Their associated nonradiative recombination processes eventually quenched laser action. However, exceptions do occur, and the early GaN films that replaced ZnSe as candidates for blue-green laser applications did not eliminate defect structures. In their case, however, the defects did not grow with use, and they formed a defect acceptor band that still allowed radiative transitions from the conduction band. The result was a decrease in bandgap energy and lasing frequency, but lasing action was maintained.

The two major film growth techniques are chemical vapor deposition (CVD) and molecular beam epitaxy (MBE). Some materials, like GaAs laser diodes can be grown by liquid-phase epitaxy (LPE), which consists of sliding liquid Ga saturated with As over a GaAs substrate. GaAs precipitates upon contact with the substrate.

In CVD, gases are reacted in a hot reactor to decompose and precipitate as the desired compound on a substrate. Heating is accomplished by RF induction heating. The gases decompose in contact with the hot substrate. Sometimes, photons are used to trigger the precipitation. This is photo-assisted CVD. One typical reaction for the formation of Si consists of the decomposition of silane (SiH_4) into Si and hydrogen. If the starting materials used are organometallic compounds, the method is called MOCVD. The formation of GaAs from trimethyl gallium and arsine follows this reaction:

$$Ga(CH_3)_3 + AsH_3 \rightarrow GaAs + 3CH_4 \qquad (5.90)$$

One of the critical factors in reactor design is the temperature profile above the substrate. In collaboration with T. Anderson at the University of Florida, we designed a system whereby the gas region above the substrate could be monitored by Raman scattering. The method allowed a quantitative measure of the concentration of reactor components (arsine and trimethyl gallium) as well as GaAs as a function of height over the substrate. At the same time, by using the rotational bands of the carrier nitrogen gas, we could monitor accurately the temperature dependence of the region probed.

Chemical vapor deposition has also been used very successfully for nonsemiconductor applications, such as the formation of high-purity optical fibers for communications. Here, the serendipitous result that silicon tetrachloride ($SiCl_4$), germanium tetrachloride, boron trichloride BCl_3, and phosphorus pentachloride PCl_5 vaporize at very low temperatures (12–162 °C) compared to $FeCl_2$ (670 °C) and $CuCl_2$ (620 °C) causes a very large difference in vapor pressures between them at the boiling temperature of $SiCl_4$. Consequently, the glass components can be precipitated from the fluoride vapors with high purity and an Fe and Cu content below a few parts per billion, which is necessary for long-distance optical fibers.

The MBE method allows the greatest control over the purity of components and deposition conditions. The process is conducted under ultrahigh vacuum. The gas is provided as a well-collimated, molecular beam from an effusion cell. By combining molecular beams in predetermined proportions from several cells, compound semiconductors may be grown with high control over growth rates, composition, and other conditions. In situ testing of the beam and the deposited films allows for monitoring and controlling of the deposition process. Among the many methods used to determine the quality of film growth, reflective high-energy electron diffraction (RHEED) provides a structural analysis of the film in real time during growth.

5.9.2 Color

Semiconductors are low-pass filters. Thus, their color formation is subtractive. White light incident on the semiconductor has its upper frequencies absorbed and its lower frequencies transmitted, with the bandgap energy serving as the cutoff as follows:

$$\text{light absorbed:} \quad h\nu \geq \xi_G$$
$$\text{light transmitted:} \quad h\nu < \xi_G$$

(5.91)

Consequently, the color of the semiconductor is made up of the transmitted frequencies only and should follow Table 5.4. Colored semiconductors are found in paints [cadmium yellow (CdS), cadmium orange (CdS_xSe_{1-x}), vermillion (CdHgS)], and they are used to form colored glass filters in photography.

The actual color of semiconductors is best observed in thin platelets or films deposited on a transparent substrate. In high-ξ_G materials, the discrepancy between the subtractive and the actual colors comes from defect-band absorption. For example, diamonds are found with several colors. From its high bandgap energy, diamond should display no color, since it absorbs in the UV. Most diamonds follow this rule, and blue, green, and yellow diamonds are rare. Such colors result from the formation of deep donor (yellow) and deep acceptor (blue and green) bands resulting from nitrogen and boron impurities, respectively.

A discrepancy which occurs in low-ξ_G materials, as they exhibit essentially a silver color instead of black, results from the thermal excitation of electrons into the conduction band. These free electrons act to shield the interior of the semiconductor from the incident electromagnetic wave up to the plasma frequency, producing a high reflectivity coefficient. The plasma frequency is temperature dependent, since it is proportional to the total number of free electrons in the semiconductor, as follows:

Semiconductor	Bandgap energy, ξ_G	Actual color
Diamond	5.5 eV	Clear
GaN	3.4 eV	Green
ZnSe	2.6 eV	Yellow
CdS	2.5 eV	Yellow
CdSe	1.75 eV	Red
CdTe	1.5 eV	Red
Si	1.11 eV	Black
Ge	0.9 eV	Black
HgS	-0.3 eV	Black

Table 5.4: Bandgap and subtractive colors of semiconductors.

$$\omega_p = \frac{q_e^2}{m\varepsilon} N_f = \frac{q_e^2}{m\varepsilon} \int_{\xi_G}^{\infty} \frac{1}{1 + e^{(\xi-\xi_F)/k_B T}} d\xi \qquad (5.92)$$

5.9.3   Properties

What are important characteristics of semiconductors? Such a question could easily take several books to answer. Many of the important characteristics will vary depending on the application of the material. Here, we will discuss only a few basics.

We have seen how lattice constant, crystal structure, and the ease of propagation of extended defects can affect semiconductor applications. The bandgap energy is important in determining the wavelength of emission of semiconductor diodes and lasers. Relative bandgap energies are important for quantum confinement of free carriers in quantum well structures. Free-carrier dynamics and scattering processes affect the behavior of photoexcited carriers. Finally, contact potentials between layers and with metal electrodes are important. We will now examine these.

5.9.3.1   Bandgap energies

The first electron affinity $(e\psi)$ is the energy required to add an electron to a neutral atom. It is the energy difference between the vacuum level and the bottom of the conduction band. The first ionization potential (ξ_{ie}) is the energy required to remove the highest-energy valence electron from a neutral atom. It is the energy difference between the vacuum level and the top of the uppermost valence band. Consequently, the energy gap is the difference between them:

$$\xi_G = \xi_{ie} - e\psi \qquad (5.93)$$

5.9.3.2   Contact potentials

Contacts between metals and semiconductors and between different semiconductors involve the flow of free carriers across the interface, since the electron affinity varies from material to material. In a metal, the free-electron states reach up to the Fermi energy, ξ_{Fm}. The work function, $e\phi_m$, is the difference in energy between the vacuum level and the Fermi energy. Figure 5.21 shows an energy diagram of the bands in the metal and an intrinsic semiconductor when they are far apart, and the resultant change from making a contact between them. Far from each other, the two

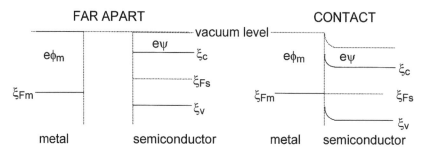

Figure 5.21: Contact potential between a metal and an intrinsic semiconductor.

materials have the same vacuum level and there is no charge on either.
When contact is established, current will flow between them until the
Fermi energies of the two materials across the interface equalizes by the
formation of a space charge near and across the interface. This will cause
a contact voltage to develop as the two vacuum levels are displaced with
respect to each other. The semiconductor Fermi energy occurs near the
middle of the gap in intrinsic (undoped) semiconductors, as predicted by
Eq. 5.36. In the sketch, since the semiconductor Fermi energy, ξ_{Fs}, is
higher than the metal's, ξ_{Fm}, the vacuum level of the semiconductor must
be lowered. This is accomplished by the buildup of a space charge of holes
on the interface side of the semiconductor and electrons on the interface
side of the metal. A positive voltage develops between them as follows:

$$\Delta V = \phi_m - (\psi + V_s) = \phi_m - \left(\psi + \frac{1}{2e}\xi_G\right) \qquad (5.94)$$

Note that this voltage is positive only, because the example showed a
metal work function larger than the sum of the electron affinity and half
the bandgap energy of the semiconductor. The contact potential can in
general have either sign, depending on the relative magnitudes of these
properties.

Figure 5.22 shows similar sketches for the contact between a metal and
either an n-type or a p-type doped semiconductor. The voltages developed
are:

$$\text{n}-\text{type:} \qquad \Delta V_{mn} = \phi_m - (\psi + V_n) > 0$$

$$\text{p}-\text{type:} \qquad \Delta V_{mp} = \phi_m - \left(\frac{1}{e}\xi_G - V_p\right) < 0 \qquad (5.95)$$

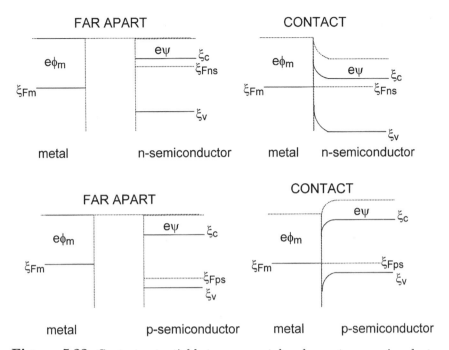

Figure 5.22: Contact potential between a metal and an n-type semiconductor and between a metal and a p-type semiconductor.

Contact between oppositely doped semiconductors, p–n junctions, are of great technical importance. This is shown in Fig. 5.23. The voltage developed in a p–n junction is calculated as:

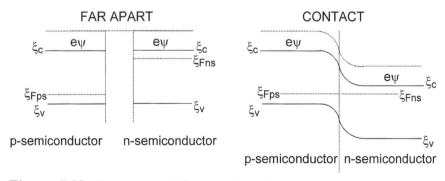

Figure 5.23: Contact potential at a p–n junction.

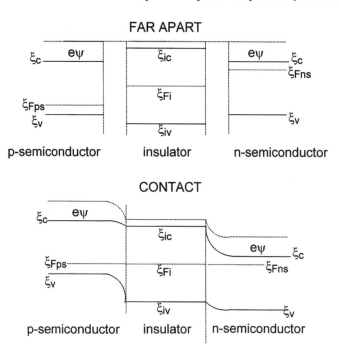

Figure 5.24: Contact potential at a p–i–n junction.

p–n junction: $\Delta V_{p-n} = (\psi_p - \psi_n) + \dfrac{1}{e}\xi_{Gp} - (V_p + V_n)$ (5.96)

Note that it depends on the bandgap energy of the p-type semiconductor.

Another important contact involves a p–n sandwich around an insulator: a p–i–n junction, as sketched in Fig. 5.24 and whose voltage is written as

p – i – n junction: $\Delta V_{p-i-n} = (\psi_p - \psi_n) + \dfrac{1}{e}\xi_{Gp} - (V_p + V_n)$ (5.97)

Note that the voltage developed is independent of the bandgap of the insulator!

A p–i–n junction sandwiched between the same metal with the simplifying assumptions that $V_n = V_p = 0$ and $\psi_i = \psi_n = \psi_p$ gives the following voltage, again independent of the insulator bandgap energy:

m – p – i – n – m junction: $\Delta V_{m-p-i-n-m} = \dfrac{1}{2e}\left(\xi_{Gp} + \xi_{Gn}\right)$ (5.98)

The general rules to follow are:

1. The Fermi energy must be the same for all materials in contact. The bias voltage developed is equal to the shift in vacuum level.
2. As charges flow across the interface to drive the Fermi energy up or down, positive charges drive the Fermi energy away from the conduction band, and negative charges drive the Fermi energy away from the valence band.
3. A forward bias lowers the Fermi energy of the p-type semiconductor below that of the n-type semiconductor; a reverse bias does the opposite.

5.10 Quantum well structures, quantum wires, and quantum dots

In the parabolic approximation, the wave functions representing the nearly free electron in a bulk semiconductor are formed from a sum of plane waves that use the effective mass concept to exhibit dispersion (k-space dependence). The same approximation will hold for the dynamic behavior of a free exciton. If boundaries are created over distances of the order of the free carrier or exciton Bohr radii, then confinement effects must be taken into account. Boundaries are formed in one dimension in multiple quantum wells that are stacked thin films, in two dimensions in quantum wires, and in three dimensions in quantum dots.

Looking first at quantum wells, let's examine the effect of forming a sandwich of a semiconductor between two insulators if one stacks thin films of different materials. To a first approximation, the insulator will be represented by a material with a much larger bandgap energy than the semiconductor. If the bandgap energy of the insulator is infinite, then the wave functions must vanish at the semiconductor–insulator interfaces. This reduces to the problem of a free electron confined in one dimension. The vanishing of the wave function at the interfaces $[\Psi(0) = \Psi(\ell) = 0]$ forces the energy states of the semiconductor to become quantized according to the following equation:

$$\xi_n = \frac{\hbar^2 n^2 \pi^2}{2m^* \ell^2} \tag{5.99}$$

The same argument may be used to examine the behavior of quantum dots. Here, the Schroedinger equation must be solved in three dimensions, with the wave function vanishing at the radius of the dot $[\Psi(a_0) = 0]$. In its simplest form (parabolic approximation), the Schroedinger equation may be written as follows for an electron and hole in a quantum dot in three dimensions:

$$-\left[\frac{\hbar^2}{2m_e^*}\nabla_e^2\Psi + \frac{\hbar^2}{2m_h^*}\nabla_h^2\Psi\right] - \frac{q_e^2}{4\pi\varepsilon_0 r} = \xi\Psi \qquad (5.100)$$

The equation is written in terms of spatial coordinates for the electron and the hole separately. It is possible to solve this equation by separating variables into center-of-mass coordinates, which includes the coordinate for the center of mass of the correlated electron and hole (R, Θ, Φ) and relative coordinates (r, θ, ϕ). This allows breaking up the Schroedinger equation into a center of mass with no potential energy term to represent the dynamic motion of the correlated electron hole and the more complex relative coordinates equation.

These equations are also subject to the boundary condition that $\Psi(a_0) = 0$. Under the condition that the dot radius is much larger than the exciton Bohr radius, $a_0 >> a_x$, an approximate solution may be obtained by assuming that the electron and hole remain correlated and the confinement of the boundary condition is acting only on the translational motion of the correlated exciton. In three dimensions, the wave function for the translation motion of the correlated exciton is composed of spherical harmonics, as discussed in Chapter 1. The requirement that the wave function vanish at the radius of the dot, a_0, forces the product of a_0 with the wave vector to equal the n zeroes of the spherical Bessel functions of order ℓ, which we will denote by $\alpha_{n\ell}$, to yield the following energy quantization:

$$\Delta\xi_{n\ell} = \frac{\hbar^2\alpha_{n\ell}^2}{2M^*a_0^2} \qquad (5.101)$$

where $\Delta\xi_{nl}$ is the change in binding energy from that of the exciton in the bulk solid and M^* is the effective mass of the exciton $[M^* = m_s^* + m_h^*]$. The correlated electron–hole solution holds well in the range $10a_x \geq a_0 \geq 3a_x$. Shifts in bandgap energy have been observed in many confined semiconductors. An example is seen in Fig. 5.14, where a blue shift in the absorption edge is apparent for decreasing particle size. The data for this material and for CdS follow Eq. (5.101) closely (Potter and Simmons 1988).

When the quantum dot radius approaches the exciton Bohr radius, the approximation breaks down and numerical solutions are possible. Efros and Efros (1982) have suggested a regime where the dot radius is smaller than the exciton Bohr radius, in which the correlation energy becomes negligible when compared to the electron and hole kinetic energies. In this case, the electron and hole are fully decoupled and have independent spherical Bessel wave functions that yield an energy shift of

$$\Delta \xi_{n\ell} = \frac{\hbar^2 \alpha_{n\ell}^2}{2\mu^* a_0^2} \qquad (5.102)$$

where μ^* is the effective reduced mass of the electron and hole $[1/\mu^* = 1/m_e^* + 1/m_h^*]$. Measurements show that this approximation is valid for quantum dots with radii smaller than the exciton Bohr radius.

These shifts in exciton energy (blue shifts) can be approximated by measuring the absorption edge or the position of the lowest exciton transition in semiconductors of various sizes. Results from our own studies showed that, in general, the data in the literature reporting absorption edge measurements suffers from an inability to estimate the effect of inhomogenous broadening of the exciton transition due to distributions in dot sizes in any real sample. Our studies showed that if the inhomogeous broadening effect is accounted for, then the blue shift in exciton energy with decreasing quantum dot sizes below the exciton Bohr radius follows closely the Efros–Efros approximation of decoupled electron and hole. These measurements were conducted on CdTe quantum dots varying in size between 200 and 12 Å, compared to an exciton Bohr radius of 75 Å. As mentioned earlier in the chapter, CdTe has two two-dimensional L-point excitons. These have Bohr radii of 12 and 15 Å, so, though crystals of 50 Å are smaller than the exciton Bohr radius at the Γ point, the same crystals are much larger than the L-point excitons. Thus the same crystals that exhibited decoupled carrier behavior at the Γ point [Eq. (5.102)] also exhibited translational confinement of coupled L-point excitons [Eq. (5.101)], at the same time supporting both extremes of the Efros–Efros model.

Measurements on the same materials showed an unexpected saturation of the blue shift at very small quantum dot sizes. An analysis of this behavior with 20-band $k \cdot p$ calculations showed that a saturation in the

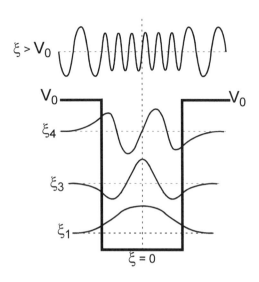

Free particle wave functions
in and near a potential well

Figure 5.25: Illustration of quantum confinement of the wave functions of a particle in a potential well for different energy levels. Note that due to the finite value of V_0, there is penetration of the wave function into the high potential region [as described by Eq. (5.103)]. The penetration increases with energy: $\Delta x(\xi_4) > \Delta x(\xi_3) > \Delta x(\xi_1)$. The fully free function $\xi > V_0$ still shows the effect of the underlying potential well.

blue shift arises at small sizes from a flattening of the conduction band due to mixing of the central valley states with the side valley (L and X points) conduction-band states.

Wave functions for a single particle in a quantum well are shown in Fig. 5.25. If the insulator does not have a very much larger bandgap energy than the semiconductor, then penetration of the semiconductor wave function is allowed into the insulator. Note that since the excited states of the particle have progressively higher energy, the tunneling distance into the insulator increases according to Eq. (5.103). This reduces the blue shift and can allow tunneling between neighboring semiconductor structures if there is a small separation between them. This condition occurs in multiple quantum wells, where the insulator is just another semiconductor with a small increase in bandgap energy and the insulator films are of the order of tunneling distances. The latter can be estimated using the following equation:

$$\Delta x = \frac{1}{2} \frac{\hbar}{\sqrt{2m(V_0 - \xi)}} \tag{5.103}$$

where Δx is the distance of penetration before the probability of finding the particle drops to $1/e$ of its value, V_0 is the potential energy barrier, and ξ is the energy of the particle. Note that this expression depends on the carrier mass, so electrons, since they have a smaller mass, can tunnel much further than excitons. Tunnel barriers in quantum wells can be designed as partial filters to pass electrons and not excitons. This is a clever method for increasing the relative concentration of excitons in a film.

Appendix 5A

Derivation of the Carrier Concentration Equation

The general carrier concentration equation [Eq. (5.32)] is as follows:

$$n = \frac{(2m_e^*)^{3/2}}{2\pi^2\hbar^3} \int_{\xi_G}^{\infty} \sqrt{\xi - \xi_G}\ \frac{1}{e^{(\xi-\xi_F)/k_BT} + 1} d\xi \qquad (5.104)$$

The solution is obtained by letting $y = \xi - \xi_G$ and $dy = d\xi$, with $\Delta = \xi_G - \xi_F$.

$$n = \frac{(2m_e^*)^{3/2}}{2\pi^2\hbar^3} \int_{0}^{\infty} dy\sqrt{y}\ \frac{1}{1 + e^{(y+\Delta)/k_BT}} \qquad (5.105)$$

Now, letting $x = y/k_BT$, $dx = dy/k_BT$, and $\eta = -\Delta/k_BT = (\xi_F - \xi_G)/k_BT$ yields:

$$n = \frac{(2m_e^*)^{3/2}}{2\pi^2\hbar^3}(k_BT)^{3/2} \int_{0}^{\infty} dx\sqrt{x}\frac{1}{1 + e^{x-\eta}}$$

$$= \frac{(2m_e^* k_BT)^{3/2}}{2\pi^2\hbar^3} F_{1/2}(\eta) \qquad (5.106)$$

$$F_{1/2}(\eta) = \exp[-0.32881 + 0.74041\eta - 0.045417\eta^2 - 8.797 \times 10^{-4}\eta^3$$

$$+ 1.5117 \times 10^{-4}\eta^4] \quad \text{for } -10 < \eta < +10$$

Appendix 5B

Derivation of Absorption from Direct Interband Transitions

We begin with an equation for a transition rate probability for a transition between states Ψ_i and Ψ_f, written in terms of the quantum mechanical expectation value of the dot product between the wave polarization vector, $\mathbf{e}$, and the momentum operator, $\mathbf{p}$, as follows:

$$W_{fi} = \frac{2\pi}{\hbar} \left(\frac{q_e E_0}{m_0 \omega}\right)^2 |<f|\mathbf{e} \cdot \mathbf{p}|i>|^2 \delta(\xi_f - \xi_i - \hbar\omega) \tag{5.107}$$

where

$$\mathbf{e} \cdot \mathbf{p} = -i\hbar(e_x \frac{\partial}{\partial x} + e_y \frac{\partial}{\partial y} + e_z \frac{\partial}{\partial z}) \tag{5.108}$$

The overall absorption is the difference between the power absorption, P_+, and the power emission, P_-, rates:

$$P_+ = \frac{2\pi}{\hbar}(\hbar\omega)\left(\frac{q_e E_0}{m_0 \omega}\right)^2 2\sum_{i,f} |<f|\mathbf{e} \cdot \mathbf{p}|i>|^2 f(\xi_i)(1 - f(\xi_f))\delta(\xi_f - \xi_i - \hbar\omega)$$

$$P_- = -\frac{2\pi}{\hbar}(\hbar\omega)\left(\frac{q_e E_0}{m_0 \omega}\right)^2 2\sum_{i,f} |<f|\mathbf{e} \cdot \mathbf{p}|i>|^2 f(\xi_i)(1 - f(\xi_f))\delta(\xi_f - \xi_i + \hbar\omega)$$

$$= 1/2(P_+ + P_-) = \frac{2\pi}{\hbar}(\hbar\omega)\left(\frac{q_e E_0}{m_0 \omega}\right)^2 2\sum_{if} |<f|\mathbf{e}\cdot\mathbf{p}|i>|^2 [f(\xi_i)$$

$$- f(\xi_f)]\delta(\xi_f - \xi_i - \hbar\omega)$$

(5.109)

Consequently, for a transition between states in the valence and conduction bands with the same wave vector (k), the imaginary part of the dielectric function ϵ_2 may be written as

$$\epsilon_2 = \left(\frac{1}{\omega\epsilon_0}\right)\frac{\pi q_e^2}{m_0^2\omega} 2\sum_k |P_{cv}(k)|^2 \delta(\xi_c(k) - \xi_v(k) - h\nu)$$

(5.110)

where

$$P_{cv}(k) = \int_{\text{cell}} U_{ck}^*(r)(\mathbf{e}\cdot\mathbf{p})U_{vk}(r)d^3r$$

(5.111)

Since the Bloch functions, $U_{ck}(k)$ and $U_{vk}(k)$, are weakly dependent on k, we can approximate the integral with $k = 0$. The delta function reduces to the optical joint density of states, and the energies may be approximated as:

$$\xi_v(k) = -\frac{\hbar^2 k^2}{2m_h^*}$$

$$\xi_c(k) = \xi_G + \frac{\hbar^2 k^2}{2m_e^*}$$

(5.112)

$$\xi_c(k) - \xi_v(k) = \xi_G + \frac{\hbar^2 k^2}{2}\left(\frac{1}{m_e^*} + \frac{1}{m_h^*}\right) = \xi_G + \frac{\hbar^2 k^2}{2\mu^*} = h\nu$$

This allows solution of the equation for k and substitution into the optical joint density of states:

$$k^2 = \frac{2\mu^*}{\hbar^2}(h\nu - \xi_G)$$

$$n(\nu) = \frac{(2\mu^*)^{3/2}}{\pi^2\hbar^3}(h\nu - \xi_G)^{1/2}$$

(5.113)

This leads to the absorption coefficient:

$$\alpha = \frac{\omega \epsilon_2}{cn} = \frac{\pi q_e^2}{cn\omega} |P_{cv}(o)|^2 n_{\text{opt}}(h\nu)$$

$$= \frac{\pi q_e^2}{nch^2} \frac{(2\mu^*)^{3/2}}{h\nu} |P_{cv}(0)|^2 \qquad (5.114)$$

$$\approx \frac{q_e^2}{nch^2} \frac{(2\mu^*)^{3/2}}{m_e^* h\nu} (h\nu - \xi_G)^{1/2}$$

Appendix 5C

Band Structure of Semiconductors

The outermost valence electrons of the elements making up semiconductors are in either s or p orbital states. The major elementary semiconductors, Si, Ge, and diamond, crystallize in the *diamond cubic* structure, in which all atoms are tetrahedrally coordinated with the O_h point group symmetry. The band structure of these crystals can be thought to be made from a hybridization of the elemental electronic states. As discussed in the introduction of Chapter 2 for diamond, the lower levels of carbon's electronic structure, $(1s)^2 (2s)^2$, are filled. The extra two electrons are in the $(2p)^2$ states, leaving four unfilled $(2p)$ states. As the atoms coalesce to form a crystal, the wave functions overlap to form symmetric and antisymmetric hybrid (sp) states. The electrons in the same states become spin-paired ($\uparrow \downarrow$). The hybrid states are formed by mixing two states from the $(2p)$ levels with two states from the $(2s)$ level to form the valence band, which is filled with the four valence electrons from each atom. These orbitals exhibit tetrahedral symmetry, which leads to tetrahedral atomic coordination. This leaves four unfilled $(2p)$ states to form the conduction band. The hybrid $(sp)^3$ states [(sp) valence $+ (p)^2$ conduction] decrease the energy of the crystal and form its cohesive, or binding, energy. Figure 5.9 shows the energy levels for Si and Ge.

The diamond cubic crystals have the same basic structure, yet they exhibit far different behavior from each other:

1. Diamond has six electrons per atom with electronic states [He] $(2s)^2$ $(2p)^2$. It is a direct-gap semiconductor with a bandgap energy of $5.5\,\mathrm{eV}$ at the Γ point.

268

2. Germanium has 32 electrons per atom with electronic states [Ar] $(4s)^2$ $(4p)^2$. It is an indirect-gap semiconductor (Fig. 5.9) with a Γ-point energy gap of 1.1 eV, an X-point gap of 1.01 eV, and a minimum energy gap near the L point at 0.9 eV at 0K and 0.67eV at 300K.

3. Silicon has 14 electrons per atom with electronic states [Ne] $(3s)^2$ $(3p)^2$. It is an indirect-gap semiconductor like Ge, except that it displays no minimum in the conduction-band energy at the Γ point. Its minimum occurs in the Δ direction near, but not, at the X point with a value of 1.1 eV (Fig. 5.9).

Compound semiconductors are formed from either column 3-5, column 2-6, or column 1-7 elements. In the cubic form they exhibit the zinc blende structure, in which the larger anions form the FCC cell while the cations occupy half the tetrahedral holes. These materials have point group symmetry T_d and include GaAs, GaP, ZnSe, CdTe, CuCl, and CuBr. Some have direct gaps (GaAs and CdTe—see Fig. 5.8), and others have indirect gaps (GaP). In the hexagonal form, the crystals exhibit the wurtzite structure, in which the anions form an HCP lattice and the cations fill two-thirds of the tetrahedral holes. This structure has the C_{6v} symmetry with a polar c-axis. Column 2-6 and some column 3-5 compounds exhibit this structure, including ZnO, CdS, and GaN.

In the cubic structures, the valence bands are degenerate at $k = 0$, as the heavy-hole and light-hole bands coincide. The hexagonal structures have all three distinct bands.

Semiconductor band structures may be calculated using a variety of methods, among which the principal ones are tight binding, pseudopotential, and $k \cdot p$ methods. All methods use expansion solutions made from the sums of Bloch functions or basis sets. The tight-binding method is empirical, but it provides a reasonably good description of the band structure and eigenstates and the effect of defects near the band edges. In the tight-binding method, the core electrons are insensitive to neighboring atoms, and only the valence electron states are perturbed by the neighboring atoms. The s and p orbitals making up the basis sets are localized over the periodic lattice. Spin-orbit coupling can be included in the method. The heavy-hole band is described by $J = 3/2$ and $m_j = \pm 3/2$ orbitals, the light-hole band by $J = 3/2$ and $m_j = \pm 1/2$ orbitals, and the split off band by $J = 1/2$ and $m_j = \pm 1/2$ orbitals. The spin orbit coupling between the $J = 3/2$ and the $J = 1/2$ states is Δ, and varies between 0.04 for Si to 0.03 for Ge, 0.34 for GaAs, and 0.38 for InAs.

The pseudopotential method uses the fact that valence-and conduction band states are orthogonal to core states. This is used in the formulation of the pseudopotential in the Hamiltonian. This has the effect of subtracting the effect of the core levels from the pseudopotential, which results in a smooth spatial dependence. The solution is expanded in terms of plane waves.

The $k \cdot p$ method uses the results of approaches like tight-binding to calculate the band structure at the edges of the Brillouin zone. Then the method uses perturbation theory to describe the electronic structure away from the high-symmetry points. The Hamiltonian is broken up into three components when the Bloch solution is substituted into the Schroedinger equation:

$$\psi_{nk}(r) = e^{ik \cdot r} U_{nk}(r) \qquad (5.115)$$

$$
\begin{aligned}
H_0 &= p^2/2m \\
H_1 &= \left(\hbar/m\right)k \cdot p \qquad (5.116) \\
H_2 &= \hbar^2 k^2/2m
\end{aligned}
$$

The H_0 term is the zeroth-order term and corresponds to a free particle. The H_1 term is the first-order perturbation, and the H_2 term is the second-order perturbation. The results are as follows

Zeroth order:

$$
\begin{aligned}
U_{nk} &= U_{n0} \\
\xi_{nk} &= \xi_n(0)
\end{aligned}
\qquad (5.117)
$$

First order:

$$U_{nk} = U_{n0} + \frac{\hbar}{m}\sum k \cdot \frac{\langle n_0|p|n_0\rangle}{\xi_n(0) - \xi_{nk}}$$

$$\xi_{nk} = \xi_n(0) + \frac{\hbar}{m}k \cdot \langle n_0'|p|n_0\rangle \qquad (5.118)$$

Sanders, Stanton and Chang (1989) have used tight-binding and 20-band $k \cdot p$ methods to calculate successfully the band structures of bulk and quantum confined (wells, wires, and dots) semiconductors.

References

Anderson, P. W. 1958. *Phys. Rev.* 109:1492.

Ashcroft, N. W., and N. D. Mermin. 1976. *Solid State Physics*. Philadelphia: Holt Rinehart and Winston.

Bailey, D. W., C. J. Stanton, and K. Hess. 1990. *Phys. Rev.* B42:3423.

Campbell, S. A. 1996. *The Science and Engineering of Microelectronic Fabrication*, Oxford NY.

Chiang, Y.-M., D. Birnie, III, and W. D. Kingery. 1997. *Physical Ceramics*. New York: Wiley.

Connell, G. A. N. 1985. Optical properties of amorphous semiconductors. In *Amorphous Semiconductors*, edited by M. H. Brodsky. Berlin: Springer.

Efros, Al. L., and A. L. Efros. 1982. Interband absorption of light in a semiconductor sphere. *Soviet Phys. Semicond.* 16:772–775.

Harrison, W. 1989. *Electronic Structures and the Properties of Solids*. New York: Dover.

Ibach, Harald, and Hans Luth. 1996. *Solid State Physics*. Berlin: Springer.

Kane, E. O. 1996. The $k \cdot p$ Method. In *Semiconductors and Semimetals 1*, 75–100. New York: Academic Press.

Kittel, C. 1996. *Introduction to Solid State Physics*. New York: Wiley.

Klingshirn, C. F. 1997. *Semiconductor Optics*. Berlin: Springer.

Loudon, R. 1983. *The Quantum Theory of Light*. Oxford, UK: Clarendon Press.

Mahajan, S., and K. S. Sree Horsha. 1999. *Principles of Growth and Processing of semiconductors*. Basin: McGraw Hill.

Mott, N. F. 1990. *Metal–Insulator Transitions*. London: Taylor and Francis.

Mott, N. F., and A. E. Davis. 1971. *Electronic Processes in Non-Crystalline Solids*. London: Oxford University Press.

Pankove, Jacques I. 1971. *Optical Processes in Semiconductors*. New York: Dover.

Peyghambarian, N., S. W. Koch, and A. Mysyrowicz. 1993. *Introduction to Semiconductor Optics*. Englewood, NJ: Prentice Hall.

Pollack, F. H., C. W. Higginbotham, and M. Cardona. 1996. *J. Phys. Soc. Japan*, Supplement 20.

Potter, B. G., and J. H. Simmons. 1988. Quantum-size effects in CdS–glass composites. *Phys. Rev.* B37, 838–845.

— 1990. Quantum-confinement effects in CdTe–glass composite thin films, made using RF magnetron sputtering. *J. Appl. Phys.* 68:1218–1224.

— 1991. Quantum-confinement effects at the L-point of CdTe. *Phys. Rev.* B43:2234–2238.

Potter, B. G., J. H. Simmons, P. Kumar, and C. J. Stanton. 1994. Quantum-size effects on the band edge of CdTe clusters in glass. *J. Appl. Phys.* 75:8039–8045.

Rosen, D. L., A. G. Donkas, A. Katz, Y. Budansky, and R. R. Alfano. 1984. Techniques in time resolved luminescence spectroscopy. In *Semiconductors Probed by Ultrafast Laser Spectroscopy*, Vol.2, edited by R. R. Alfano. Orlando, FL: Academic Press.

Sanders, G. D., C. J. Stanton, and Y. C. Chang. 1993. Theory of electronic and optical transport properties in Si quantum wires. *Phys. Rev.* B48:11067.

Singh, J., 1993. *Physics of Semiconductors and Their Heterostructures*. New York: McGraw-Hill.

Solymar, L., and D. Walsh. 1989. *Lectures on the Electrical Properties of Materials.* 4th ed. Oxford, UK: Oxford Science.

Stanton, C. J., and D. W. Bailey. 1991. Evaluating photoexcitation experiments using Monte Carlo simulations. In *Monte Carlo Device Simulations: Full Band and Beyond*, edited by K. Hess. Boston: Kluwer Academic.

Tauc, J., R. Grigorovici, and A. Vancu. 1966. Optical properties and electronic structure of amorphous germanium. *Phys. Stat. Sol.* 15:627.

Wang, L., and J. H. Simmons. 1995. Observation of exciton and biexciton processes in $Cd_xZn_{1-x}Se/ZnSe$ $(x = 0.2)$. *Appl. Phys. Lett.* 67:1450.

Weisbuch C., and B. Vinter. 1991. *Quantum Semiconductor Structures*. Boston: Academic Press.

Yu, P. Y., and M. Cardona. 1996. *Fundamentals of Semiconductors.* Berlin: Springer.

Chapter 6

Optical Gain and Lasers

6.1 Introduction

Now that we understand the source of optical behavior for various classes of materials, it is time to examine how electromagnetic waves interact with matter and how this interaction leads to optical devices. The next two chapters are devoted to this goal. In this chapter, we will examine the features of spontaneous and stimulated emission, population inversion, and gain. This knowledge will allow us to examine the operation of lasers. In effect, this chapter is intermediate between the study of materials on the basis of type of material and the study of materials on the basis of property. Thus, in the subsequent chapter we will concentrate on discussing materials that have common properties or behaviors or materials that can be used for the same kind of application.

When an electromagnetic wave propagates through a material, there are numerous interaction mechanisms that can take place. We examined scattering and absorption in Chapter 3 and reflection in Chapter 2. If the propagating light has a frequency that corresponds to the energy difference between two levels in the energy structure of the material, a strong interaction develops. These levels can be electronic states, atomic states, or vibrational states. The interaction that develops can take the form of absorption or emission of light.

6.2   Spontaneous emission

For simplicity, we will assume that we have only a two-level atom and examine the transitions between these two levels. Spontaneous emission occurs with or without the presence of an electromagnetic field. Its behavior is the same under either condition, so let's look at the case without an electromagnetic field. We shall assume that the lower level has energy ξ_1 and that the upper level has energy ξ_2 as shown in Fig. 6.1. If level 2 is occupied and level 1 has available states in its density of states, then atoms in level 2 will decay to level 1 spontaneously, at a rate, $(X_{21})_{\text{spont}}$, proportional to the occupation density of level 2, N_2, as shown in Eq. (6.1):

$$(X_{21})_{\text{spont}} = -\frac{dN_2}{dt} = A_{21}N_2 = \frac{N_2}{\tau_s} \tag{6.1}$$

where A_{21} is the transition rate coefficient and τ_s is the spontaneous lifetime of state 2.

6.3   Line shapes

Measurements of spontaneous transitions show line shapes with measurable width, which indicate a source of line broadening. Examination of the problem reveals that there are two separate conditions that will lead to line broadening. In the first, the atoms are

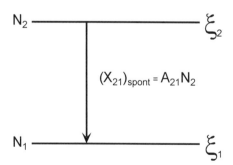

Figure 6.1: Spontaneous emission in a two-level system. X_{21} is the spontaneous emission rate.

all assumed to be identical, so the energy difference between the two levels is the same for all atoms. This leads to homogeneous line broadening. In the second, inhomogeneous line broadening, the atoms are assumed to be in slightly different environments, so the energy difference between their two states is not all the same.

6.3.1 Homogeneous line broadening

Homogeneous line broadening arises despite having a system of identical atoms. Among the causes of homogeneous line broadening are the following: (a) a finite spontaneous lifetime of the excited state, (b) inelastic collisions of the excited atom with phonons or other atoms, (c) transitions to other levels, (d) elastic collisions that destroy phase relationships, (f) pressure-induced line broadening.

In general, the major source of homogeneous line broadening is the finite lifetime of the excited state. As the population of the excited state decreases with time, the intensity of spontaneous emission also decreases in time. This decrease appears as a broadening of the transition frequency. This is shown by examining the decaying optical emission, $I(t)$, in terms of an oscillatory function, $\cos \omega t$, and a decaying term, $e^{-t/\tau}$:

$$I(t) = I_0 e^{-t/\tau} \cos \omega_0 t = {}^1\!/_2\, I_0 \left[e^{i(\omega_0 + i\gamma)t} + e^{-i(\omega_0 - i\gamma)t} \right] \qquad (6.2)$$

where $\gamma = \tau^{-1}$. The spectral shape, $g(\nu)$, of the emission function is found by calculating the Fourier transform of $I(t)$ as follows:

$$I(\omega) = \int_0^\infty I(t) e^{-i\omega x} dt = \frac{I_0}{2} \int_0^\infty e^{i(\omega_0 - \omega)t - \gamma t} dt + \frac{I_0}{2} \int_0^\infty e^{i(\omega_0 + \omega)t - \gamma t} dt$$

$$= \frac{I_0}{2} \left(\frac{1}{\omega_0 - \omega + i\gamma} - \frac{1}{\omega_0 + \omega - i\gamma} \right) \qquad (6.3)$$

This function has a spectral density of

$$|I(\omega)|^2 = \frac{1}{(\omega_0 - \omega)^2 + \gamma^2} \qquad (6.4)$$

which gives a Lorentzian line shape in frequency, ν, written as

$$g(\nu) = \frac{2\Delta\nu}{2\pi[(\nu_0 - \nu)^2 + (\Delta\nu)^2]} \qquad (6.5)$$

Consequently, if the spontaneous emission line shape is Lorentzian, it is a clear sign of homogeneous broadening. (The Lorentzian line shape is drawn in Fig. 6.2.)

Inelastic collisions also broaden the width of the emission line. They do this by adding or subtracting a small amount of kinetic energy (from the collision) to either the upper or the lower state or to both states of the transition. They can take place between atoms, between electrons, between atoms and electrons, and between all of these and phonons.

6.3.2   Inhomogeneous line broadening

Inhomogeneous broadening occurs if the atoms are distinguishable. This can result from the strong dependence of the energy of the excited and ground states, and consequently the emission or absorption frequency, on the local environment of the atom and on structural differences between atoms. Such differences can be caused by structural defects, impurities, random strain, lattice imperfections, or the lack of crystalline order (glasses and amorphous films). Here, it is reasonable to assume that the transition energies $h\nu_i$, with homogeneous line widths, $\Delta\nu_i$, are randomly distributed about some average energy, $h\nu_0$, with a distribution width $\Delta\nu_0$, giving rise to a random or Gaussian distribution of homogeneous line shapes. Thus, the resulting emission takes the form of a Voigt function (e.g., a convolution of Gaussian and Lorentzian functions):

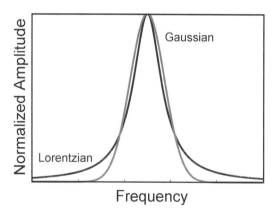

Figure 6.2: Line shapes corresponding to homogeneous (Lorentzian) and inhomogeneous (Gaussian) broadening. The Gaussian is the lighter shade. It is broader at the peak but narrower in the wings.

$$g(\nu) = \int_0^{\infty} \frac{e^{-1/2[(\nu_{0i}-\nu_0)/\Delta\nu_0]^2}}{[(\nu_{0i}-\nu)^2 + (\Delta\nu_i)^2]} d\nu_{0i} \qquad (6.6)$$

Generally, the inhomogeneous line width will be much larger than the homogeneous line width. In that case, the line shape can be simplified to the Gaussian envelope:

$$g(\nu) = \frac{1}{\sqrt{2\pi}\,\Delta\nu} e^{-(\nu-\nu_0)^2/2\Delta\nu^2} \qquad (6.7)$$

A Gaussian line shape is shown in Fig. 6.2. Differences are evident in the wings of the line. The Lorentzian line is narrower at the peak, but it has higher wings. If the distribution of environments is not random, then different line shapes are possible. For example, if there is simply a bimodal structure, in which the atoms sit in only two different environments, then the emission line will consist of the addition of two Lorentzian lines.

Inhomogeneous broadening can also result from Doppler-shifted transition energies due to the Maxwellian distribution of velocities of atoms in a gas. The result is also a Gaussian line shape. The well-known helium–neon laser will exhibit such a Maxwellian Doppler-shifted emission line with Gaussian half width $\Delta\nu = 1.5 \times 10^9$ Hz on a central frequency of $\nu_0 = 4.7 \times 10^{14}$ Hz. The CO_2 laser has the following characteristics ($\Delta\nu = 6 \times 10^7$ Hz, $\nu_0 = 2.8 \times 10^{13}$ Hz).

6.4   Stimulated emission and absorption

If an external field is present, then the two-level atom can absorb radiation or emit radiation of the same frequency. Absorption is easily understood, since the population of the lower level, N_1 is available for transitions to the upper level in the presence of an external field. Spontaneous emission, discussed earlier, consists of the transition from the excited to the ground state in the absence of an external field if the ground state is not fully occupied. Stimulated, or induced, emission is less intuitive and occurs only in the presence of an external field. It results from a resonance between the atom and the external field that causes the emission of a photon of the same frequency as the external electromagnetic field. This *induced* emission occurs in addition to the spontaneous emission process and has the same probability for occurrence as the absorptive process, which is also induced (Fig. 6.3).

Denoting the transition rate per population as W (such that $X = NW$) and the energy density of the electromagnetic field as $\rho(\nu)$, we may compare the following:

$$(W_{21})_{\text{induced}} = B_{21}\rho(\nu) \qquad \text{and} \qquad (W_{12})_{\text{induced}} = B_{12}\rho(\nu)$$

$$(W_{21})_{\text{total}} = A_{21} + B_{21}\rho(\nu) \quad \text{and} \quad (W_{12})_{\text{total}} = B_{12}\rho(\nu) \qquad (6.8)$$

where A_{21} is the spontaneous emission rate, B_{12} is the induced absorption rate, and B_{21} is the stimulated/induced emission rate. At thermal equilibrium, the populations do not change with time, and: $N_2 W_{21} = N_1 W_{12}$, which gives

$$N_2[B_{21}\rho(\nu) + A_{21}] = N_1 B_{12}\rho(\nu) \qquad (6.9)$$

Assuming a Boltzmann population distribution for the two level populations at thermal equilibrium, and assuming the black-body radiation law for the energy density $\rho(\nu)$, we can calculate the relationship between the spontaneous and induced rates:

$$\frac{N_2}{N_1} = e^{-h\nu/k_B T}$$

$$\rho(\nu) = \frac{8\pi n^3 h\nu^3}{c^3} \frac{1}{e^{h\nu/k_B T} - 1} \qquad (6.10)$$

$$\frac{A_{21}}{B_{12}e^{h\nu/k_B T} - B_{21}} = \frac{8\pi n^3 h\nu^3}{c^3(e^{h\nu/k_B T} - 1)}$$

where c is the speed of light in vacuum and n is the refractive index. The Einstein model predicts that the induced absorption and induced

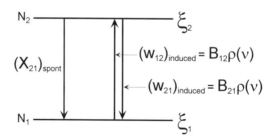

Figure 6.3: Two-level system demonstrating induced absorption (W_{12}) and induced emission (W_{21}).

emission probabilities must be equal, so $B_{12} = B_{21}$. This leads to a relationship between the induced emission probability and the spontaneous emission probability at thermal equilibrium:

$$\frac{A_{21}}{B_{21}} = \frac{8\pi n^3 h \nu^3}{c^3} \tag{6.11}$$

From this, we may now calculate the induced transition rate for any incident optical field $\rho(\nu)$:

$$(W_{21})_{\text{induced}} = (W_{12})_{\text{induced}} = B_{12}\rho(\nu) = A_{21}\frac{c^3}{8\pi n^3 h \nu^3}\rho(\nu) = \frac{c^3}{8\pi n^3 h \nu^3 \tau_{\text{s}}}\rho(\nu) \tag{6.12}$$

If the incident electromagnetic field is monochromatic, its energy density may be written in terms of its intensity $I(\nu_0)$ as follows:

$$\rho(\nu) = \rho(\nu_0)g(\nu) = \frac{nI(\nu_0)}{c} \tag{6.13}$$

and the transition rate for the monochromatic field becomes:

$$W_{\nu_0}(\nu)_{\text{induced}} = \frac{c^3\rho(\nu_0)}{8\pi n^3 h \nu^3 \tau_{\text{s}}}g(\nu) = \frac{c^2 I(\nu_0)}{8\pi n^2 h \nu^3 \tau_{\text{s}}}g(\nu) = \frac{\lambda^2 I(\nu_0)}{8\pi n^2 h \nu \tau_{\text{s}}}g(\nu) \tag{6.14}$$

6.5 Absorption and amplification (gain)

If we abandon thermal equilibrium and allow the population densities to change, conditions may be developed by which the electromagnetic wave can gain energy by propagating through a medium. In this case, the population densities must be inverted by an external means. This process therefore does not violate the second law of thermodynamics, since more energy is supplied to the system to cause the population inversion than is gained by the propagating wave.

The problem is best treated by considering a monochromatic plane wave propagating through a medium with N_2 atoms per unit volume in state 2 and N_1 atoms/volume in state 1. Neglecting spontaneous emission, the transition rate from 2 to 1 is $N_2 W_i$ and from 1 to 2 is $N_1 W_i$, where W_i is the induced transition rate. Consequently, we may write the net power lost or gained by the propagating wave as

$$\frac{\text{net optical power}}{\text{volume}} = (N_2 - N_1)W_i h\nu \tag{6.15}$$

If we assume that the radiation is added coherently (e.g., phase matching), then, using Eq. (6.14), the variation in intensity with propagation distance in the medium is:

$$\frac{dI(\nu_0)}{dz} = (N_2 - N_1)\frac{c^2 g(\nu)}{8\pi n^2 \nu^2 \tau_s}I(\nu_0) = (N_2 - N_1)\frac{\lambda^2 g(\nu)}{8\pi n^2 \tau_s}I(\nu_0)$$

$$I_{\nu_0}(z) = I_{\nu_0}(0)e^{\Gamma(\nu)z} \tag{6.16}$$

The gain, $\Gamma(\nu)$, is thus written as

$$\Gamma(\nu) = (N_2 - N_1)\frac{c^2}{8\pi n^2 \nu^2 \tau_s}g(\nu) \tag{6.17}$$

It is now clear that if there is population inversion, the medium has gain; if not, the medium has loss, depending on the sign of $\Gamma(\nu)$:

$$N_2 > N_1 \rightarrow \text{Amplifying medium}$$
$$N_2 < N_1 \rightarrow \text{Absorbing medium}$$

In real media, this is not strictly true, because the material also has intrinsic losses that must be balanced by an additional amount of population inversion. Consequently, there is a threshold population inversion above which there will be a net gain. In laser cavities, one can calculate this threshold population inversion, based on the properties of the cavity. We will examine this shortly.

6.6   Operational characteristics of lasers

6.6.1   Three-level, four-level, and n-level lasers

Lasers, in theory, operate through optical transitions between two levels. In practice, many more levels take part in their operation. In general, however, they are described through simplifications to either three-level or four-level systems. These two types of structures are shown in Fig. 6.4. In both systems, the lasing transition is between ξ_2 and ξ_1. In the three-level laser, the system is externally pumped to a pump state with energy ξ_p, that is higher than the upper level of the lasing transition, ξ_2. A rapid nonradiative decay occurs and lowers the excitation from ξ_p to ξ_2. The

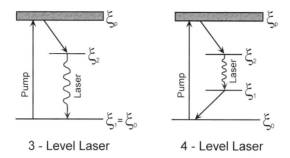

Figure 6.4: Comparison of the states of a three-level and a four-level laser.

population of state 2 then increases until lasing occurs, at which point the excited state radiatively transfers population to ξ_1, which is also the ground state of the medium. In the four-level laser, pumping to get population inversion in state 2 occurs in the same way. The difference is that state 1 is not the ground state of the medium, and there is an additional, rapid and nonradiative transition required to the ground state, ξ_0, after lasing. The basic difference between the two laser structures is in the population density of the lower lasing state, ξ_1. The population density of that state in a three-level laser is always thermal. By contrast, the thermal population in the four-level laser is in the ground state, ξ_0. If the difference in energies between the lower lasing state and the ground state is larger than thermal energy $(\xi_1 - \xi_0 > k_B T)$, then the occupation density of level 1 can be very small. Once we have determined the type of laser cavity of interest, later, we will compare the threshold population inversions of the two lasers. Laser systems with more levels can essentially be reduced to either three-level or four-level types, depending on the thermal population of the lower energy level of the transition, ξ_1, as just described.

6.6.2 Gain and gain saturation

If we consider a four-level laser with a degenerate upper state, it is possible to examine gain saturation and compare its value for homogeneously and inhomogeneously broadened media.

In a four-level laser (Fig. 6.5), the relevant transitions include the pump rate $P_2 = R_2$, from the ground state to the upper lasing level ξ_2, which is desirable, the pump rate $P_1 = R_1$, from the ground state to the

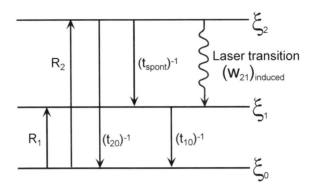

Figure 6.5: Schematic of a four-level laser system, defining the various transition rates. R_1 and R_2 are the pump rates to level ξ_2, so they include the pump rate to the pump state plus the decay down to the upper laser level.

lower lasing level, ξ_1, which is undesirable, the spontaneous emission transition from level 2 to level 1 with lifetime $\tau_s = t_{spont}$, the induced (lasing) transition from level 2 to level 1 with rate $(W_{21})_{induced}$, the time for decay of state 2 directly to state 0, t_{20}, which is spontaneous, and the time for decay of state 1 to state 0, t_{10}, which is also spontaneous.

The equations giving the induced transition rate and the rates of change of the lasing level populations are:

$$(W_{12})_{induced} = W_i(\nu) = \frac{\lambda^2 g(\nu)}{8\pi n^2 h\nu\tau_s} I(\nu_0)$$

$$\frac{dN_2}{dt} = P_2 - \frac{N_2}{t_2} - (N_2 - N_1)W_i(\nu) \qquad (6.18)$$

$$\frac{dN_1}{dt} = P_1 - \frac{N_1}{t_1} + \frac{N_2}{\tau_s} + (N_2 - N_1)W_i(\nu)$$

The total lifetime of state 2 is t_2, so $t_2^{-1} = t_{20}^{-1} + \tau_s^{-1}$. The total lifetime of state 1 is t_1, so $t_1 = t_{10}$.

At steady state, the population is constant, so the time derivatives go to zero. In that case, it is possible to solve for the population inversion:

$$N_2 - N_1 = \frac{[P_2 t_2 - (P_1 + P_2\bar{\tau})t_1]}{1 + [t_2 + t_1(1 - \bar{\tau})]W_i(\nu)} \qquad (6.19)$$

where $\bar{\tau} = t_2/\tau_s$. In the absence of stimulated emission, but still with pumping, the maximum population inversion, ΔN^0, is expressed as follows:

$$\Delta N^0 = P_2 t_2 - (P_1 + P_2 \bar{\tau}) t_1 \qquad (6.20)$$

and the population inversion expression for a homogeneously broadened system simplifies to the following:

$$N_2 - N_1 = \frac{\Delta N^0}{1 + Q \tau_s W_i}$$

$$N_2 - N_1 = \frac{\Delta N^0}{1 + \left[Q \lambda^2 g(\nu)/8\pi n^2 h\nu \right] I(\nu_0)} = \frac{\Delta N^0}{1 + I(\nu_0)/I_s} \qquad (6.21)$$

where

$$I_s = \frac{8\pi n^2 h\nu}{Q\lambda^2 g(\nu)} \quad \text{and} \quad Q = [t_2 + t_1(1 - \bar{\tau})]$$

In efficient laser systems, $t_2 \approx \tau_s$, $\bar{\tau} \approx 1$, $t_1 \ll t_2$, and $Q \approx 1$. This reduces the population inversion equation to the following:

$$N_2 - N_1 = \frac{P_2 \tau_s}{1 + \tau_s W_i} \qquad (6.22)$$

This equation shows that the population inversion is proportional to the product of the pump rate into the upper excited state, and the spontaneous lifetime of the state, divided, essentially, by the transition rate for the lasing transition.

Going back to the general equation [Eq. (6.21)], the gain for a homogeneously broadened medium can be calculated by combination with Eq. (6.17) to obtain the following:

$$\Gamma(\nu)_h = \frac{\Delta N^0 \lambda^2 g(\nu)}{8\pi n^2 \tau_s} \frac{1}{1 + I(\nu_0)/I_s} = \frac{\Gamma_0(\nu)}{1 + I(\nu_0)/I_s} \qquad (6.23)$$

The same calculation can be conducted for an inhomogeneously broadened medium. In this case, the medium is treated as composed of a sum of homogeneous packets, each with a slightly different transition energy. The calculation reveals a different gain function:

$$\Gamma(\nu)_i = \frac{\Gamma_0(\nu)}{\sqrt{1 + I(\nu_0)/I_s}} \quad \text{with } I_s = \frac{2\pi^2 n^2 h\nu \Delta\nu}{Q\lambda^2} \qquad (6.24)$$

A comparison of the decrease in gain with increasing light intensity between the homogenous and the inhomogeneous cases is shown in Fig. 6.6. The equations show that the homogeneous gain drops to $0.5\Gamma_0$ when the light intensity reaches the saturation value, I_s. The inhomogeneous gain, however, drops to only $0.7\Gamma_0$ by saturation intensity. Gain saturation sets in more slowly in the inhomogeneous system because, for each group of ν_0 values, as the population decreases in the excited state, the increase in light intensity causes more groups with different ν_0 to be excited, partly compensating for the saturation. In addition, the saturation intensity, I_s, does not depend on the line shape but rather on the line width.

6.6.3   Hole burning

An interesting experiment consists of measuring the gain as a function of frequency if monochromatic light of high intensity is incident on the material with frequency ν different from the center frequency, ν_0. As shown in Fig. 6.7, the homogeneous gain saturates over all frequencies. In contrast, the inhomogeneous gain saturates only around the ν frequency,

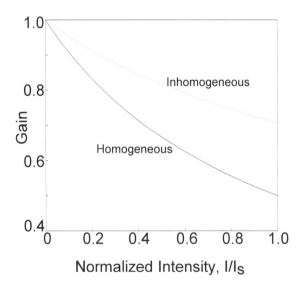

Figure 6.6: Gain in a laser cavity as a function of pump intensity. Note that as the intensity approaches the saturation value, the gain drops to about half its initial value for a homogeneous system. The inhomogeneous system drops only to 70% of the initial gain.

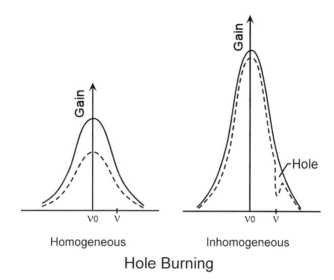

Hole Burning

Figure 6.7: Hole burning in an inhomogeneously broadened system. The saturation of the resonance at ν affects the population only within a homogeneous linewidth of that resonance value. In contrast, a homogeneously broadened system loses gain over the entire population.

and the remainder of the gain curve remains essentially unchanged. The hole in the gain curve at ν is referred to as *hole burning*. The width of the hole is the homogeneous line width for the packet with center frequency at ν. By scanning the ν frequency over the inhomogeneous line, it is possible to measure the homogeneous line width at each frequency through the hole-burning spectra.

6.7 Laser cavity characteristics

6.7.1 Fabry–Perot laser cavities

The Fabry–Perot cavity consists of a slab of material between two highly reflective surfaces. As shown in Fig. 6.8, multiple reflections from the surface and multiple internal reflections within the cavity result in transmission of only the cavity resonant wavelength. All other wavelengths are eliminated by destructive interference. As the reflection coefficient of the two reflectors is increased, the range of transmitted

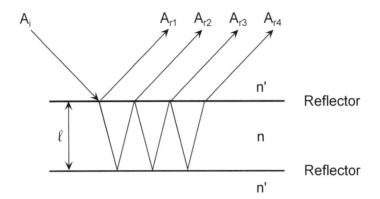

Figure 6.8: Sketch of the reflection interference in a Fabry–Perot cavity.

frequencies narrows drastically (see Fig. 6.9). Consequently, when the end mirror reflectivity is very high, the cavity propagates only the resonant wavelength. This cavity serves well as a laser cavity, since the depopulation of the excited upper level by stimulated emission can be stimulated only by the resonant wavelength. Thus, all the population inversion achieved goes to the laser wavelength transition, and only that wavelength is amplified. The only departure from this behavior is caused

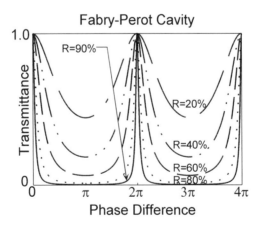

Figure 6.9: Plot of the calculated transmittance for a Fabry–Perot cavity as a function of the cavity length and for different levels of mirror reflectance (20, 40, 60, 80, and 90%).

by any spontaneous emission in the cavity from the excited state to other states, but this radiation makes only one trip through the cavity before undergoing destructive interference, so it is not amplified.

Mathematically, these conditions are treated as follows. The optical path length δ through a cavity of thickness ℓ and refractive index n for a beam of wavelength λ that is incident at an angle θ is written as

$$\delta = \frac{4\pi n \ell \cos \theta}{\lambda} \tag{6.25}$$

The reflectance and transmittance (Section 3.4.2) of the cavity are:

$$\frac{A_r}{A_0} = \frac{(1 - e^{i\delta})\sqrt{r}}{1 - re^{i\delta}} \qquad \frac{A_t}{A_0} = \frac{t}{1 - re^{i\delta}} \tag{6.26}$$

where r and t are, respectively, the mirror reflectance and transmittance amplitude coefficients.

Now, let's look at a laser oscillator with front mirror (r_1 and t_1) and rear mirror (r_2 and t_2) reflectivity and transmittance, containing a medium of dielectric constant ϵ. If the cavity has both gain (from a population inversion in the medium filling the Fabry–Perot cavity) and propagation losses, α, the electric field after an infinite number of paths through the medium is written as

$$\frac{E_t}{E_0} = \frac{t_1 t_2 e^{i(k + \Delta k)\ell} e^{(\Gamma - \alpha)\ell/2}}{1 - r_1 r_2 e^{-2i(k + \Delta k)\ell} e^{(\Gamma - \alpha)\ell}} \quad \text{with } \Delta k = \frac{k\epsilon}{2n^2} \tag{6.27}$$

From Eq. (6.17) for the gain of the system, it is clear that a population inversion will cause the gain to be positive, so the second term in the denominator may grow to be near 1. At some point, the transmitted wave will be larger than the incident wave. Under that condition, the cavity becomes an oscillator. When the denominator is zero, the transmitted intensity will diverge, indicating the existence of a transmitted wave without any incident wave ($E_0 = 0$). The wave therefore travels a roundtrip through the cavity without any change of intensity or phase. These conditions are expressed as follows:

no intensity change: $\quad r_1 r_2 e^{-2i(k + \Delta k)\ell} e^{(\Gamma - \alpha)\ell} = 1 \tag{6.28a}$

no phase change: $\quad 2(k + \Delta k)\ell = 2\pi m \quad \text{where } m = \text{integer} \tag{6.28b}$

The phase-change equation sets up a condition on the cavity length and gives a set of resonant wavelengths for the cavity:

$$\lambda_m = \frac{2\pi}{m}\left(1+\frac{\varepsilon}{2n^2}\right) \tag{6.29}$$

The gain value that produces no intensity change is called the threshold gain:

$$\Gamma(\nu) = \alpha - \frac{1}{\ell}\ln r_1 r_2 \tag{6.30}$$

which requires a threshold population inversion of

$$N_t = (N_2 - N_1)_t = \frac{8\pi n^2 \tau_s}{g(\nu)\lambda^2}\left(\alpha - \frac{1}{\ell}\ln r_1 r_2\right) \tag{6.31}$$

For a more detailed coverage of this subject, we refer the readers to Yariv (1976), from which the following example of the He-Ne laser is reproduced.

The helium-neon laser operates at $\lambda = 6.328 \times 10^{-5}$ cm. A typical cavity length is $\ell = 10$ cm.

$$\tau_s = 10^{-7}\,\text{sec}$$

$$\frac{1}{g(\nu)} \approx 10^9\,\text{Hz}$$

$$r_1^2 = r_2^2 = 0.98 \quad \text{and} \quad \alpha \approx 0$$

$$N_t \approx 10^9\,\text{cm}^{-3}$$

6.7.2 Population inversion in three-level and four-level lasers

In a four-level laser, $\xi_1 - \xi_0 \gg k_B T$, so the population of the lower transition level is negligible ($N_1 \approx 0$). Also, the lifetime of atoms in state 1 is short compared to the lifetime of the upper transition level (state 2); consequently, atoms that have made the induced or spontaneous transition to state 1 do not linger in that state and quickly decay to the ground state (ξ_0). This allows threshold population inversion to be reached essentially as soon as the population of state 2 reaches the threshold value:

$$N_t = N_2 - N_1 \approx N_2 \tag{6.32}$$

Laser oscillation thus begins as soon as the population of the upper excited state reaches the threshold value.

In a three-level laser, the lower transition level is also the ground state. Consequently, it has a thermal population. When the laser is pumped, N_2 must be raised above the thermal population of N_1 in order to obtain population inversion. The point where $N_2 = N_1$ gives zero gain. Thus, if the number of active atoms per unit volume is N_0, then zero gain is achieved when $N_2 = N_1 = {}^1/_2 N_0$. Population inversion is achieved by pumping half the threshold population into N_2, since they will also leave N_1 . This leads to the following threshold condition for cavity oscillation:

$$N_2 = \frac{N_0}{2} + \frac{N_t}{2} \quad \text{and} \quad N_1 = \frac{N_0}{2} - \frac{N_1}{2}$$

(6.33)

$$N_2 - N_1 = N_t$$

Since in most lasers, $N_0 \gg N_t$, we can calculate the ratios of pump rate for three- and four-level lasers to be

$$\frac{(N_2)_{3-\text{level}}}{(N_2)_{4-\text{level}}} = \frac{{}^1/_2 N_0}{N_t}$$

(6.34)

In ruby lasers, this ratio can be as high as 100!

6.7.3   Issues of output power coupling in laser cavities

Let us review what happens inside a laser cavity.

1. The pump induces a population inversion.
2. The presence of the optical output induces transitions between the upper and lower lasing levels, and vice versa. But since there is a population inversion, $N_2 > N_1$, and more emission (down) transitions occur than absorption (up) transitions. This causes $(N_2 - N_1)$ to be less than ΔN^0, corresponding to gain saturation.
3. When the gain balances or just exceeds the losses in the cavity, steady oscillation is achieved.
4. There are two major sources of loss in the laser cavity:
    a. True losses that result from absorption and scattering in all the cavity components, including the laser medium, the inert matrix or gas, and the mirrors.

b. Loss from the cavity due to output of the laser energy. This is done through a coupling mirror that transfers lasing energy above a given level outside the cavity. This is the useful component of the laser and brings up complex issues.

The output coupler in a laser cavity must perform two opposite functions. By exhibiting a high reflection coefficient, it extracts maximum power from the inverted population of the medium and it narrows the cavity emission line, improving the quality and power of the laser. But this also prevents power from escaping the cavity and providing external power. In order to deliver useful power from the laser, the output mirror transmission must be increased. However, as the transmission is increased, the inversion threshold must also be increased. Eventually, as transmission is further increased, the inversion threshold necessary for oscillation will be larger than the ability to pump population inversion and the laser will cease to operate. The transmission level for the output coupler is clearly very important for laser operation, and some optimum setting must be established.

It is possible to calculate and measure the output power for a given laser as a function of output coupler transmission for different unavoidable internal losses. Figure 6.10 shows the variation in output power as a function of output coupler transmission from 0–15%, for a set of internal loss values of 0, 1.7, 3.5, and 6% for a typical 6328 He–Ne laser. The unsaturated gain of the laser corresponds to the point where the laser stops oscillating with no internal losses. In this example, the value is 12%, the point where the 0% internal loss curve crosses the abscissa. The curves show that for each given value of internal loss, there is a transmission coefficient for the output coupler that yields maximum power output from the laser. The power output, P_0, may be written in terms of the output coupler transmittance, T_{oc}, and the internal residual losses, α_i, as follows:

$$P_0 = \frac{8\pi n^2 h\nu \, \Delta\nu \, A_m}{\lambda^2 (t_2/\tau_s)} \left(\frac{\Gamma_0}{T_{oc} + \alpha_i} - 1 \right) T_{oc} \qquad (6.35)$$

where A_m is the mode cross-sectional area. The optimum value of output power is found by differentiating Eq. (6.35) with respect to the output coupler transmittance, T_{oc}:

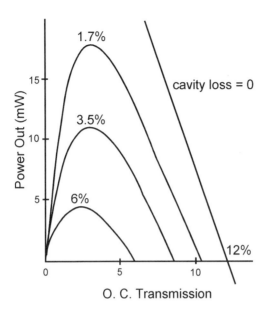

Figure 6.10: Variation in output power from a laser cavity for various output coupler transmission values for different cavity losses (6, 3.5, 1.7, and 0%). The maximum output coupler transmission possible is 12% for no cavity loss before the gain drops to zero (after Laures 1964).

$$(P_0)_{\text{opt}} = \frac{8\pi n^2 h \nu \,\Delta\nu \, A_m}{\lambda^2 (t_2/\tau_s)} \left(\sqrt{\Gamma_0} - \sqrt{\alpha_i}\right)^2 = I_s A_m \left(\sqrt{\Gamma_0} - \sqrt{\alpha_i}\right)^2 \qquad (6.36)$$

6.7.4   Issues of mode locking in laser cavities

In a system under steady-state oscillation, the gain at the oscillation frequency, ν_0, is clamped at the total cavity loss:

$$\Gamma_t(\nu_0) = \alpha - \frac{1}{\ell}\ln r_1 r_2 \qquad (6.37)$$

However, the gain curve is sufficiently broad to cover several cavity resonances [Eq. (6.29)], as shown in Fig. 6.11, and these super-and sub-harmonics of ν_0 would oscillate in the resonator if they could achieve sufficient gain to overcome the cavity losses. They are related to each other in frequency according to the following equation:

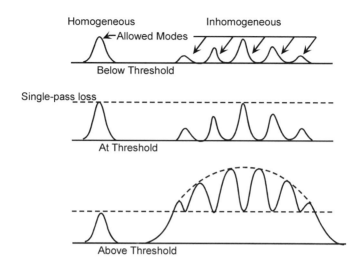

Single-Pass Gain for Homogeneously and Inhomogeneously
Broadened Media

Figure 6.11: Comparison of the gain saturation behavior of a homogeneously broadened and an inhomogeneously broadened medium. The cavity of the inhomogeneous medium is assumed to contain five modes. Note the clamping of the gain at the cavity resonance but not in between.

$$\nu_{q\pm 1} = \nu_q \pm \frac{c}{2n\ell} \qquad (6.38)$$

where n is the refractive index of the cavity medium. In the case of a homogeneously broadened medium, all the atoms react in the same fashion to the optical stimulus. The threshold gain at ν_0 is clamped at the single-pass loss, and any increase in population inversion cannot increase the gain because of saturation. Consequently, the threshold gain is the maximum achievable over the entire transition width, and it reaches its peak at ν_0. Since the peak gain is at ν_0 and it just matches the single pass cavity loss, none of the other resonances can produce a net gain and the cavity resonator can oscillate only at the single frequency ν_0.

In the inhomogeneously broadened case, the atoms are sufficiently different that they can act independently. This causes a major difference in the gain behavior. Below threshold, the gain curve is essentially the same as in the homogeneous medium. As the gain reaches threshold, it becomes clamped at ν_0 to the value of the single-pass cavity loss, as in the homogeneous case. However, if the population inversion is increased beyond this point, we find that while the gain at ν_0 remains clamped, the gain at the other resonances do not, since they have not yet reached the clamping value. Consequently, we see the gain increase at the other resonances until each one reaches the single pass cavity loss. In between, the gain actually increases further, as shown in Fig. 6.11. Further increases in population inversion produce a curve that shows hole burning in the gain at the clamped resonant frequencies. Thus the inhomogeneously broadened medium produces multiple modes in its oscillations. These modes are emitted from the cavity with no phase coherence and produce marked oscillations in the output power. The behavior of a typical laser with five modes in a cavity with an inhomogeneously broadened medium is shown in Fig. 6.11. The behavior is compared to that of a homogenous medium. Notice how the gain is clamped at the single-pass loss at the resonances, but rises far above that value in between for the inhomogeneous medium.

There are two solutions to the problem of multiple incoherent mode generation in an inhomogeneously broadened laser medium:

1. Reduce the cavity length, which increases the mode spacing ($\Delta\nu = c/2n\ell$) sufficiently that the gain can reach threshold only at the single mode of the system, ν_0.
2. Use mode-locking methods to maintain all modes in relative phase with respect to one another.

A useful method for achieving mode locking is to make the phases zero for all modes. Actual methods include changing the ratio of modulating wavelength to intermode spacing, and using masks that remove the out-of-phase components (acoustic Bragg shutters, saturable absorbers, and Q-switches)

6.7.4.1 *Acoustic Bragg diffraction*

This approach consists of modulating the gain or losses of the laser at a modulation frequency $\omega_m = \pi c/n\ell$, which is the intermode frequency spacing.

Consider a thin shutter within the laser resonator. The shutter is closed most of the time except for a brief opening every $2n\ell/c$ sec. A single random laser mode will not oscillate because of the high loss. (The open time is too short to allow the oscillation to build up.) Arbitrary phase oscillations will not develop for the same reason. Only the mode that happens to coincide with the opening of the shutter will pass and have a chance for amplification. Eventually, this mode will propagate amplified, and the others will decay quickly.

If the phases are locked, the short pulse of light arrives at the shutter when it is open; and if $\tau_{\text{open}} > \tau_{\text{pulse}}$, the mode-locked pulse will be unaffected by the shutter (e.g., only modes satisfying this temporally periodic condition develop and the shutter continuously restores the phase coherence).

The shutter forms a temporal periodic loss condition. This condition can be induced with an acoustic Bragg diffraction of the laser intensity by a standing acoustic wave. The refractive index is modulated by the acoustic strain in the material at frequency ω_m. In turn, the index modulation forms a phase diffraction grating with spatial period of $2\pi/k_m$. The diffraction loss reaches a peak twice in the acoustic period. Thus the loss modulation frequency is $2\omega_m$, and mode locking occurs when $\Delta\omega = 2\omega_m = 2\pi c/n\ell$.

6.7.4.2   Saturable absorber (bleachable dye)

This method is used in high-power solid-state lasers and continuous dye lasers. The absorber has an absorption α that decreases with light intensity in the cavity (e.g., it absorbs while the pulse builds up). This high absorption forces the gain threshold to high population-inversion values. Near maximum intensity, the saturable absorber intensity threshold is passed and it becomes transparent. This sudden decrease in total cavity loss causes the cavity's population inversion to exceed the threshold, which triggers stimulated emission that grows rapidly. This fires the pulse through the cavity, depleting all the population inversion. Then as the light intensity drops, the saturable absorber absorbs again and rebuilds the energy in the cavity.

In very fast systems, the timing of acoustic Bragg diffraction becomes very difficult. Therefore, since saturable absorbers are self-timed (i.e., they switch when the signal is there), they are preferable for fast lasers (Ti-sapphire femtosecond-pulse lasers).

6.7.4.3   Giant pulse — Q-switch (intense short burst of oscillations)

The Q of a cavity is the inverse of the loss of that cavity. Initially the loss is increased (Q is degraded), so the gain builds up well above its steady-state saturation value without oscillation. When the inversion reaches its peak, Q is restored abruptly. Since the gain is well above threshold, this causes a rapid buildup of oscillations. This is the same process as in the saturable absorber, and in fact saturable absorbers are a subset of Q-switches. However, the most used Q-switchings are induced electronically. Here are three examples of Q-switching in a laser cavity.

1. *Rotating end mirror:* Mount one of the reflecting mirrors on a rotating shaft so that optical losses are very high except for a brief period when the mirrors are nearly parallel.
2. *Saturable absorber:* Saturable absorbers are essentially Q-switches. However, they cannot be modulated as rapidly as with the other methods.
3. *Electro-optic crystals (liquid Kerr cell):* Here, one places a voltage-controlled gate in the resonant cavity, the electro-optic (EEO) crystal. Its presence causes $\pi / 2$ retardation (phase shift).

When the *voltage is on* the following condition develops (see Fig. 6.12): Vertically polarized, laser light $\rightarrow$ passes a vertically polarized analyzer $\rightarrow$ EEO crystal (causes circular polarization) $\rightarrow$ mirror (reflects the light back to the EEO crystal still circularly polarized) $\rightarrow$ EEO crystal phase-shifts the light to horizontal polarization $\rightarrow$ vertically polarized analyzer blocks the beam $\rightarrow$ **large cavity loss.**

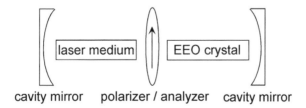

Laser Cavity

Figure 6.12: Q-switching using an electro-optic crystal. When the voltage is on, the rotation of polarization in the EEO crystal causes a large cavity loss. When the voltage is turned off, the reduction in loss stimulates a large laser pulse.

When inversion reaches its peak the *voltage is turned off* and the
following condition develops: Laser light is vertically polarized and it
travels in vertical polarization in both directions across the cavity
→ **low cavity loss** → giant laser pulse.

6.7.5   Laser efficiency

The most efficient lasers are the semiconductor lasers. All others are
relatively inefficient, and most laser users know that laser cooling is a
critical requirement for high-power lasers.

The total efficiency of a laser is the product of the factors that enter into
producing the light output. These are divided into the following three
classes.

Quantum efficiency: This factor represents the atomic or electronic
efficiency of the laser transition. It is defined as the ratio of the laser
output optical energy to the laser optical pump energy:

$$\eta_{qe} = \frac{\xi_2 - \xi_1}{\xi_p} = \frac{h\nu_{out}}{\xi_p} \tag{6.39}$$

where ξ_p is the pump energy (i.e., the energy difference between the
pump state and the ground state of the lasing system) and lasing takes
place between ξ_2 and ξ_1. This is essentially the efficiency of inducing the
population inversion.

Coupling efficiency: Once the stimulated emission is generated by the
population inversion, the laser light must be coupled out of the laser
cavity. This was discussed in section 6.7.3, on output couplers. The
coupling efficiency cannot be perfect if the electromagnetic emission
makes several gain trips through the cavity.

Pumping efficiency: This factor measures the fraction of the pump
energy that is transformed into excitation of the system to the pump
level. It includes the amount of flashlamp energy that is transformed to
light or the electric current energy that is required to excite the system
to the pump level.

Typical total efficiencies are ~ 1% for solid-state lasers, ~ 30% for CO_2
lasers, and ~ 100% for GaAs junction lasers.

6.8   Examples of laser systems

There are four primary categories of lasers in use today. These include solid-state lasers, gas lasers, liquid lasers, and semiconductor lasers. In the following sections we will discuss a number of examples of particular laser systems that fall into these categories. A more detailed description of the operation of these systems may be found in Yariv (1976), Verdeyen (1995), Weber (1991), and Hecht (1992). Instruction manuals for various lasers also contain much useful information about their operational characteristics.

6.8.1   Ruby lasers

The ruby laser is the first visible light laser to follow the early microwave masers. It produces pink light at a wavelength of $\lambda_0 = 694.3$ nm and is an example of a solid-state laser.

The active components of the ruby laser are Cr^{3+} ions, which are present as impurities in single crystal Al_2O_3. The chromium ion, Cr^{+3}, is introduced as a substitutional impurity replacing the aluminum ion Al^{+3} in the corundum structure of alumina. The corundum structure is hexagonal close packed (HCP), with its Al^{+3} ions in octahedral coordination filling two-thirds of the available sites in the HCP structure of the O^{-2} ions. The Cr^{3+} concentration is about 0.05% by weight. Ruby is typically brilliant red in gemstones, where the chromium concentration is about 1%. Chromium has an outer shell configuration of $(3d)^3$ with five empty d orbitals. The color results from interlevel transitions between different chromium d orbitals.

Ruby is a three-level laser. Figure 6.13 shows a rough sketch of the optical transitions of the ruby laser. The optical radiation of a flashlamp between 400 and 600 nm is absorbed by the Cr^{+3} ions, pumping them into the $^4F_2$ band. This is followed by a rapid nonradiative transition (50 nsec) to the ^2E levels, which are split into a 2A level and an E level separated by 29 cm^{-1}. The lasing transition is from the E level to the ground state, with a wavelength of 694.3 nm and a linewidth of 11 cm^{-1} (0.3 GHz, or 5.3Å). This makes ruby lasers three-level lasers. Ruby has a quantum efficiency of about 80%. The excited-state lifetime is 3 msec, and $t_2 \simeq t_{spont}$. Assuming a threshold pump energy of 3 J/cm^2 for a population inversion of 2×10^{19} atoms/cm^3 and a 1% pumping efficiency requires a flashlamp energy density of 300 J/cm^3.

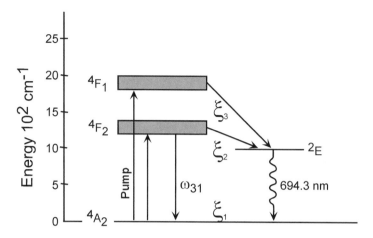

Figure 6.13: Sketch of the operation of a three-level ruby laser.

$$^* \nu = 1\,\mathrm{cm}^{-1} = 0.03\,\mathrm{GHz} = 1.24 \times 10^{-4}\,\mathrm{eV}$$

6.8.2   Nd^{3+}:YAG lasers

The most used solid-state laser employs the neodynium, Nd^{3+}, rare earth as an impurity in crystalline Y$_3$Al$_5$O$_{12}$ yttrium aluminum garnet (YAG). Its operating wavelength is $\lambda = 1.0641\,\mu\mathrm{m}$ at 300 K. The Nd^{+3} ion substitutes for Y^{+3} in the cubic oxygen lattice. Because of the size difference between the two ions (Nd is 3% larger), only about 1% doping can be achieved, otherwise the crystal becomes severely strained. To increase the doping level, the crystalline YAG host is replaced by glass hosts of various compositions.

The Nd^{3+} laser is a four-level laser. Figures 6.14 and 6.15 show a rough sketch of the optical transitions of the Nd-YAG laser. A flashlamp pumps the ions into the ^{4}F$_{5/2}$-^{2}H$_{9/2}$ excited band (12,400–13,000 cm^{-1}), and they quickly drop nonradiatively to the ^{4}F$_{3/2}$ level (11,507 and 11,423 spin-orbit split states). The optical lasing transition is between the ^{4}F$_{3/2}$ and ^{4}I$_{11/2}$ states of neodynium. The lower-level energies, ξ_1, are at 2,111 and 2,028 cm^{-1} above the ground state, so their Maxwell–Boltzmann population is 5×10^{-5} that of the ground-state population. The quantum efficiency of the laser is 75%.

The lasing transitions $\xi_2 - \xi_1$ are 1.06415 and 1.0644 µm, with a width

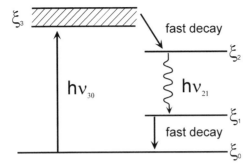

Figure 6.14: Sketch of the operation of a four-level Nd-YAG laser.

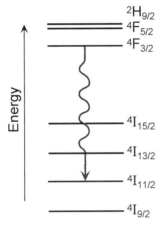

Figure 6.15: Energy diagram for the Nd-YAG laser.

of $6\,\mathrm{cm}^{-1}$ (0.2 GHz). The excited-state lifetime is 0.255 msec. The lower-state lifetime is 30 ns. For the same population inversion, the optical gain of Nd-YAG is 75 times that of ruby.

Assuming a flashlamp pumping efficiency of 5%, pulse operation takes place at a flashlamp energy of about $0.3\,\mathrm{J/cm}^3$.

Continuous wave (CW) operation is possible in these lasers as power is given off by spontaneous emission (critical fluorescence) just below threshold. The broad fluorescence emission spectrum turns to laser activity in the tuned Fabry–Perot cavity. The flashlamp energy required for CW operation is 80 watts.

Neodynium-YAG lasers today are efficiently pumped by semiconductor lasers instead of flashlamps. Gallium arsenide semiconductor lasers have an output at 808 nm (12,376 cm^{-1}), which matches reasonably well the $^4F_{5/2}$ energy level. Since semiconductor lasers are more efficient than flash lamps, a great improvement in efficiency and reliability is achieved.

6.8.3 Neodynium-glass lasers

When a need for large high-power solid-state lasers developed, Nd-glass lasers quickly became popular, since they are not limited by crystal growth and strain criteria. Today Nd-glass laser rods several feet in diameter are produced. The lasing frequency of the Nd^{+3} ion in the glass host is broader (inhomogeneously broadened) and lower in energy: $\lambda_0 = 1.059\,\mu m$ (see Fig. 6.16). The fluorescence line width from the glass host is broadened by a factor of 50 times that of the YAG crystal, due to variations in the local structure and ligand field around the Nd^{+3} ion.

Continuous wave (CW) operation requires a flashlamp power of 5 kW compared to only 80 W for the YAG host. This high value makes the glass lasers useful only in the pulsed state, where a minimum flashlamp pump energy of 8.5 J/cm^3 is required.

Neodynium-glass rods have desirable uses in amplification as well as for lasers. As amplifiers, the property of interest is the amount of energy storage achievable through population inversion. As discussed in Section 6.6, inhomogeneously broadened media have the ability to reach much

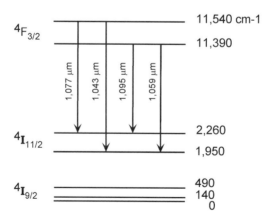

Figure 6.16: Energy diagram for a Nd-glass laser.

larger population-inversion and saturation levels than homogeneously broadened media. Along with the higher doping densities possible in the glass, the high saturation values make glass amplifiers very desirable.

A variety of glass hosts have been used for the Nd-glass lasers, including a number of alkali borosilicates. Recently, fluorophosphate glasses have been used preferentially for high power, due to the lower linear and nonlinear refractive indices of these glasses compared to those of other hosts. The latter increases the threshold for self-focusing damage of the laser rod. This damage mechanism was discovered to be the principal limitation to the use of Nd-glass lasers and amplifiers at high power. The mechanism results from the tendency of glasses to increase their refractive index as a function of light intensity. This is a nonlinear optical process that causes self-focusing of the beam in the laser rod. The self-focusing is a runaway process that feeds upon itself as it increases the local power density, which further increases the refractive index and causes more severe self-focusing. Eventually, the energy density of the beam will vaporize the glass along thin parallel paths through the rod. Fluorophosphate glasses, with their low non-linear index coefficient, minimize this effect.

Here is a comparison of some of the characteristics of YAG and glass hosts for Nd^{+3} lasers.

YAG
$\lambda_0 = 1{,}064\,nm$
$n = 1.5$
$\Delta\nu = 6\,cm^{-1}$
$t_{spont} = 5.5 \times 10^{-4}\,sec$
$N_t = 1.0 \times 10^{15}\,cm^{-3}$

Glass
$\lambda_0 = 1{,}059\,nm$
$n = 1.5$
$\Delta\nu = 200\,cm^{-1}$
$t_{spont} = 3.5 \times 10^{-4}\,sec$
$N_t = 9.05 \times 10^{15}\,cm^{-3}$

6.8.4 Tunable titanium-sapphire lasers

Titanium-sapphire lasers are solid-state lasers in which the laser rod is the Al_2O_3 crystal (sapphire) doped with about 1% Ti^{+4} as a substitutional impurity for Al^{+3}. The substitution is in an octahedral site; but since the ion is Ti^{+4}, there is an additional hole associated with the titanium. Titanium-sapphire has the ability to produce gain due to population inversion over the range 680–1,100 nm.

Figure 6.17 shows a schematic of the Coherent Mira® laser (Coherent Laser Corp.). Its output is about 1 W over the range 720–790 nm with a

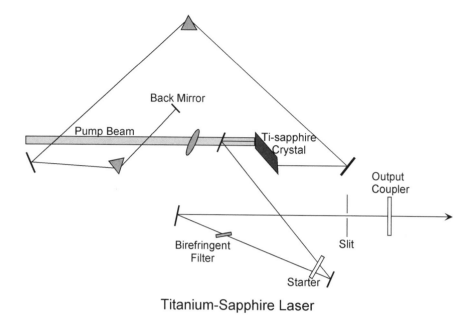

Titanium-Sapphire Laser

Figure 6.17: Diagram of a titanium-sapphire laser (from Coherent, Inc.).

pump from either an argon laser (8 W, all lines) or a doubled semiconductor laser. This laser is interesting in its design because it uses Kerr-lens mode locking, which is a self-focusing process from the optical Kerr effect (Chapter 7) that increases the refractive index of the Ti-sapphire crystal. This intensity-dependent refractive index is a nonlinear optical process typical of most materials. It is a product of the nonlinear refractive index, n_2, and the light intensity, I, and it is found to be depend strongly on the low-intensity or linear refractive index, n_0, increasing with an increase in the index. The total index, n, is written as

$$n = n_0 + n_2(n_0)I$$

The Ti-sapphire crystal has a lower-threshold output for CW operation; therefore, all the gain will be depleted by the CW mode. The Kerr-lens mode-locking process reduces the beam waist at a manually operated aperture. Thus, the operator closes the aperture and introduces a large loss in the CW mode, but does not affect the pulsed mode due to its smaller waist. This operation causes the pulsed modes to have the lower threshold. As gain builds up in the pulsed modes, increased Kerr self-focusing causes a better overlap of modes in the cavity, which further increases the gain and narrows the waist (see Fig. 6.18).

Active mode locking is obtained by an electronic or acousto-optic shutter that opens and closes at a regular interval. In ring lasers, the mode-locking frequency must match exactly the pulse roundtrip frequency. If the cavity length is changed, so must be the mode-locking frequency. (This is the case for YAG lasers.) In passive mode locking, one uses a saturable absorber. The Kerr-lens mode locking causes an intensity-dependent beam profile. The introduction of the shutter or slit selects the most intense beam, since that is the one with the narrowest waist. Gain is produced only for this mode as the others suffer large losses. In the design shown in Fig. 6.17, a pulse roundtrip takes 13.2 ns, causing a pulse repetition rate of 76 MHz.

The pulsed operation is caused by the introduction of a vibrating glass bar in the cavity, which causes two longitudinal modes to form simultaneously due to spatial hole burning. By rapidly sweeping the thickness of the bars across the optical path, the frequency and phases of the tuned cavity resonances are changed periodically. This transient condition causes more modes to form than can be carried by the cavity. These modes act incoherently until some coherence develops. This forms a pulse that satisfies the Kerr mode-locking constraint and an amplified pulse emerges. Once pulsed operation is started, the vibrating bars are no longer needed.

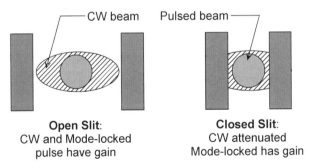

Open Slit:
CW and Mode-locked
pulse have gain

Closed Slit:
CW attenuated
Mode-locked has gain

Passive Mode-Locking

Figure 6.18: Passive mode locking in a titanium-sapphire laser using a slit that can be closed to attenuate the CW beam while passing the pulsed beam. The narrower waist of the pulsed beam results from non-linear self-focusing of the beam in the sapphire crystal (Coherent Laser Corp.).

Pulse width is determined by the group velocity dispersion of the cavity. This arises from the fact that all materials have dispersion in their refractive index. The velocity of a pulse is determined by the group velocity of its Fourier components ($v_g = dn/d\lambda$). The chirp spectrum of a pulse defines the frequency distribution of the wave packet. Normal dispersion is a decrease in refractive index (increase in propagation velocity) with increasing wavelength. Thus a normally dispersed pulse will have its blue components retarded with respect to its red components. This is called *positive* chirp. *Negative* chirp has red components retarded with respect to blue components.

Self-phase modulation causes focusing and a positive change in the refractive index of the beam, with a spatial and temporal dependence. The wavelength of the peak intensity of the pulse has the highest refractive index and is retarded the most, since the nonlinear refractive index is proportional to the linear refractive index. This will happen at the blue end of the spectrum, since the refractive index is highest there. Therefore, self-phase modulation will add to the positive chirp of a cavity. The remaining positive chirp results from all the dispersive elements of the cavity.

Dispersion compensation is therefore required to reduce the positive chirp of the pulses in the cavity. This is accomplished with either a set of gratings or a pair of intercavity prisms. Figure 6.19 shows how the prisms

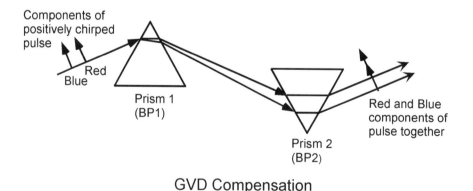

GVD Compensation

Figure 6.19: Since most materials exhibit normal dispersion, the blue component of a pulse will travel more slowly than the red component. The use of two prisms reverses the dispersion by forcing the red beam to travel further in the second prism and thus brings the spectral components back in phase.

can be adjusted to reverse positive chirp. This adjustment can be controlled by the amount of prism material traversed by the beam. Note, however, that since the prisms also add positive chirp, which increases with the amount of prism material traversed by the beam, there is a saturation condition, beyond which negative chirp cannot be produced and only positive chirp results. Group velocity dispersion (GVD) compensation can also be obtained by careful selection of materials, according to the following equation: $GVD = d^2 n / d\lambda^2$. Commercial short-pulse Ti-sapphire lasers produce about 100-fsec pulses. Laboratory lasers with pulse widths of 15 fsec have been produced due to selection of improved materials.

6.8.5 Tunable alexandrite lasers

As in the ruby laser, the active ion of this tunable solid-state laser is Cr^{+3}. The crystal host for the dopant is chrysoberyl ($BeAl_2O_4$). What makes the alexandrite laser tunable is the presence of phonon-separated levels in the lower level ($1940 \, cm^{-1}$) of the laser transition. This phonon-broadened band allows temperature-dependent tuning over a wide range of energies (680–800 nm).

6.8.6 Color center lasers

Color center lasers are a class of solid-state lasers that is tunable in the near infrared (generally over wavelengths from 1.4 to 3 μm). These lasers rely on optical pumping of F-center, or color center, crystals, which are thin crystals into which point defects have deliberately been introduced (through exposure to electron-beam irradiation, doping, or thermal annealing). The introduction of microscopic defects into the crystal structure results in coloration of the normally colorless materials, giving the laser its name. Some typical F-center lasers are formed by doping potassium chloride or rubidium chloride with either lithium (F_A center) or with sodium (F_B center). Although numerous materials have been used as host crystals for such lasers, alkali halides are the most common. Pure alkali halides have the advantage of being optically isotropic, transparent over broad wavelength bands in the visible and near infrared, and polishable, with careful processing, to an acceptable optical quality for use in a laser cavity. Alkali halide crystals may be doped, for example, with thallium to produce neutral thallium defects. Such defects consist of a thallium impurity adjacent to an anion vacancy, with an electron bound to the complex, and have emission wavelengths in the 1.4–1.7-µm range. This system has particular technological relevance since it covers one of the wavelength bands of interest for optical telecommunications.

Color center lasers are pumped by an external laser that has an output wavelength that is absorbed by the crystal (for the Tl color center laser just described, the laser may be pumped at 1.06 µm by a Nd:YAG laser, for example). Absorption of the pump photon excites the atoms at the defect site in the color center crystal and causes them to rearrange. Such rearrangement is accompanied by a radiationless relaxation to a lower energy state, which places the defect in the upper lasing level for the system. Relaxation from this level to the lower lasing level occurs via stimulated emission while the defect site is in the upper excited state. This system, thus, mimics a standard four-level laser. Since numerous vibrational states exist in the crystal, color center crystals exhibit homogeneous broadening of their emission spectra. This broadening means that through careful adjustment of the laser cavity length it is possible to tune the output of these lasers continuously over a significant wavelength band.

6.8.7   Fiber lasers and erbium-doped fiber amplifiers

Fiber lasers consist of Nd-doped germanosilicate optical fibers optically pumped. In the past, the cavity was formed by cleaving the ends of the fibers. Coupling to the transmitting fibers was accomplished through a splice, with some associated intensity loss. Today, photosensitive gratings (see Chapter 7) are formed at resonant intervals in a continuous fiber. These gratings act as mirrors and form the Fabry–Perot laser cavity. Since the gratings can be formed without cleaving the fibers, there is no coupling loss between the laser and the transmission medium. The laser operation is similar to the Nd-glass solid state lasers covered earlier.

Optical signal transmission for communications has been conducted at 1.3 and 1.55 microns, which are the loss minima in germanosilicate fibers. These minima occur on either side of a water absorption overtone. The loss is lower at 1.55 µm, but the dispersion is lower at 1.3 m. Long communication distances require optical amplifiers. Erbium amplifiers have made convenient and efficient repeaters for long-distance communications.

The fiber amplifiers work like a laser. They are formed into cavities at various intervals in a continuous fiber by use of photosensitive gratings. The selected region of the fiber is doped with erbium. Pumping of the erbium electronic states is accomplished with 980-nm light from a solid-state or InGaAs semiconductor laser, between the $^4I_{15/2}$ and $^4I_{11/2}$ states. Pumping is also possible between the $^4I_{15/2}$ ground state and the $^4I_{13/2}$ excited state. Laser emission occurs between the excited $^4I_{11/2}$ and the $^4I_{15/2}$ ground state. The emission wavelength is 1.55 μm, with an inhomogeneous width of about 0.0003 μm, which is considerable. The broadening is due to the many m_j levels in the manifolds of each j state.

Erbium fiber amplifiers are among the key elements that have made global communications possible via optical fiber, at a reasonable price.

6.8.8   Helium–neon laser

This was the first CW laser. It is a gas laser that operates with a mixture of He and Ne in a ratio of between 5/1 and 20/1. The gas pressure, p, is determined in relation to the tube bore diameter, d, according to the equation $p \cdot d \approx 0.36$ torr-cm. A typical tube is 10–100 cm long and 2–8 mm in diameter.

Unlike the solid-state lasers that operate by transitions between electronic levels of impurities or dopants, the gas lasers operate through transitions between electronic states of the atoms of the gas in the laser. The gases are contained inside a glass tube with end flats that are inclined at Brewster's angle (37°) in order to minimize reflection of the p-polarized mode. The cavity mirrors are outside the glass tube, so any loss at the end plates of the tube are included in the total cavity loss. This means that the laser output is highly polarized in the vertical direction. The gas atoms are excited by either a DC or an RF current discharge. The energetic electrons of the current discharge excite the atoms into numerous excited states. As they cascade down to the ground state, they collect in the long-lived $2\,^1S$ state and $2\,^3S$ states of helium, which have lifetimes of 100 μsec and 5 μsec, respectively. These excited atoms in turn transfer energy to the 2S and 3S levels of neon. The excited neon atoms then undergo several electronic de-excitation transitions that produce laser light (Fig. 6.20).

There are three major lasing transitions in neon:

632.8 nm: $2s_2 \rightarrow 2p_4$. This is the most commonly known and used laser transition. $t_1 = 10^{-8}$ sec and $t_2 = 10^{-7}$ sec. ($t_1 < t_2$ satisfies population inversion between 3s and 2p.)

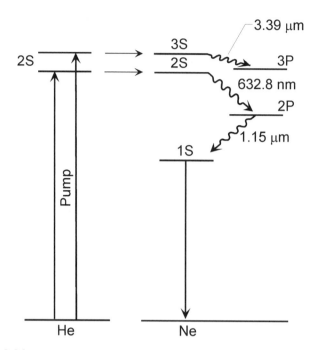

Figure 6.20: Sketch of the operation of He–Ne lasers showing the sources of its three emissions.

1.1523 μm: $2p_4 \rightarrow 1s_2$. This was the first demonstrated laser transition in a gas laser.

3.3913 μm: $3s_2 \rightarrow 3p_4$.

Many other transitions are possible, because the m states of the atoms allow a variety of closely spaced transitions.

The 3.39-μm oscillation has a much lower threshold than the 632.8-nm oscillation. Therefore, only the 3.39-μm laser should work. In order to enhance the 632.8-nm oscillation, the gas is contained inside a glass tube with glass windows in the optical path. Since glass absorbs at 3.39 μm, the loss is so high that it raises the threshold above that of the 632.8-nm oscillation.

The Brewster's windows (Fig. 6.21) do the same thing with the different polarizations. They reflect the beam with s-polarization, causing a loss of energy and raising the threshold for oscillation of the s-polarization. Consequently, the laser output is p-polarized.

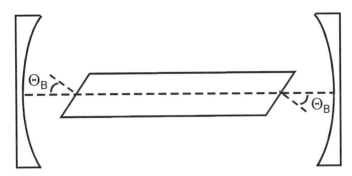

Figure 6.21: Use of Brewster's angle on the windows of the gas container in gas lasers. The light from the cavity encounters no reflective loss in the *p*-polarization.

6.8.9 Argon-ion and krypton-ion lasers

These are important gas lasers because of their range of operation, between 0.34 and 0.80µm. Argon-ion and krypton-ion lasers do not need other gases. The transition states are developed by electronic excitations of the gas ions. Excitation involves energetic collisions between electrons and the gas ions (4–5 eV). The singly ionized gas ions are formed by excitation of the ground state, and the first ionization for Ar^+ costs about 15.75 eV of energy. In order to reach the upper pump states, an additional 19.68 eV is required. Although the ionization and excitation processes can be separated, they still both require numerous simultaneous collisions to pump the laser, since the ion–electron collisions provide only about 4–5 eV per collision. This demands a very high current density through a narrow-bore tube containing the gas, causing a great deal of heat. Details of the collisions are still not clearly understood. The interaction of the current with the gas atoms is enhanced by a large longitudinal magnetic field that causes the electrons to spiral in their travel down the tube. Consequently, the laser is very delicate, having a heavy magnet wound around a glass or ceramic tube. The assembly also gets very hot, requiring a large amount of cooling.

Argon-ion and Krypton-ion lasers are very desirable because of the numerous intense laser transitions that they can support.

These lasers operate in CW mode with a quantum efficiency of 7–13% and an overall efficiency of 0.03%. Despite these shortcomings, they are very useful lasers because of the wavelengths and range of wavelengths available in CW mode, including UV lines (see Table 6.1). The latter

Ar$^+$-ion laser λ(nm)	Kr$^+$-ion laser λ(nm)
351.1–363.8 multiple lines	337.4–356.4 multiple lines
457.9–514.5 multiple lines	406.7–415.4 multiple lines
514.5	647.1–676.4 multiple lines
501.7	799.3
496.5	752.5
488.0	676.4
476.5	647.1
472.7	568.2
465.8	530.9
457.9	520.8
454.5	482.5
	476.5
	413.1

Table 6.1: Wavelengths of laser lines for Argon and Krypton (TEM$_{\infty}$ lines).

require an increase in magnetic field over operation in the visible, in order to increase the energy exchange between the electron current and the gas molecules to reach the upper states of the UV transition. The output can be cleaned up with a Fabry–Perot in-line cavity filter to produce single-mode oscillations. Argon lasers are also commonly used as pumps for dye lasers (discussed later).

All the transitions obey the standard selection rules: $\Delta J = \pm 1, 0$ and $\Delta L = \pm 1, 0$. The $\Delta S = 0$ rule is broken, as in the transition for the 514.5-nm line.

6.8.10   Helium–cadmium laser

The helium-cadmium laser is another gas laser. It is a reliable source of UV light and has been used for luminescence studies in wide-gap semiconductors. Its oscillation wavelength is $\lambda = 325$ nm. It works like the He–Ne laser through electronic excitations of vaporized Cd ions in helium.

6.8.11 Excimer lasers

Excimer lasers are also gas lasers. The atomic excitation to the upper level of the lasing transition requires an even larger energy discharge than do the argon-ion lasers. This requires either a spark gap or a thyratron, which is far more efficient and reliable. Different lasing transitions are produced with different fills of rare gas. These active rare-gas–halogen compounds are formed by exciting an electron from the outer closed shell of the rare gas to the higher s-state, thus making it behave as an alkali metal. In this condition, it can be made to react with a halogen element, such as F^{-1} and Cl^{-1}, to form rare-gas–halide compounds:

$$Ar + F_2 \rightarrow Ar^* + F_2 \rightarrow Ar^+ F^- + F$$

The lowest ionically bound (excimer) state of $Ar^+ F^-$ is 6.7 eV. This is formed under high excitation energy, in which an electron jumps from the rare earth to the halogen gas and the two ions become bound to each other. The excess energy of the reaction is distributed in the gas (e.g., kinetic energy picked up by the free fluorine and other collisions) to stabilize the excimer molecule. The radiative lifetime of the excited molecule is short and produces the laser gain. The emission wavelengths are shorter for lighter rare gases, due to the higher ionization potentials, and longer for the lighter halogen gases, due to the higher electron affinities. The most common wavelengths produced are listed in the following table.

Typical wavelength	Gas	Radiative lifetime
157 nm	F_2	—
175 nm	ArCl	—
193 nm	ArF	4.2 ns
222 nm	KrCl	
248 nm	KrF	~ 8 ns
308 nm	XeCl	11 ns
351 nm	XeF	~ 15 ns

The lasers use carrier gases of He or Ne. Output energies are about 50–800 mJ per pulse at a repetition rate of 1–100 Hz, with pulse widths of 8–30 nsec. In general, the output is not well collimated. Often an energy scan across the beam reveals dark regions where there is destructive interference between many interfering modes. Consequently, some spatial filtering and refocusing are beneficial.

6.8.12   Nitrogen gas lasers

Nitrogen gas lasers are another example of gaseous discharge lasers. They have an output wavelength of $\lambda = 308\,\text{nm}$, with a repetition rate of 1–10 Hz, and pulse durations of 300 psec. Nitrogen lasers are often sold with dye lasers, in which the dye is contained in a cuvette and can be used both for the laser light generation and amplification. The low repetition rate makes them difficult to use for spectroscopy with weak signals, but the UV output and short pulse length are beneficial for studies that require tunable wavelengths.

6.8.13   Carbon dioxide (CO_2) Lasers

In essence, the CO_2 laser is also a gas discharge laser that gains its excitation from kinetic energy transfers from elastic and inelastic collisions between a large current of electrons and gas molecules. Unlike the others, however, it is a molecular laser; that is, it does not work by means of electronic transitions of atoms, but rather by means of kinetic excitations of the gas molecules. The lasing transitions take place between phonon vibrational states of the molecule.

The gas mixture in the laser tube consists of a mixture of CO_2 and N_2. Excitation of the gas takes place in both gases. Nitrogen has only one normal mode of vibration. The N_2 molecule is a simple harmonic oscillator with energy levels ξ_n represented as

$$\xi_n = h\nu_0(n + {}^1/_2)$$

The ground state for N_2 is $\xi_0 = {}^1/_2 h\nu_0$, and the excited state is $\xi_1 = {}^3/_2 h\nu_0 = 2326\,\text{cm}^{-1}$.

Carbon dioxide is more complex, having three distinct modes of vibration:

$\nu_1 = $ symmetric stretch $= (100)$ state $= 1388.3\,\text{cm}^{-1}$

$\nu_2 = $ bending (degenerate) $= (010) = 667.3\,\text{cm}^{-1}$ and $(020) = 1285.5\,\text{cm}^{-1}$

$\nu_3 = $ asymmetric stretch $= (001) = 2349.3\,\text{cm}^{-1}$

The vibrational states of the carbon dioxide and nitrogen gases are also divided into rotational states.

It is useful to find that the excitation for the nitrogen gas and the (000) → (001) transition for the carbon dioxide gas are within 23 cm^{-1} of each other. Thus the energy gained by the nitrogen molecules is readily converted into excitation of the (001) state of carbon dioxide, with only a slight addition of kinetic energy.

The laser is pumped with a plasma discharge to populate the excited energy levels of both the CO_2 and the N_2 molecules: (2349cm^{-1}) for (001) CO_2 (v_3) and 2326 cm^{-1} for N_2 (v_1). Laser emission takes place between the odd rotational levels of the (001) state and the even rotational levels of the (100) and (020) states. The 10.6-μm radiation of the laser is a transition from the (001) state to the (100) state [e.g., 2349.3 cm^{-1}→1288.3 cm^{-1}]. The 9.6-μm emission occurs between the (001) and the (020) states. These are tunable emissions due to the variations in the rotational levels (Fig. 6.22).

The CO_2 laser exhibits a 30% working efficiency for the following reasons.

1. The laser levels are near the ground state → high atomic quantum efficiency $v_{21}/v_{30} \approx 45\%$

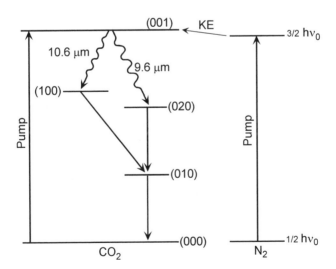

Figure 6.22: Energy diagram for a CO_2 laser.

2. A large fraction of CO_2 molecules are excited by the electron impact, and they cascade down the energy ladder to the long-lived v_3 (001) vibrational state.

3. Because of the large energy level separation of N_2 in the upper levels, most of the excited atoms go only to the $n = 1$ state. Their energy is easily transferred to CO_2 molecules through collisions, with the difference of 18–23 cm^{-1} made up from kinetic energy (e.g., the collision has a high cross section). In fact, most of the N_2 molecules excited to $n = 1$ lose their energy through collisions with CO_2 molecules rather than transitions to $n = 0$. This means that the excited lifetime of the v_3 state of CO_2 is much less than the collision time. (Otherwise CO_2 molecules would also excite N_2 molecules.)

Carbon dioxide lasers can emit very high power, on the order of kilowatts. The pulse rate is determined by the plasma discharge source. Also, CW operation at lower power is possible.

6.8.14 Copper vapor lasers

Copper vapor lasers (CVLs) are neutral metal vapor lasers with high efficiency and very high power output. Typically these lasers are operated at kilohertz repetition rates and have output powers averaging 100 W. In a typical CVL, pieces of copper metal are placed in a laser discharge tube and the tube is heated to about 1500 °C in order to vaporize the metal. Inert gas is added to the tube, most commonly Ne, He, or Ar, to improve lasing efficiency, in part by depopulating the lower lasing level through collisions. The vaporized metal gas is pumped by a fast, high-voltage electronic discharge that passes longitudinally along the plasma tube. A thyratron is used to obtain the high pump voltages (10–20 kV) and repetition rates necessary for laser operation. Electron collisions excite the copper atoms to the upper lasing state, from which they decay. Decay to the lower, metastable laser levels is possible via two transitions that provide emission either at 510.6 nm, in the green, or at 578.2 nm, in the yellow. Thus, CVLs represent a typical three-level laser scheme with two possible routes available for radiative decay to the lower states. The high gain of these lasers, on the order of 10–30% per cm, means that the lasers can operate without need for mirrors in the cavity. In fact, a sealed plasma tube with polished flat glass faces is adequate for efficient operation.

6.8.15 Organic dye lasers

Dye lasers are a type of liquid laser that relies on the emission properties of organic dyes to supply lasing wavelengths. Dyes absorb strongly in the visible and exhibit broad luminescence that often spans a wide range of wavelengths. The energy level diagram of dye molecules consists of vibrational states with groups of rotational levels. Among these states, the singlet-to-triplet-state transitions are generally undesirable and less likely than the singlet-to-singlet transitions. Stimulated emission of the dye molecules occurs when light energy used to excite the molecules and pump the population from the singlet ground state to one of the singlet excited states quickly decays down the rotational states to the lowest level in each vibrational band (Fig. 6.23). Luminescence occurs through transitions to all states (rotational) of the lower vibrational band. Cavity resonance conditions cause a single transition to be selected for amplification of stimulated emission, and that produces the laser output. The laser output wavelength can be tuned by changing the cavity length.

The upper-state lifetime is short-lived (typically 10 nsec), so pumping to obtain population inversion is difficult, and high-power lasers are used to pump dye lasers. The pump energies are usually in the green or UV wavelength range. Typical pumping lasers include doubled Nd, argon-ion, and nitrogen lasers.

Due to the high pump energies required, typical dye lasers consist of flowing liquids. This allows thermal equilibrium to be maintained,

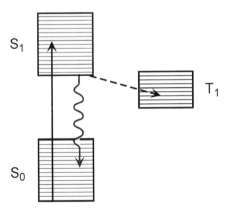

Figure 6.23: Typical transitions between the rotational levels of various vibrational states for a dye laser. The luminescence comes from singlet-to-singlet transitions, which are faster than singlet-to-triplet transitions.

because the exposed liquid is heated for only a short time. The nozzles develop hypersonic fluid velocities in the dye jet. However, this is still long compared to the upper-state or luminescence decay time.

Dye lasers can be simple to operate, depending on the cavity configuration of the laser. Their primary benefits are the range of fully tunable wavelengths that can be accessed with a suite of well-selected dyes and the fact that the dyes themselves are relatively inexpensive (certainly less expensive than a series of gas or solid-state lasers capable of the same range of wavelength tunability would be). The main drawbacks of dye lasers are that the dyes themselves are messy and the flow jets tend to splatter occasionally when air bubbles or dirt get into the dye recirculating system. This frequently results in dye droplets falling on relevant cavity optics (such as the lenses used to focus pump laser light onto a narrow portion of the dye jet), which then necessitates careful cleaning of the cavity optics. In addition, most dyes are carcinogenic, and dye spills are difficult to clean up since the dyes permanently stain many surfaces (tables, floors, clothes, etc.). Table 6.2 shows the range of wavelengths accessible with typical commercial dyes.

In the past, colliding-pulse-mode-locked lasers were very popular for providing short pulses (60 fsec). These are ring lasers in which two counterpropagating modes travel. The gain medium is either a dye or

Organic dye*	Wavelength range when pumped with argon laser
Stilbene 420	420–480 nm
Coumarin 480	480–530 nm
Coumarin 540	520–580 nm
Rhodamine 560	550–600 nm
Rhodamine 590	560–650 nm
DCM	610–730 nm
LD 700	710–850 nm
LDS 821	800–940 nm
IR140	920–980 nm

*Data from Exciton, Inc., Catalog.

Table 6.2: Laser dyes.

copper vapor (CPM), and a saturable absorber provides a high cavity loss, except when the two pulses arrive simultaneously at the absorber from the two opposite directions. This superposition produces a short, intense pulse of light that depends only on the high intensity portion of the colliding modes.

6.8.16 Free-electron laser

The free-electron laser (FEL) (like the x-ray laser, which will not be discussed here) is a unique laser that fits into none of the dominant classes of laser systems. It is designed to generate electromagnetic waves over a broad range of wavelengths, from the microwave range to the UV range. The "laser" does not operate through transitions between well-defined levels. Instead, the radiation comes from the accelerated motion of free electrons in a periodic magnetic field. High-energy electrons emit synchrotron radiation when a magnetic field bends their path or accelerates or decelerates them. This emission of radiation can be continuous, since free electrons lack the discrete energy levels of atoms and molecules. The FEL is designed to produce free electrons that travel down a magnetic guide equipped with an array of magnets of periodically alternating polarity. This chamber is called a wiggler or undulator, since it causes the motion of the electrons to wobble periodically as they traverse the magnetic field. The phenomenon of electron bunching causes the electrons to clump in packets separated by one period of the magnet. Electrons with the same energy will emit radiation at the same wavelength, and the light emitted from different clumps will add together in phase. Light emitted along the axis of the magnet will be amplified if it satisfies the resonance conditions of the cavity. In general, wavelength tunability is obtained by keeping the undulator spacing fixed and varying the electron energy.

The FEL can serve as either an oscillator or an amplifier. The oscillator configuration requires that the emitted synchrotron radiation be confined inside a resonance cavity with mirrors at both ends. In the amplifier configuration, an optical signal is injected into the undulator and the high-current electron beam provides gain without the need for optical feedback. Free-electron lasers have a tremendous application potential because of their continuous tunability. However, such lasers are extremely large and expensive and are not generally commercially available.

6.9   Semiconductor lasers

Stimulated coherent emission was demonstrated in GaAs and GaAsP in 1962. The structures formed were p–n junctions. Semiconductor lasers do not operate much like other lasers. The laser transitions are between the conduction and valence bands of the semiconductor, and pumping is by forward bias current. The interband transitions involve two *distributions* of energy levels rather than only two levels (e.g., for independent atoms), so the emission can be much broader. The propagation characteristics of electromagnetic waves in the optical cavity are totally determined by the material. In fact, the material makes up the optical cavity and cleaving, the end faces on the semiconductor structure is an important aspect of semiconductor laser fabrication. The simplest lasers are made by p–n junctions. But more complex, heterojunction lasers are more often used. They are both described next.

6.9.1   p–n Junctions

In an n-type semiconductor, the Fermi level is raised above the bottom of the conduction band, so there are occupied states in the conduction band due to the high concentration of donor impurities. In a p-type semiconductor, the Fermi level is depressed into the valence band, so there are empty states in the top of the valence band, due to the high concentration of acceptor impurities. In both cases, the semiconductor behaves like a metal and the conductivity remains nonzero even at 0K. Now, as electrons in the conduction band on the n-side decay to the empty states in the valence band on the p-side, there is spontaneous emission of radiation (Fig. 6.24).

If we recall that interband absorption is written as

$$\alpha = \frac{\omega e^2 x_{cv}(2\mu)^{3/2}}{2\pi\epsilon\eta c\hbar^3}(\hbar\omega - \xi_g)^{1/2}$$

it is possible to calculate the absorption and amplification due to free carrier excitation and recombination at 0 K:

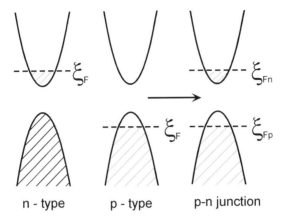

Figure 6.24: Formation of a p–n junction brings together materials with free electrons and free holes.

$$\alpha_{cv} = 0 \qquad\qquad\qquad \text{for } \hbar\omega < \xi_g$$

Absorption: $\quad \alpha = +K(\hbar\omega - \xi_g)^{1/2} \quad$ for $\xi_g < \hbar\omega < \xi_{Fn} - \xi_{Fp}$

Gain: $\quad\quad \gamma = -K(\hbar\omega - \xi_g)^{1/2} \quad$ for $\hbar\omega > \xi_{Fn} - \xi_{Fp}$

Above 0 K, we must worry about the occupancy factor, $f(\xi_c)$ or $f(\xi_v)$, for the appropriate states:

$$\Delta N = N_{c\to v} - N_{v\to c} = f_v(\xi_k) - f_c(\xi_k)$$

$$\alpha = -K(\hbar\omega - \xi_g)^{1/2}[f_v(\xi_k) - f_c(\xi_k)]$$

$$\gamma = -\alpha(\omega)$$

Note that we use the same k-values for direct transitions. The maximum gain is $\gamma_0(\omega) = +K(\hbar\omega - \xi g)^{1/2}$. For frequencies $\hbar\omega$ between ξ_g and $\xi_{Fn} - \xi_{Fp}$, the gain becomes

$$\gamma_0(\omega) = K(\xi_{Fn} - \xi_{Fp} - \xi_g)^{1/2}$$

6.9.2   Degenerate p–n junction lasers

In order to have a recombination, an inverted population is required. High current is passed through the junction to maintain the inverted population, and the forward bias ξ_g/e. Consequently, there exists a region near the depletion layer that contains simultaneously a degenerate population of electrons and holes, under forward bias (Fig. 6.25). Gain is present only in the region (active region) where there are both electrons and holes. The thickness of the active layer is simply the distance that electrons injected into the p region can diffuse before recombining with a hole: $t_{\mathrm{recomb}} \approx 10^{-9}$ sec in GaAs. The diffusion coefficient is $D \approx 10\,\mathrm{cm}^2/\mathrm{sec} \Rightarrow \Delta\ell = \sqrt{Dt_{\mathrm{recomb}}} \cong 10^{-4}\,\mathrm{cm}$, So the penetration distance is quite large.

In GaAs lasers, Zn atoms are doped as acceptors and give: $N_A = 10^{20}\,\mathrm{cm}^{-3}$; Te atoms are doped as donors and give: $N_D = 10^{18}\,\mathrm{cm}^{-3}$.

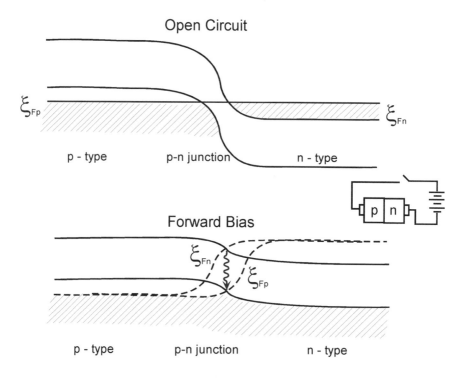

Homo Junction Laser

Figure 6.25: Under forward bias, the free electrons and free holes are driven toward the junction, where recombination leads to intense light emission (luminescence). In a cavity, the gain produces a laser.

The optical resonator is formed by polishing a pair of opposite crystal faces or, in GaAs, by cleaving the (110) faces.

These junctions, however, when used as amplifiers must account for the effect of impurities. These can cause major changes in the band structure. At high concentrations of donors ($> 10^{18}$ cm^{-3}) the donor tail merges with the conduction band and the optical transitions terminate on the acceptor band. This reduces the operational frequency of the laser. At even higher impurity concentrations, the transition becomes too broad to lase.

Also of importance is the refractive index profile. We can change the width of the active region by suitable index profiles in such a way as to vary the refractive index perpendicular to the direction of the layer. This variation in index in the lateral direction, *[n(z)]*, has the effect of containing the emission laterally while the end mirrors contain it longitudinally.

Threshold condition: Not all the energy travels within the active region where it is amplified. Lateral emission in the p and n regions leads to attenuation (α). Also, there is loss through the end reflectors. This leads to a threshold condition and a threshold current. For a layer of thickness d and a cavity of length ℓ, *the threshold current is calculated to be*

$$J_t = \frac{8\pi \nu^2 e d n^2 \Delta\nu}{c^2} \left(\alpha - \frac{1}{\ell} \ell \mathrm{n}R \right)$$

In GaAs, $\Delta\nu = 200$ cm^{-1}, $\alpha - (1/\ell)\ell u R = 20$ cm^{-1}, $\lambda = 0.84$ μm, $n = 3.35$, and $d = 1$ μm. This gives a threshold current of: $J_t \approx 75$ A/cm^2, at low temperatures.

6.9.3 Heterojunction lasers

In p–n junction lasers, it is difficult to confine the inversion layer to a small region; therefore, the threshold currents are high. But electromagnetic confinement can be provided by waveguiding the optical wave by forming an index profile. Waveguiding is achieved by having regions of high index surrounded by regions of low index. The electromagnetic wave is then confined to the region of higher refractive index.

Using the fact that GaAs has a higher index than Ga$_{1-x}$Al$_x$As ($\Delta_n = -0.4x$), structures can be constructed to confine the optical mode in a limited active region. The alloy Ga$_{1-x}$Al$_x$As also has a larger gap than GaAs. This provides an energy barrier to prevent carriers from diffusing out of the active region. Combined, these two effects lead to an active region whose size can be controlled by design of the structure. A typical double-heterostructure laser is formed with GaAs–AlGaAs. Conditions for the formation of a double-heterostructure laser are (in going from GaAs to $Al_xGa_{1-x}As$): (1) the index decreases, (2) the bandgap increases, (3) there is lattice matching between the two crystals, (4) the Fermi level offset is small, and (5) the coefficient of expansion of the two crystals match. Table 6.3 lists some repesentative laser materials and corresponding wavelengths.

Material	Wavelength (nm)
GaAs	838 (4.2 K)
	843 (77 K)
InP	910 (77 K)
InAs	3,100 (77 K)
InSb	526 (10 K)
PbSe	8,500 (4.2 K)
PbTe	6,500 (12 K)
$Ga(As_xP_{1-x})$	650–840
$(Ga_xIn_{1-x})As$	840–3,500
$In(As_xP_{1-x})$	910–3,500
GaSb	1,600 (77 K)
$Pb_{1-x}Sn_xTe$	(9.5–28 μm) (12 K)
$Ga_{1-x}Al_xAs$	690–850
InGaAsP	1,200–1,600
InGaP	500–700
ZnSe	490–500 (4.2 K)
$In_xGa_{1-x}N$	450–650

Source; Yariv 1976

Table 6.3: Oscillation wavelength and operating temperature of some semiconductor p-n junction lasers.

References

Bennett, W. R. 1962. Gaseous optical masers. *Appl. Optics, Suppl. 1, Optical Masers* 1:24.

Coherent Lasers, 5100 Patrick Henry Dr., Santa Clara, CA 95054

Exciton, Inc., P.O.Box 31126, Dayton, OH 45437

Geusic, J. E., H. M. Marcos, and L. G. Van Uitert. 1964. Laser oscillations in Nd-doped yttrium aluminum, yttrium gallium, and gadolinium garnets. *Appl. Phys. Lett.* 4:182.

Hecht, J. 1992. *The Laser Guidebook.* 2d ed. Blue Ridge Summit, PA: McGraw-Hill.

Laures, P. 1964. Variation of the 6328 Å laser output power, with mirror transmission. *Phys. Lett.* 10:61.

Maiman, T. H. 1964. Optical and microwave experiments in ruby. *Phys. Rev. Lett.* 4:182.

Snitzer, E., and C. G. Young. 1968. Glass lasers. In *Lasers*, Vol. 2, edited by A. K. Levine. New York: Marcel Dekker.

Verdeyen, J. T. 1995. *Laser Electronics.* Englewood Cliffs, NJ: Prentice Hall.

Weber, M. J., ed. 1991. *CRC Handbook of Laser Science and Technology.* Volumes I–V and Supplements. Boca Raton, FL: CRC Press.

Yariv, A. 1975. *Quantum Electronics.* New York: Wiley.

— 1976. *Introduction to Optical Electronics.* New York: Holt, Rinehart Winston.

Yariv, A. 1996. *Optical Electronics in Modern Communications.* 5th ed. New York: Oxford University Press.

Chapter 7

Nonlinear Optical Processes in Materials

7.1 Introduction

The nonlinear optical response of materials is utilized in many applications, from optical coupling between fibers to laser light generation and modulation; it is critical to photonic computing and logic operations. Optical nonlinearity is defined simply as any deviation from the linear relationship between a material's polarization response and an applied electromagnetic field. A nonlinear polarization can result from the application of a DC electric field, from a modulation in the intensity of the applied electromagnetic field, or from a combination of both. The type of nonlinear optical behavior exhibited by each material depends upon its atomic structure and composition. Many different processes can result in an optical nonlinearity in the material response. Most materials, in fact, exhibit more than one of these processes. The field of nonlinear optics (NLO) is much more extensive than can be covered in a single chapter or even a single book. Here, we will examine the predominant processes that lead to nonlinear behavior and give examples of technologically important materials systems notable for their NLO properties.

In studying optical processes in materials, it is customary to use the semiclassical approach in which photons are treated by the classical form of Maxwell's equations and electrons and holes are treated quantum mechanically.

7.1.1   Linear materials

First, let us review the propagation of plane waves in an isotropic homogeneous medium. The electric and magnetic permittivities, ϵ and μ, are constant:

$$E_x = E_{0x}e^{i(kz-\omega t)}$$

$$H_y = H_{0y}e^{i(kz-\omega t)} \quad \text{and} \quad H_{0y} = \frac{E_{0x}}{\sqrt{\mu/\epsilon}} = \frac{E_{0x}}{\mu c} \tag{7.1}$$

The velocity of the wave is

$$\upsilon = \omega/k = \omega/\sqrt{\mu\epsilon} = c/n$$

where n is the refractive index. Since μ is essentially μ_0, then $n = \sqrt{\epsilon/\epsilon_0} = c/\upsilon$, where c is the speed of light in a vacuum.

Reviewing the polarization relations in a linear medium, we have the polarization, P

$$P = \epsilon_0 \chi E$$

where χ is the electric susceptibility. The electric displacement field is

$$D = \epsilon E = \epsilon_0 E + P = \epsilon_0(1 + \chi)E$$

with $\epsilon = \epsilon_0(1 + \chi)$ and $n = \sqrt{\epsilon/\epsilon_0} = \sqrt{1 + \chi}$.

7.1.2   Nonlinear processes

We next look at the refractive index and the physical processes that will change it. The refractive index is a measure of the velocity of the electromagnetic wave in the medium. This velocity is affected by the local polarization processes. There are many kinds of polarization processes; some are important to nonlinear optical behavior, and other are not. Let us now examine the important ones.

We saw in Fig. 3.1 that there are several polarization processes in materials that can affect electromagnetic waves. Since, however, we are interested in visible, IR, and UV light, we see that only the ionic polarizability (phonon density of states) and the electronic excitations play a role at these frequencies.

Let us first consider transparent and isotropic materials, i.e., materials with no permanent dipoles. Transparent materials exhibit their transparency in a frequency or wavelength range between the ionic polarizability and the electronic excitation processes. The region of transparency, therefore, occurs at electromagnetic wave frequencies too high for the ionic polarizability to follow and too low to resonate with the electronic excitation polarizabilities. In the transparency region, the ionic relaxation resonances are too slow to occur during the period of oscillation of the wave. Therefore, the ionic polarizabilities play a role only if the optical wave frequency is near or below one of the fundamental or harmonic (multiphonon) vibrations of the structure. The electronic excitation processes are faster than the wave oscillation, so they can easily rearrange the polarization during the period of the optical wave. Consequently, the existence of electronic excitations at frequencies higher than that of the optical wave will affect the wave velocity and refractive index. From Fig. 3.1, as the frequency of the electromagnetic wave is reduced, the wave period will become long enough to include the electronic polarization, then the ionic polarization, then the molecular polarization, and so on. Each time another polarization mechanism is included, the dielectric constant far from resonances will increase. Thus at low frequencies, the dielectric constant will be larger than at higher frequencies.

Returning to the influence of the higher-frequency electronic excitations on the polarization in the transparency region, one might wonder how such an interaction could take place, since the electronic excitations require more photon energy than is carried by the wave (i.e., $h\nu < \xi_G$). Quantum mechanically, the photon does not have enough energy to cause an excitation of valence electrons across the bandgap. For this interaction, it is necessary to invoke multiphoton processes. Therefore the simultaneous coincidence of several photons in an electronic absorption volume would supply enough energy to excite an electron across the bandgap of a transparent material if $h\nu < \xi_G$. For example, a two-photon process could take place in the region: $^1/_2\xi_G \leq h\nu \leq \xi_G$. The probability of such a coincidence in time and space increases linearly with intensity for each photon required. Therefore the likelihood that an n-photon process will occur is proportional to the wave intensity to the nth power (i.e., the oscillator strength of the two-photon process will depend on the *square* of the intensity).

Three-photon, four-photon, and n-photon processes work in the same way to supply energy to the bound electron for excitation, and the probability for coincidence in time and space varies more sharply with intensity. Thus the polarization processes due to interaction of the wave with the *outer, loosely bound, valence electrons* of a dielectric material affect the transparency of a material and its refractive index. This intensity dependence in the absorption and index of refraction of the material is a nonlinear optical process.

Each time that the optical wave frequency matches a multiphoton transition energy, optical resonances occur with the oscillators of the material, because a new polarization process is enabled. These resonances are very weak at frequencies or photon energies far below the bandgap energy of the material, since they require the coincidence of many photons. However, they become significant at the two-photon absorption frequency ($^1/_2\xi_G$), and they dominate all other processes at the bandgap energy (ξ_G), where the induced change in polarization is very large. These processes will be treated later.

Nonlinear optical behavior can also be observed in materials that have permanent polarization anisotropies (i.e., permanent dipoles, as would be found in noncentrosymmetric crystals). Here, one can visualize a change in polarization behavior resulting from the application of a DC electric field. The net coherent force from the interaction of the applied field with the aligned dipoles in the material causes a directional stress in the exposed region of the material and consequently changes the optical propagation characteristic and the polarization of the material in the presence of the electromagnetic wave.

The processes that we have sketched in this section are designed to give the reader a sampling of what is to come. They are only a beginning. Next we describe some of the principal concepts and mathematics of nonlinear optics.

7.2   Mathematical treatment

Application of Maxwell's equations to nonlinear media:

$$\nabla \times E = -\frac{\partial B}{\partial t} \qquad \nabla \times B = \mu_0 \left(J + \varepsilon_0 \frac{\partial E}{\partial t} \right) \qquad (7.2)$$

$$\nabla \cdot E = \frac{\rho}{\varepsilon_0} \qquad \nabla \cdot B = 0 \qquad (7.3)$$

where *J(r,t)* and *ρ(r,t)* are current and charge densities. The charge conservation law gives

$$\nabla \cdot J = -\frac{\partial \rho}{\partial t} = 0 \tag{7.4}$$

J and ρ could be expanded into multipoles, but these expansions are not useful, since they are unphysical. Instead, we let

$$J = J_0 + \frac{\partial P}{\partial t} \tag{7.5}$$

where $J_0 = $ DC DC current density and P = generalized electric polarization, which is a function of the field.

$$\nabla \times B = \mu_0 \left(\frac{\partial}{\partial t} (\varepsilon_0 E + P) + J_0 \right) \tag{7.6}$$

$$\nabla \cdot (\varepsilon_0 E + P) = \frac{\rho_{\text{free}}}{\varepsilon} \quad \text{and} \quad \nabla \cdot P = -\rho_{\text{bound}} \tag{7.7}$$

Linear optics:

$$P(r,t) = \varepsilon_0 \int_{-\infty}^{\infty} \chi^{(1)}(r - r', t - t') \cdot E(r', t') dr' dt' \tag{7.8}$$

If $E = $ monochromatic plane wave:

$$P(r,t) = \varepsilon_0 \int_{-\infty}^{\infty} \chi^{(1)}(k, \omega) \cdot E(k, \omega) dk d\omega e^{i(k \cdot r - \omega t)} \tag{7.9}$$

$$\varepsilon_D(k, \omega) = 1 + \chi^{(1)}(k, \omega) \tag{7.10}$$

Nonlinear optics: At small E values, P can be expanded in powers of E.

$$
\begin{aligned}
P(r,t) = &\; \varepsilon_0 \int_{-\infty}^{\infty} \chi^{(1)}(r - r_1, t - t_1) \cdot E(r_1, t_1) \, dr_1 \, dt_1 \\
&+ \varepsilon_0 \int_{-\infty}^{\infty} \chi^{(2)}(r - r_1, t - t_1; r - r_2, t - t_2) : E(r_1, t_1) E(r_2, t_2) \, dr_1 \, dr_2 \, dt_1 dt_2 \\
&+ \varepsilon_0 \int_{-\infty}^{\infty} \chi^{(3)}(r - r_1, t - t_1; r - r_2, t - t_2; r - r_3, t - t_3) : \\
&\quad E(r_1, t_1) E(r_2, t_2) E(r_3, t_3) \, dr_1 \, dr_2 \, dr_3 \, dt_1 \, dt_2 \, dt_3 \\
&+ \dots
\end{aligned}
\tag{7.11}
$$

If E can be expressed as a group of monochromatic plane waves:

$$E(r,t) = \sum_i E(k_i, \omega_i) e^{i(\vec{k}_i \cdot \vec{r} - \omega_i t)} \tag{7.12}$$

then in Fourier transform space:

$$P(k,\omega) = P^{(1)}(k,\omega) + P^{(2)}(k,\omega) + P^{(3)}(k,\omega) + \ldots \qquad (7.13)$$

$$P^{(1)}(k,\omega) = \varepsilon_0 \chi^{(1)}(k,\omega) \cdot E(k,\omega)$$

$$P^{(2)}(k,\omega) = \varepsilon_0 \chi^{(2)}(k = k_i + k_j, \omega = \omega_i + \omega_j) : E(k_i,\omega_i)E(k_j,\omega_j)$$

$$P^{(3)}(k,\omega) = \varepsilon_0 \chi^{(3)}(k = k_i + k_j + k_\ell, \omega = \omega_i + \omega_j + \omega_\ell) :$$
$$E(k_i,\omega_i)E(k_j,\omega_j)E(k_\ell,\omega_\ell)$$

such that

$$\chi^{(n)}(k_1 + \cdots + k_n, \omega_1 + \cdots + \omega_n) = \int_{-\infty}^{\infty} \chi^{(n)}(r - r_1, t - t_1; \cdots; r - r_n, t - t_n)$$
$$e^{-i[k_1(r-r_1)-\omega_1(t-t_1)+\cdots+k_n(r-r_n)-\omega_n(t-t_n)]} \, dr_1 \, dr_2 \, dr_3 \ldots dr_n \, dt_1 \, dt_2 \ldots dt_n$$
$$(7.14)$$

If $\chi^{(n)}$ is known for a given medium, then, in principle the nth-order optical effects can be predicted in the medium, from Maxwell's equations. Physically, the higher-order susceptibility terms become important only when the electric field is large, since $\chi^{(n)}$ is proportional to $[E(r)]^n$. Each term in the expansion of the susceptibility, $\chi^{(n)}$ is related to the microscopic structure of the medium and can be properly evaluated only through a full quantum mechanical calculation of the electronic states of the system. However, simple classical models are very instructive and can illustrate the origin of nonlinear optical behavior in simple materials. The first example is the anharmonic oscillator model.

7.2.1   The anharmonic oscillator

Consider a medium composed of N oscillators per unit volume. Physically each oscillator describes an electron bound to a core or an IR active molecular vibration. The potential at x near x_0 (x_0 = interatomic distance) can be written as a Taylor series expansion:

$$V(x - x_0) = V(x_0) + k_i(x - x_0) + \frac{1}{2}k_2(x - x_0)^2 + \frac{1}{3!}k_3(x - x_0)^3$$
$$+ \frac{1}{4!}k_4(x - x_0)^4 + \ldots$$

where the k_n values correspond to the nth derivative of the potential at x_0. The first term vanishes, since we assume that the potential at x_0 is at equilibrium. It is also simpler to assume that $x_0 = 0$. The first remaining term in the expansion, as we have seen in Chapter 3, gives the simple harmonic oscillator:

$$V(x) = {}^1/_2 kx^2 \tag{7.15}$$

7.2.1.1 Optical linearity—Simple harmonic oscillator

The force for a simple harmonic oscillator is expressed as follows:

$$F = -\nabla V = -kx \tag{7.16}$$

Free oscillator: For a free oscillator, there is no driving force, and Newton's second law is simply the following:

$$m\frac{d^2x}{dt^2} + 2\Gamma m \frac{dx}{dt} = -kx \tag{7.17}$$

with a free oscillator with damping constant $2m\Gamma$. It simplifies to the form

$$m\left[\frac{d^2x}{dt^2} + 2\Gamma\frac{dx}{dt} + \omega_0{}^2x\right] = 0 \quad \text{with} \quad \omega_0{}^2 = \frac{k}{m} \tag{7.18}$$

The solution for the displacement from equilibrium is

$$x_i = \frac{-2\Gamma \pm \sqrt{4\Gamma^2 - 4\omega_0{}^2}}{2} = -\Gamma \pm i\sqrt{\omega_0{}^2 - \Gamma^2} = -\Gamma \pm i\mu \tag{7.19}$$

with $\mu = \sqrt{\omega_0{}^2 - \Gamma^2} \approx \omega_0$. For a free oscillator the displacement has the following familiar form:

$$x(t) = x(0)e^{-\Gamma t}e^{\pm i\mu t} \tag{7.20}$$

Driven oscillator: If the oscillator is driven, we can assume an external field *E(t)* and a corresponding force *F*:

$$E(t) = E_0 e^{-i\omega t}$$
$$F(t) = qE(t) = qE_0 e^{-i\omega t} \tag{7.21}$$

to alter the differential equation to the following form:

$$m\left[\frac{d^2x}{dt^2} + 2\Gamma\frac{dx}{dt} + \omega_0{}^2x\right] = qE_0 e^{-i\omega t} \tag{7.22}$$

The resulting solution must include a homogenous solution, x_c, and a particular solution, x_p:

$$x(t) = x_c + x_p$$

where

$$x_c = \begin{cases} \text{homogeneous solution with no driving force} \\ x_0 e^{-\Gamma t} e^{\pm i\omega_0 t} \quad \text{(decays with time)} \end{cases}$$

x_p = steady-state, particular or inhomogeneous solution

After an initial period, $x(t) \to x_p$.

Assume an inhomogeneous solution of the form $x_p = Ae^{-i\omega t}$. Substitute into the differential equation:

$$\frac{dx_p}{dt} = -i\omega Ae^{-i\omega t} \qquad \frac{d^2 x_p}{dt^2} = -\omega^2 Ae^{-i\omega t} \tag{7.23}$$

$$m[-\omega^2 A - 2i\Gamma kA\omega + A\omega_0^2] = qE_0 \Rightarrow A = \frac{qE_0/m}{(\omega_0^2 - \omega^2) - 2i\Gamma\omega} \tag{7.24}$$

$$x(t) = \frac{qE_0}{m} \frac{e^{-i\omega t}}{(\omega_0^2 - \omega^2) - 2i\Gamma\omega} + x(0)e^{-\Gamma t}e^{\pm i\omega_0 t} \tag{7.25}$$

For oscillating electrons, $q = q_e$ = electronic charge. Assuming that the homogeneous solution decays rapidly, the optical susceptibility becomes

$$\chi(t) = Nqx(t) = \frac{Nq^2 E_0}{m} \frac{e^{-i\omega t}}{(\omega_0^2 - \omega^2) - 2i\Gamma\omega} \tag{7.26}$$

The phase difference between $\chi(t)$ and the applied field $E(t)$ is

$$\Phi = \tan^{-1} \frac{2\omega\Gamma}{\omega_0^2 - \omega^2} \tag{7.27}$$

In a medium composed of N oscillators per unit volume, the polarization is then

$$P(t) = -Nqx(t) = \frac{Nq^2 E_0}{m} \frac{e^{-i\omega t}}{(\omega_0^2 - \omega^2) - 2i\Gamma\omega} \tag{7.28}$$

7.2.1.2   Second-order optical Nonlinearity—Anharmonic oscillator (Shen 1984)

Now we examine the effect of having an interatomic potential with anharmonic terms:

$$V(x) = \frac{1}{2}m\omega_0{}^2x^2 + \frac{1}{3}\max^3 + \frac{1}{4}mBx^4 + \dots \qquad (7.29)$$

and

$$F(x) = -m\omega_0{}^2x - \max^2 - mBx^3 - \dots \qquad (7.30)$$

The asymmetry in the potential field $(1/3\max^3)$ can reflect a lack of symmetry in the crystal field itself. If we examine the effect of adding just this term to the simple harmonic oscillator, we begin to appreciate the effects of nonlinear polarization processes.

$$m\frac{d^2x}{dt^2} + 2m\Gamma\frac{dx}{dt} + m\omega_0{}^2x + \max^2 = F = driving\ force \qquad (7.31)$$

Consider a driving electric field with component frequencies ω_1 and ω_2:

$$F = q[E_1(e^{-i\omega_1 t} + e^{+i\omega_1 t}) + E_2(e^{-i\omega_2 t} + e^{+i\omega_2 t})] \qquad (7.32)$$

$$\frac{d^2x}{dt^2} + 2\Gamma\frac{dx}{dt} + \omega_0{}^2x + ax^2 = \frac{q}{m}[E_1(e^{-i\omega_1 t} + e^{+i\omega_1 t}) + E_2(e^{-i\omega_2 t} + e^{+i\omega_2 t})] \qquad (7.33)$$

The anharmonic term ax^2 is assumed to be small, so it can be treated as a perturbation in the successive approximation of finding a solution.

$$x(t) = x^{(1)}(t) + x^{(2)}(t) + x^{(3)}(t) + \cdots \qquad (7.34)$$

where each term is a successively smaller correction.

Consider the following linear problem:
$x^{(1)} = x^{(1)}(\omega_1) + x^{(1)}(\omega_2) + $ complex conjugates:

$$\frac{d^2x^{(1)}}{dt^2} + \Gamma\frac{dx^{(1)}}{dt} + \omega_0{}^2x^{(1)} = \frac{q}{m}E_1e^{-i\omega_1 t} \qquad (7.35)$$

$$x^{(1)}(\omega_j) = \frac{(q/m)E_j}{(\omega_0{}^2 - \omega_j{}^2 - 2i\omega_j\Gamma)}e^{-i\omega_j t} \qquad (7.36)$$

This is the zeroth approximation and constitutes the *linear optical approximation*. This is an excellent approximation in most materials at low light intensities.

With the advent of lasers, where the optical power density is very high, smaller approximation terms that show an intensity dependence become important. In this case, the optical susceptibility $\chi^{(2)}$, $\chi^{(3)}$ and higher-order terms are all dependent upon the intensity of the applied field, $E(t)$, to the same power as the order of the susceptibility term.

The second-order solution is found by approximating the anharmonic term with the harmonic solution, as follows:

$$ax^2 = a(x^{(1)})^2$$

$$\frac{d^2x^{(2)}}{dt^2} + 2\Gamma\frac{dx^{(2)}}{dt} + \omega_0{}^2 x^{(2)} + ax^{(1)^2} = \frac{q}{m}[E_1(e^{-i\omega_1 t} + e^{+i\omega_1 t})$$

$$+ E_2(e^{-i\omega_2 t} + e^{+i\omega_2 t})] \tag{7.37}$$

This gives a second order solution of the form

$$x^{(2)} = x^{(2)}(\omega_1 + \omega_2) + x^{(2)}(\omega_1 - \omega_2) + x^{(2)}(2\omega_1) + x^{(2)}(2\omega_2) + x^{(2)}(0) \tag{7.38}$$

where:

$$
\begin{aligned}
x^{(2)}(\omega_1 + \omega_2) &= x_{12}e^{-i(\omega_1+\omega_2)t} + cc \\
x^{(2)}(\omega_1 - \omega_2) &= x_{21}e^{-i(\omega_1-\omega_2)t} + cc \\
x^{(2)}(2\omega_1) &= x_{11}e^{-2i\omega_1 t} + cc \\
x^{(2)}(2\omega_2) &= x_{22}e^{-2i\omega_2 t} + cc
\end{aligned}
\tag{7.39}
$$

where cc is the complex conjugate of the form $x_{ij}{}^* e^{+i(\omega_1+\omega_2)t}$ Substituting this into the differential equation yields the following results, which give the terms of the second-order solution:

$$x^{(2)}(\omega_1 \pm \omega_2) = \frac{-2a(q/m)^2 E_1 E_2}{(\omega_0{}^2 - \omega_1{}^2 - i\omega_1\Gamma)(\omega_0{}^2 - \omega_2{}^2 \mp i\omega_2\Gamma)} \frac{e^{-i(\omega_1 \pm \omega_2)t}}{[\omega_0{}^2 - (\omega_1 \pm \omega_2)^2 - i(\omega_1 \pm \omega_2)\Gamma]}$$

$$x^{(2)}(2w_j) = \frac{-a(q/m)^2 E_j{}^2}{(\omega_0{}^2 - \omega_j{}^2 - i\omega_j\Gamma)^2(\omega_0{}^2 - 4\omega_j{}^2 - 2i\omega_j\Gamma)}e^{-2i\omega_j t} \tag{7.40}$$

$$x^{(2)}(0) = a\left(\frac{q}{m}\right)^2 \frac{1}{\omega_0{}^2}\left[\frac{1}{\omega_0{}^2 - \omega_1{}^2 - i\omega_1\Gamma} + \frac{1}{\omega_0{}^2 - \omega_2{}^2 - i\omega_2\Gamma}\right]$$

By successive iterations, the higher-order solutions can be obtained. As seen in the 2nd-order solution, new frequency components appear in the polarization at $\omega_1 + \omega_2$, $\omega_1 - \omega_2$, $2\omega_1$, $2\omega_2$, and 0. These appeared through a quadratic interaction of the field with the oscillator via the anharmonic term of the force.

The oscillating polarization components will radiate and generate new

EM waves at $\omega_1 \pm \omega_2$, $2\omega_1$, $2\omega_2$. The zero-frequency polarization component is known as *optical rectification*.

Treating ax^2 as a small term is equivalent to expanding P in a power series of E.

Assuming nonresonance, (i.e., $\omega_0 \gg \omega_1$, ω_2):

$$\left| \frac{P^{(2)}}{P^{(1)}} \right| \sim \left| \frac{qaE}{m\omega_0{}^4} \right| \tag{7.41}$$

For an electron bound to the core of an ion, $qE_{at} =$ total binding force on the electron:

$$|qE_{at}| \sim m\omega_0{}^2 x - max^2$$
$$\sim \frac{m\omega_0{}^4}{a} \tag{7.42}$$

Therefore:

$$\left| \frac{P^{(2)}}{P^{(1)}} \right| \sim \left| \frac{E}{E_{at}} \right|$$

and in general:

$$\left| \frac{P^{(n+1)}}{P^{(n)}} \right| \sim \left| \frac{E}{E_{at}} \right| \tag{7.43}$$

Typically $E_{at} \sim 3 \times 10^8$/Vcm. The electromagnetic wave field for a 2.5 W/cm^2 laser beam is about 80 V/cm$\to |E/E_{at}| \sim 10^{-7}$. *In general, the nonlinear polarization is much smaller than the linear polarization.*

For an incident wave with frequency ω_1 or ω_2, we now have several new physical processes:

2nd harmonic generation	Output waves at $2\omega_1$, or $2\omega_2$.
Parametric oscillation	Waves in: $[\omega_1$ and $\omega_2]$
	Waves out: $[(\omega_1 - \omega_2)$ and $(\omega_1 + \omega_2)]$
Rectification	DC output ($\omega = 0$)
Spontaneous parametric fluorescence	Waves in: $[\omega_3]$
	Waves out: $[\omega_1$ and $\omega_2]$

where, clearly, parametric upconversion is a subset of parametric oscillation. The susceptibility may now be written as follows.

Linear susceptibility:

$$\chi^{(1)}(\omega_j) = \frac{N(q^2/m)E_j}{(\omega_0{}^2 - \omega_j{}^2) - i\omega_j \Gamma} e^{-i\omega_j t} \tag{7.44}$$

The second-order susceptibility terms associated with the foregoing processes are as follows:

Parametric oscillation:

$$\chi^{(2)}(\omega_1 \pm \omega_2) = \frac{-2aN(q^3/m^2)E_1 E_2}{[(\omega_0^2 - \omega_1^2) - i\omega_1\Gamma][(\omega_0^2 - \omega_2^2) \mp i\omega_2\Gamma]} \frac{e^{-i(\omega_1 \pm \omega_2)t}}{[\omega_0^2 - (\omega_1 \pm \omega_2)^2 - i(\omega_1 \pm \omega_2)\Gamma]}$$

(7.45)

Second harmonic generation (SHG):

$$\chi^{(2)}(2\omega_j) = \frac{-aN(q^3/m^2)E_j^2}{[(\omega_0^2 - \omega_j^2 - i\omega_j\Gamma)^2(\omega_0^2 - 4\omega_j^2 - 2i\omega_j\Gamma)]} e^{-2i\omega_j t}$$

(7.46)

Optical rectification:

$$\chi^{(2)}(0) = -aN(q^3/m^2)\frac{1}{\omega_0^2}\left[\frac{1}{\omega_0^2 - \omega_1^2 - i\omega_1\Gamma} + \frac{1}{\omega_0^2 - \omega_2^2 - i\omega_2\Gamma}\right]$$

(7.47)

One may ask, "What material shows this behavior?" The asymmetry in the potential field at very low disturbances arises from a lack of symmetry in the crystal field itself:

$$V(x) = \frac{1}{2}m\omega_0^2 x^2 + \frac{1}{3}\max^3 + \frac{1}{4}mBx^4 + \cdots$$

(7.48)

The second term is the nonsymmetric term in the potential and corresponds to the case of crystals with noncentrosymmetric unit cells! This means that if the crystal structure has no center of inversion symmetry, then local dipoles exist. In a crystal, the dipoles will be aligned and the material will exhibit a nonlinear polarization and directional index change under the application of a field. The polarization can be affected nonlinearly by an applied DC field, since it acts on the aligned dipoles of the structure. This is the underlying mechanism of $\chi^{(2)}$ electric susceptibility, or second-order optical nonlinearity.

Interestingly, macroscopically isotropic materials, such as polymers and noncrystalline materials, can still show second-order optical nonlinearity. This results from the formation of dipoles due to charge separation in side groups of conjugated polymers or due to charged defects or impurities in glass. These dipoles are aligned by an applied *external* field (poling). Once the dipoles are aligned, the materials respond as do noncentrosymmetric crystals, and they can exhibit parametric oscillation, second harmonic generation, and optical rectification.

These effects can also take place in crystals with center of inversion symmetry when exposed to a very large field. Under this condition, the displacement of the electron cloud from equilibrium can be so large that anharmonic components of the potential may become evident far from equilibrium. Since electron clouds are bound by Coulomb forces to the ionic cores, the restoring forces are not strictly proportional to the displacement from equilibrium for large deviations. Consequently, all materials can become nonlinear at sufficiently high fields. Of these, some experience an anharmonic potential, and they exhibit second-order optical nonlinearity. Others still experience a symmetric potential even at large deviations. These materials exhibit third-order optical non-linearity.

7.2.2 Third-order optical Nonlinearity

The third-order optical nonlinearity arises from the next term in the multipole expansion of the polarizability. If a crystal has center of inversion symmetry, the second-order susceptibility vanishes. As before, third harmonic generation is possible through $\chi^{(3)}$ as:

$$P^{(3)}(3\omega) = \varepsilon_0 \chi^{(3)}(-3\omega : \omega, \omega, \omega) : E(\omega)E(\omega)E(\omega) \qquad (7.49)$$

The third-order susceptibility is much smaller than the second-order value. For example, if $|\chi^{(2)}| \sim 10^{-7}$-$10^{-9}$ esu, the third-order susceptibility is of the order of $|\chi^{(3)}| \sim 10^{-12}$-$10^{-15}$ esu. In practice, however, frequency tripling is *not* done by $\chi^{(3)}$ because of its low value, and one uses a sum of $2\chi^{(2)}$, especially since phase matching is easier to obtain. Frequency tripling is available commercially at about 20% coupling efficiency.

If we conduct the same calculation as done for the quadratic perturbation in the potential of the simple harmonic oscillator using the cubic perturbation, Bx^3, the following terms develop in the presence of an external oscillatory electric field and a DC field:

Third-harmonic generation (THG)	$\chi^{(3)}(-3\omega; \omega, \omega, \omega)$
Nonlinear index (NLO index)	$\chi^{(3)}(-\omega; \omega, -\omega, \omega)$
Electric field–induced second harmonic generation (EFISH)	$\chi^{(3)}(-2\omega; \omega, \omega, 0)$
Quadratic electro-optic effect (QEO)	$\chi^{(3)}(-\omega; \omega, 0, 0)$
Optical rectification	$\chi^{(3)}(0; -\omega, \omega, 0)$
	$\chi^{(3)}(0; 0, 0, 0)$.

Assuming that the applied fields are written as $E_1 = E(\omega)\cos(\omega t)$ and $E_2 = E(0)$, the polarization is found to be

$$P(\omega) = \varepsilon_0 \left\{ \frac{1}{4} \chi^3(-3\omega; \omega, \omega, \omega) E^3(\omega) \cos(3\omega t) \right\}$$

$$+ \varepsilon_0 \left\{ \frac{3}{2} \chi^{(3)}(-2\omega; \omega, \omega, 0) E^2(\omega) E(0) \cos(2\omega t) \right\}$$

$$+ \varepsilon_0 \left\{ \frac{3}{4} \chi^{(3)}(-\omega; \omega, -\omega, \omega) E^3(\omega) \cos(\omega t) \right\} \qquad (7.50)$$

$$+ \varepsilon_0 \left\{ 3\chi^{(3)}(-\omega; \omega, 0, 0) E(\omega) E^2(0) \cos(\omega t) \right\}$$

$$+ \varepsilon_0 \left\{ \frac{3}{2} \chi^3(0; \omega, -\omega, 0) E^2(\omega) E(0) \right\}$$

$$+ \varepsilon_0 \left\{ \chi^{(3)}(0; 0, 0, 0) E^3(0) \right\}$$

The results of mixing a simple wave at frequency ω with a DC electrical field in a nonlinear optical material with a high third-order susceptibility can be one of the following: the emission of a wave at 3ω (THG), the emission of a wave at 2ω (EFISH), or the emission of a wave at 1ω, with a modified refractive index (NLO index, and QEO), and the formation of a DC field. If the DC field is removed, then only the terms with $E(\omega)$ alone remain: THG and NLO index. In the latter, as discussed in Chapter 3, there will be a nonlinear absorption β_2 associated with the nonlinear index n_2:

$$n = n_0 + n_2' I$$
$$\alpha = \alpha_0 + \beta_2 I \qquad (7.51)$$

where:

$$n_2' \propto \text{real} \left\{ \chi^3_{1111}(-\omega; \omega, -\omega, \omega) \right\}$$
$$\beta_2 \propto \text{imaginary} \left\{ \chi^3_{1111}(-\omega; \omega, -\omega, \omega) \right\} \qquad (7.52)$$

If the material is noncentrosymmetric, then it has a second-order susceptibility that produces optical rectification and a DC electrical field, so all the third-order susceptibility terms will be present. This happens in photorefractive crystals and SHG optical fibers.

In the next sections, we will discuss the individual terms of the polarization expansion just enumerated. We will relate these terms to relevant optical interactions and materials processes.

7.3 Second-order susceptibility

Materials that exhibit second-order susceptibility have distinct dipoles. These dipoles are usually permanent or frozen in, although a second-order effect may be developed with transient dipoles that have a lifetime longer than the period of illumination or the period of the wave. As mentioned earlier, the dipoles can result from the formation of crystals with no center of inversion symmetry or from the breakdown of center of inversion symmetry in isotropic materials by the introduction of anisotropic structures or defects that can be aligned by poling.

The presence of anisotropy in the propagation characteristics of the material means that the polarization and the applied electric field are not necessarily parallel. This means that the local electric field is no longer perpendicular to the wave propagation vector. The electric displacement vector, D, which is the sum of the local electric field and the polarization, remains normal to the propagation vector, and the wave now has two D-polarizations with different phase velocities. Mathematically, this means that the dielectric constant must become a tensor. The tensor is symmetric ($\epsilon_{ij} = \epsilon_{ji}$), so it may be diagonalized into principal dielectric constants along eigenvectors or principal directions. The refractive index thus has three values along these principal directions, and the wave propagates at different velocities for polarizations along these directions. A common occurrence is to have two of the principal refractive indices be equal ($n_x = n_y \neq n_z$). This is called *uniaxial birefringence*. In that case, the ordinary wave or ray will propagate with index $n_0 = n_x = n_y$, and the extraordinary wave or ray will propagate with index $n_e = n_z$. The optical birefringence is defined as the difference in the two indices:

$$B = \Delta n = n_e - n_0 \qquad (7.53)$$

For $\chi^{(2)}$ materials, the principal index ellipsoid is replaced by a nonlinear optic coefficient according to the following equation:

$$\Delta\left(\frac{1}{n^2}\right)_{jl} = \sum_k r_{jkl} E_k \qquad (7.54)$$

where r_{jkl} is the electro-optic tensor. The nonlinear optic coefficients d_{jkl} are obtained from the electro-optic tensor as follows:

$$d_{jkl} = -\frac{\varepsilon_j \varepsilon_i}{2\varepsilon_0} r_{jkl} \qquad (7.55)$$

with the result that the second order polarization can be written as

$$P_i^{(2)}(t) = \sum_j \sum_k d_{ijk} E_j(t) E_k(t) \qquad (7.56)$$

The d_{ijk} tensor is sometimes contracted using the piezoelectric convention ($1 = xx$, $2 = yy$, $3 = zz$, $4 = xy$, $5 = xz$, $6 = yz$). Values of the d_{ij} tensor are found in several books. The reader is referred to Yariv (1976) for an excellent coverage of this matter.

The result of a $\chi^{(2)}$ susceptibility is frequency mixing of two waves to form either second harmonic generation (SHG) and optical rectification ($\omega = 0$) from each wave, or sum frequency generation (SFG), or difference frequency generation (DFG). These are illustrated in Fig. 7.1. This mixing is a result of the modulation of the polarization and the refractive index by the incident electric field, which acts on the permanent dipoles of the material. This modulation of the dipole component of the polarization at the optical frequency causes sidebands of the incident frequency(ies) that are radiated out. This is very similar in process to the linear electro-optic effect, in which an externally applied voltage on the material causes a modulated field independent of the optical wave. The modulation is normally between DC and microwave frequencies, and this is commonly used to modulate (encode) laser beams. The Pockels coefficients that determined the electro-optic effect are proportional to the susceptibility, $\chi^{(2)}$.

In the formation of the second harmonic, sum, or difference light, it is necessary to have the conversion remain coherent with the light generated. This is called *phase matching* of the converted light to the incident light. Only materials that have a reasonable amount of phase matching will have any significant output; otherwise, the transformed output will incoherently destruct. The problem is that the incident light is a traveling wave. As it travels through a material of length L, it generates second harmonic, sum, or difference frequencies. The propagation characteristics of this light must remain in phase with the incident light beam so that the generated light remains coherent and constructive interference results. The difference in phase depends on the properties of the material, and it is expressed as a difference in wave vector amplitude and direction:

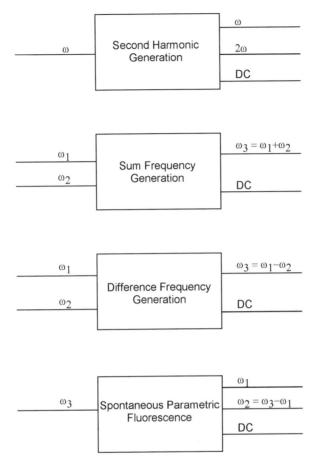

Figure 7.1: Box diagram showing the various types of processes possible with second-order susceptibility.

$$
\begin{aligned}
\text{SHG light:} \quad & \Delta k = 2k_\omega - k_{2\omega} \\
\text{SFG light:} \quad & \Delta k = k_{\omega 1} + k_{\omega 2} - k_{\omega 3} \\
\text{DFG light:} \quad & \Delta k = k_{\omega 1} - k_{\omega 2} - k_{\omega 3}
\end{aligned}
\qquad (7.57)
$$

The efficiency of generation of any of these varies as the following sinc function:

$$
\eta_{\text{eff}} \propto \frac{\sin^2(\Delta k \ L/2)}{(\Delta k \ L/2)^2}
\qquad (7.58)
$$

which is a reasonably sharp function with a maximum at $\Delta k = 0$ and a width at half max of $\Delta kL = 2.78$. This indicates that some phase mismatch is tolerated, but it reduces the length of the device and the intensity of the converted beam. A typical value of k for an optical wave is about $15,000\,\mathrm{cm}^{-1}$, so a difference in phase velocity of 0.1% will decrease the conversion efficiency to 1% of its phase matched value in a 1-cm sample.

7.3.1 Materials

7.3.1.1 Perovskites (Chiang 1997)

Crystals like barium titanate ($BaTiO_3$), $SrTiO_3$, $PbZrO_3$, $LiNbO_3$, and $KNbO_3$ have the Perovskite structure, which is cubic and made up of an oxygen cage in which the large cation has a coordination of 12, by forming an FCC lattice with the barium in the corner position and the oxygens at the face centers, while the smaller cation has a coordination of 6 at the body center. This structure is centrosymmetric. However, the structure is susceptible to a simple displacive transition. In barium titanate, for example, this occurs because the Ti ion sits in too large an oxygen octahedron. This ability of the Ti ion to rattle in its cage allows it to wander away from the central symmetric site. The room-temperature structure of $BaTiO_3$ is tetragonal, because the Ti ion has displaced 0.12Å toward one of the faces of the unit cell. The tetragonal structure is noncentrosymmetric, which gives it a permanent electric dipole. A cooperative alignment of neighboring dipoles results in regions of like dipole orientation known as *domains* . The electrical and optical response exhibited by these ferroelectric materials under applied fields is significantly influenced by the behavior of the domain ensemble present in the material. Again, in $BaTiO_3$ the transformation from cubic to tetragonal occurs at a Curie temperature of 130 °C. Pb^{+2} ions are larger than Ba^+, consequently, the instability of the Ti^{+4} is even more pronounced and the Curie temperature of $PbTiO_3$ is at 490 °C.

During the transformation, the displacement of the Ti^{+4} ion can be easily controlled with an electric field, and dipole alignment may be induced over large domains. This process, called *poling* , can lead to very high dielectric constants for capacitor applications. These materials exhibit piezoelectricity, by which the application of a field can change the size of the material. Conversely, an acoustic modulation of the material dimension can lead to generation of an electric field. For optical modulators, the application of an electric field alters the refractive index by changing the atomic density along a given direction. The application of strain has the same effect, so these materials exhibit stress birefringence as well as orientational birefringence, since the index change is not isotropic. The optical birefringence is alterable using an electric field, making these materials useful as optical light modulators and optical shutters. These effects can be enhanced by substitution of Ti for Nb in $LiNbO_3$ or by substitution of La for Ti or Zr to form PLZT materials that contain lead, lanthanum, zirconium, and titanium.

7.3.1.2   Poled polymers

Several polymer structures have also been used to produce high $\chi^{(2)}$ materials. In general, these structures are formed with chains that have polarizable ends or ends that can sustain a charged state. Thus, donor–acceptor molecules in conjugated polymers have the ability to retain opposite charges at the ends of the molecule. If the chain is made long, then the dipole moment of the molecule, which is the product of the charge and its separation, is very large. Poling before and during crosslinking allows alignment of the dipoles. Crosslinking of the chain structure retards the return to random orientations. These conjugated chain polymers have been used to produce electro-optical coefficients as high as those of $LiNbO_3$. In applications of these molecules, one of the critical parameters is the relaxation of the dipole alignment, which is difficult to control when the use temperature (room temperature) is close to the glass-transition temperature at which the material softens.

7.3.1.3   Poled glasses

Brueck and co-workers have shown that glasses can be made to form large permanent dipoles by diffusion of ions to the specimen surfaces (Alley, Brueck, and Myers 1998). Alkali-ion impurities or alkali-ion components in glasses diffuse well below the glass-transition temperature. The ions carry a positive charge while leaving behind an excess of negatively charged oxygen ions. Under a large applied voltage and at a temperature where ion diffusion is permitted, the alkali ions will diffuse to one of the glass surfaces. This generates a positively charged surface and an opposite, negatively charged surface. Cooling the glass to room temperature under the applied field freezes-in the charge segregation. The material thus has a large induced polarization. The charge segregation in the glass is not as easily reversible at room temperature as in the polymers. However, the whole charge resides near the surface (5–20 µm), so mechanical damage and corrosive processes may degrade the total polarization. Recent studies have shown that the process does not involve the alkali ion alone, and a less mobile complex (possibly involving H^+) has been suggested to explain space charge dynamics in fused silica.

7.3.1.4 Surfaces

Materials with no center of inversion symmetry have such a strong SHG signal from the bulk that the much weaker surface effects are ignored. However, if the bulk has a structure with a center of inversion symmetry, then $\chi^{(2)} = 0$ and only the surface signal is present. This signal can be thought to come from the breakdown in the center of inversion symmetry of a surface or an interface between two materials. The presence of surface-generated $\chi^{(2)}$ processes was first postulated by Bloembergen and Pershan (1962). Mizrahi and Sipe (1988) followed with a phenomenological treatment based on the assumption that the surface is represented by a very thin layer of dipoles. The SHG signal is very weak, but it is direction and polarization dependent. The result of the measurement is a calculation of the $\chi^{(2)}$ value of the specific spot of the surface under study. So the method can be used to study surface adsorption processes, and, since the laser can be scanned over the surface, spatial variations can be examined. Temporal variations in surface reactions can also be investigated by using pulsed lasers. It is difficult to calculate a priori the expected surface $\chi^{(2)}$ value for a particular process, but controlled tests have provided a guide for data interpretation, and the geometry of adsorbants may be deduced from the angular dependence of the SHG signal.

7.3.2   Systems that use the $\chi^{(2)}$ behavior of materials

7.3.2.1   Second harmonic generation

A typical frequency mixing system is shown in Figs. 7.2 for SHG. Crystals often used are KDP (KH_2PO_4) and $Ba_2NaNb_5O_{15}$. By inserting the doubling crystal into the laser cavity as shown in the figure, the output

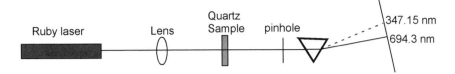

Second harmonic generation in quartz

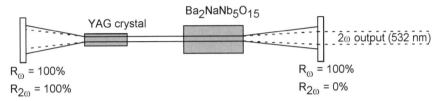

Second harmonic generation in a laser resonator

Figure 7.2: Systems set up for second harmonic generation in quartz (Franken et al. 1961) and in a laser cavity (Geusic et al. 1968).

filter can be designed to reflect fully the pump frequency so as to lower the threshold, and the only additional cavity loss is from the doubling crystal. Often the doubling crystal has a low absorption loss at the fundamental but a higher loss at the doubled frequency. This loss places an important limitation on the doubling efficiency.

7.3.2.2 Optical parametric oscillation

Optical parametric oscillators (OPOs) are the answer to an optical experimentalist's dream. They can provide laser light over a broad range of tunable frequencies. Considering the importance of measuring the optical dispersion of the various optical processes in order to understand the underlying mechanisms, it is difficult to understand how one managed without them. The optical parametric oscillator essentially takes a single frequency and separates it into two. With proper design, the majority of the energy can be transferred to a tunable output.

Optical parametric generation in a material corresponds to having an input photon with frequency ω_3 divide itself into two photons, ω_1 and ω_2, such that energy is conserved $(\omega_3 = \omega_1 + \omega_2)$. This is often called *spontaneous parametric fluorescence.* This emission is a result of the interaction of the incident photon with what are called *zero-point fluctuations* of the vacuum. These fluctuations induce spontaneous emission of an "idler" photon from the material. The idler then acts to modulate the polarization, resulting in a difference frequency generation. Any idler frequency can generate any difference beam in any direction. But the frequency and direction that dominate are those best suited for phase matching and cavity resonance.

The actual operation of an OPO requires the use of an idler beam to form the difference frequency condition. Figure 7.3 shows a typical design. A nonlinear crystal is placed into an optical cavity that provides resonances for both the idler and the output waves (ω_1 and ω_2). The input beam, ω_3, pumps the crystal to produce gain at the two frequencies. Eventually, threshold is reached and the two difference frequencies will lase. The output of the oscillator shown will consist of all three beams. In the Coherent OPO design, 820-nm 100–300-fsec pulses are focused into a BBO (β barium borate, BaB_2O_4) crystal to frequency double to 410 nm. This intense UV beam is focused into a second BBO crystal and mixed with a white-light continuum pulse. Since the idler is white light, the output difference beam can be adjusted by varying the cavity resonance over the range 450–700 nm by simply rotating the BBO crystal to maximize phase matching at each wavelength by changing the angle between the direction of propagation of the light beam and the c axis of the crystal and by compensating a delay line. The angle-tuning curve for a type I cut (31°) BBO crystal shows a range of 400 nm to 1.9 µm for a change in angle of 26–33 ° if the pumping frequency is near 400 nm. Other crystals commonly used include LBO (LiB_3O_5), $ZnGeP_2$, $LiNbO_3$, and KTP ($KTiOPO_4$).

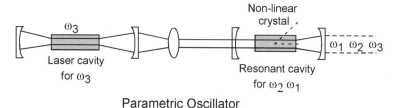

Parametric Oscillator

Figure 7.3: Parametric oscillation (Yariv 1976).

7.4 Third-order susceptibility

As seen earlier, there are many effects that result from the third-order susceptibility: (a) nonlinear index and absorption, (b) third harmonic generation, and (c) DC field–induced effects or quadratic electro-optic effect and electric field–induced second harmonic generation. These effects are also produced by several processes and occur differently in various classes of materials. Next we examine selected effects.

7.4.1 Nonlinear index and absorption

Changes in the refractive index and absorption in materials with a center of inversion symmetry lead to nonlinear optical behavior. Here, the pertinent susceptibility is the term associated with the single frequency, three-beam polarization:

$$P^{(3)}(\omega) = \frac{3}{4}\varepsilon_0\chi^{(3)}(-\omega;\omega,\omega,-\omega)E^3(\omega) \qquad (7.59)$$

Nonlinear index and absorption are tied together through Kramers–Kronig transforms. As processes induce intensity-dependent changes in polarizability, they alter both the index and the absorption. The dispersion (frequency dependence) of either one will affect the other. The processes underlying the NLO index and absorption behavior vary with frequency range. This is treated next.

7.4.1.1 Transparency region

The refractive index of a material is written as a sum of the linear part, n_0, and a nonlinear part, Δn:

$$n = n_0 + \Delta n = n_0 + 2n_2|E|^2 = n_0 + n_2'I$$

$$I = 2\varepsilon_0 n_0 c|E|^2 \quad \rightarrow \quad n_2' = \frac{n_2}{\epsilon_0 n_0 c} \qquad (7.60)$$

The NLO index processes in a material are strongly related to the dispersion behavior of that material. A crude approximation of the value of the NLO index coefficient, n_2, is obtainable from empirical equations that calculate n_2 as a function of the linear index and the dispersion (represented by the Abbe number, v_d). The Boling, Glass, Owyoung (BGO) equation (Boling, Glass, and Owyoung 1978) has been used to calculate the NLO index coefficient for various transparent materials in the region of transparency. Reasonable values are obtained if one remembers that this equation holds only far from any region of large or anomalous dispersion (in effect, well inside the range of transparency). The BGO equation is written as follows:

$$n_2 \text{ (esu)} = \frac{k(n_d - 1)(n_d^2 + 2)^2}{v_d[1.517 + (n_d + 1)(n_d^2 + 2)v_d/6n_d]^{1/2}} \qquad (7.61)$$

where $k = 6.8 \times 10^{-12}$ esu ($1\,\text{esu} = 1\,\text{cm}^3/\text{erg}$); the Abbe number $v_d = (n_d - 1)/(n_F - n_C)$; the index coefficients are: $n_d = n(587\,\text{nm})$, $n_F = n(486\,\text{nm})$, $n_C = n(656\,\text{nm})$. The NLO coefficient, n_2, associated with the electric field is commonly given in cgs units (esu), the coefficient, n_2', associated with the intensity is given in mks units (m^2/W). A useful relationship between them is

$$n_2'(m^2/W) \times 10^7 = n_2(cm^2/kW) = \frac{1}{3}\left[\frac{4\pi}{n_0}\right] n_{2\text{cgs}} \text{ (esu)} \qquad (7.62)$$

The BGO equation is an important empirical equation, but it hides the importance of the dispersion in the NLO behavior, since the latter changes character with frequency, changing drastically near resonances. Consequently, in studying the NLO index processes, it is important to keep in mind the ratios of the measuring frequency to the frequency of the electronic bandgap and to the frequency range of the phonon/multiphonon vibrations (i.e., the resonances above and below the transparency region).

Starting from the middle of the transparency region of an insulator or a semiconductor, the measuring frequency is assumed to be far from the IR absorption region and the bandgap absorption edge. In the transparency region, the optical frequency can couple weakly to some high overtones of the phonon vibration spectrum and to a fractional resonance (multiphoton absorption) of the bandgap transitions. In both cases the coupling is very weak and the intensity dependence of the polarizability is also weak. Consequently the change in refractive index with optical intensity is small and positive. This process is based on the third-order polarizability (or second-order *hyper* polarizability) of the material (often called the optical Kerr effect), and the positive value of n_2 increases gradually with frequency. This process has been measured in transparent glasses and found to be extremely fast, with response times in femtoseconds.

In the transparency region, the NLO index coefficient can be increased by raising the second hyperpolarizability of the material's constituents. For example, replacing oxygen with sulfur, selenium, or tellurium will raise n_2. Similarly, introducing components like Bi, Pb, Ba, Ga, Ti, Nb, In, and other heavy metals raises the NLO index coefficient. As a rule of thumb, heavier is better and higher valence is better (D. W. Hall et al. 1985).

The positive n_2 value obtained in this region of high transparency produces self-focusing in the medium when propagating an intense beam of light. Thus, this effect has been responsible for much self-focusing laser damage in glasses, due to the catastrophic nature of the process. If one assumes that a clean laser pulse will have a Gaussian intensity profile in the radial direction, with a maximum at the axis of a cylindrical amplifier rod, for example, then the largest value of refractive index will be found at this intensity maximum, following:

$$n = n_0 + \Delta n(r) = n_0 + n_2' I(r) = n_0 + n_2' I_0 e^{-(r^2/2\Delta r^2)} \qquad (7.63)$$

As the axial index increases above the peripheral values, a positive lens develops that focuses the beam further toward the rod axis. The resulting increase in intensity causes Δn to increase further, and the positive lens forms a tighter focus until the material vaporizes. If the beam profile is not purely Gaussian, filaments of high intensity will form at various places in the cross-sectional area of the rod where a local increase in illumination occurs by constructive interference. The solution developed to counter this effect was the choice of low-n_2 materials for lasers and amplifiers. These are the fluorophosphate glasses, since they exhibit a low linear refractive index and a low optical dispersion (large Abbe number) (see Fig. 3.7).

The index change associated with the second-order hyperpolarizability for a linearly polarized incident beam can be written as

$$\Delta n = 2n_2|E|^2 = \frac{3}{4n_0}\chi^{(3)}_{XXXX}(-\omega; \omega, \omega, -\omega)|E|^2$$

$$= \frac{3}{4n_0}[2\chi^{(3)}_{XXYY} + 2\chi^{(3)}_{XYXY}]|E|^2 \qquad (7.64)$$

For a circularly polarized incident light beam:

$$\Delta n_+ = 2n_2^+ |E_+|^2 = \frac{3}{2n_0}\chi_{XXYY}^{(3)}(-\omega;\omega,\omega,-\omega)|E_+|^2 \quad \text{for } \sigma_+ \text{ polarization}$$

$$(7.65)$$

Since the cross terms of the susceptibility are smaller than the diagonal terms, linearly polarized light causes a larger index change than circularly polarized light. (For a more detailed description of these equations and formulae for elliptically polarized light, and, in fact, for a thorough look at NLO behavior, we refer the reader to an excellent book by Sutherland (1996).)

7.4.1.2 Infrared region

At longer wavelengths, one approaches the multiphonon resonance region, and the coupling of the polarizability with the phonon resonances increases as the likelihood for an overlap with a multiphonon vibration increases. This triggers an absorptive process and causes thermal nonlinearity in the refractive index, both through the Kramers-Kronig absorption relation and the change in local temperature of the material. In the region of overlap, the light transfers energy to the phonon modes of the system through electron–phonon interactions, causing local heating. This heating can cause a shift in the phonon mode and can alter the position of the absorption band and, consequently, of the refractive index. The resulting change in refractive index is written as

$$n_2' = \frac{\alpha_T \tau}{\rho C}\left(\frac{dn}{dT}\right) \tag{7.66}$$

where α_T is the coefficient of thermal expansion, τ is a characteristic thermal diffusion time, ρ is the density of the material, and C is the specific heat. In semiconductors, the local change in temperature causes an expansion of the unit cell and, consequently, a reduction in bandgap energy. This red shift in absorption causes a positive NLO index. In glasses, as seen in Chapter 3, the temperature dependence of the refractive index can be either positive or negative. If it is positive and the index increases with temperature, the nonlinearity will be positive. If the index decreases, the NLO value will be negative. Gases and liquids will experience a decrease in refractive index with increasing temperature, leading to a negative nonlinearity.

The thermal nonlinearity is not generally desired, since it heats the sample and changes the linear index as well. However, the coupling is very fast, and investigators have observed the onset of thermal nonlinearity in subnanoseconds. When present, the thermal nonlinearity is a major problem in single-beam experiments, since its onset is so rapid. However, due to the slow propagation of thermal disturbances, two-beam experiments are less plagued by its occurrence. In many systems, and especially when an absorptive process is present, the thermal non-linearity can be very large and can dominate the other NLO processes.

The electron–phonon interaction plays a major role in the development of thermal nonlinearity. This coupling can be studied by monitoring Raman scattering and measuring the temperature of the various phonon modes through the ratio of the Stokes to the anti-Stokes scattering intensities. The latter gives:

$$\ln\left(\frac{I_{\text{Stokes}}}{I_{\text{Anti-Stokes}}}\right)_{\bar{v}_0} = \frac{h\bar{v}_0 C}{k_B T} - 4\ln\left[\frac{\bar{v}_L + \bar{v}_0}{\bar{v}_L - \bar{v}_0}\right] \qquad (7.67)$$

where $\bar{v}_0$ is the Raman shift in cm^{-1} and $\bar{v}_L$ is the excitation source wave number in cm^{-1} (Malyj and Griffiths, 1983). Experiments in our laboratories have observed variations in temperature between Raman modes that couple to an electronic excitation and those that do not. The modes that couple identify the phonon processes that provide the source for electronic excitations.

Despite some interesting characteristics, such as the interplay of the incident light with the multiphonon spectrum of a material, the thermal optical nonlinearity is considered to be a nuisance in most applications, since the process also changes the linear index, and efforts are made to avoid it.

7.4.1.3 Two-photon absorption region

If one increases the incident light frequency from the multiphonon absorption region (IR) toward the middle of the transparency region, the first resonant processes encountered are the multi-*photon* (*m* photons) absorption processes. These processes require *m* coincident photons to excite a bound electron to an excited or free state. As discussed earlier, the oscillator strength of these transitions depends on intensity to the mth power. These processes are responsible for the gradual increase in n_2 in the transparency region with increasing optical wave frequency.

The most notable multiphoton process is the two-photon absorption (TPA) process at half the bandgap energy. An excellent discussion of this process was published by Sheik-Bahae et al. (1991). They show that the value of n_2 from the TPA process for bandgap materials can be calculated to be

$$n_2 \text{ (esu)} = 3.4 \times 10^{-8} \frac{G_2(x)}{n_0 \xi_G^4} \qquad \text{with } x = h\nu/\xi_G \qquad (7.68)$$

and where

$$G_2(x) = \frac{-2 + 6x - 3x^2 - x^3 - \frac{3}{4}x^4 - \frac{3}{4}x^5 + 2(1 - 2x)^{3/2}H(1 - 2x)}{64x^6}$$

with

$$H(1 - 2x) = \left\{ \begin{array}{ll} +1 & \text{for } (1 - 2x) > 0 \\ -1 & \text{for } (1 - 2x) < 0 \end{array} \right\}$$

The significant component of this equation is the ξ_G^4 term in the denominator, which tells us that the NLO index coefficient increases with the inverse fourth power of the bandgap energy. All other factors remaining equal, a semiconductor with a 1-eV bandgap energy will have an n_2 value about 16 times greater than a semiconductor with a 2-eV bandgap energy and 256 times that of a 4-eV semiconductor. A plot of Eq. (7.68) for different bandgap energies is shown as a function of wavelength

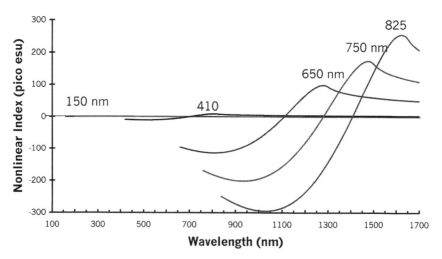

Figure 7.4: Calculation of the n_2 coefficient for materials with different bandgap energies using Eq. (7.68) (Sheik-Bahae, Hagan, and Van Stryland 1990).

in Fig. 7.4. Note the typical behavior of the nonlinear index in the presence of an intensity-dependent absorptive process. The NLO index starts positive as it approaches half the bandgap energy from longer wavelengths. Very near $^1/_2\xi_G$, the NLO index coefficient increases to a maximum and then decreases rapidly to a larger negative value, before returning asymptotically to a small negative NLO index coefficient.

The rapid change in n_2 between positive and negative values near $^1/_2\xi_G$ is a clear indication of the problems encountered by and the futility of the many early studies of NLO behavior of materials using the most convenient laser source available without concern for the dispersion of the NLO index coefficient. Say that by some stroke of bad luck, the laser wavelength happens to be slightly less than twice the wavelength of the bandgap energy; then it is possible to measure a negligible value of NLO index coefficient for a particular material at the point where the curves of Fig. 7.4 cross zero.

7.4.1.4   *Electronic processes in the band-edge region*

Near the bandgap energy, many processes can occur that affect the behavior of the NLO index coefficient through associated electronic excitation processes. The major nonlinear optical processes are associated with population redistribution processes. Next we examine a few examples.

Band filling (dynamic Burstein–Moss): If the material is illuminated within the absorption edge of the bandgap energy transition, photo excited carriers reach states in the conduction band. From there, they thermalize rapidly to the bottom of the conduction band. If the light intensity is high, and the pulse width is shorter than the electron-hole recombination time ($\tau_R \sim 10$–$50\,\mathrm{ns}$), the rapid thermalization of the photoexcited carriers fills states in the bottom of the conduction band. This results in a bleaching of absorption at the lower energies, as further electronic transitions require a higher energy to reach unfilled states. The observation is a blue shift in the bandgap energy, which decays with time at the electron-hole recombination rate. The illustration of Fig. 7.5 shows how a bleaching and blueshift in the absorption occur. This absorption bleaching process will lead to an associated change in NLO refractive index coefficient through Kramers–Kronig relations. This give a band-filling NLO coefficient calculated from a modification of Eq. 1.47, remembering that only the principal value of the integral is taken:

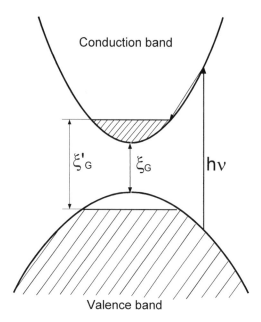

Figure 7.5: Schematic of the band-filling process that causes a transient blue shift in the bandgap absorption edge from ξ_G to ξ'_G.

$$\Delta n(\omega, I) = \frac{c}{\pi} P \int_0^\infty \frac{\Delta\alpha(\omega', I)}{\omega'^2 - \omega^2} d\omega' \qquad (7.69)$$

The result of application of this equation produces the following:

$$\begin{aligned} n_2 &< 0 && \text{for } h\nu < \xi_G \\ n_2 &> 0 && \text{for } h\nu > \xi_G \end{aligned} \qquad (7.70)$$

This process has been observed in a number of semiconductors. Figure 7.6 shows the results of an analysis of blue shift in absorption due to illumination at increasing light intensity below the bandgap energy of a filter glass made from CdS nanocrystals embedded in a borosilicate glass matrix. The semiconductor behavior observed is that of the CdS nanocrystals (Simmons, Ochoa, and Potter 1995).

Band renormalization—electron/hole plasma: In the same region as band-filling, if the light intensity is much higher, then there are sufficient free-electrons and free-holes formed to act like free-carrier plasmas. The effect of a free-electron and a free-hole plasma is to shield the Coulomb

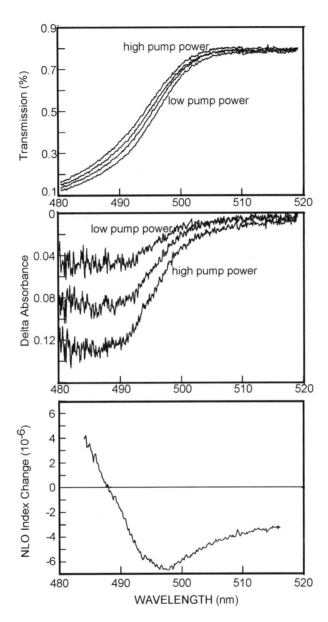

Figure 7.6: Results from measurements of the intensity-dependent absorption edge in CdS quantum dots. The upper 2 boxes show absorption bleaching. The NLO coefficient calculated results from band-filling (Simmons, Ochoa, and Potter 1995).

attraction between electrons and holes. In effect, electron–electron repulsion reduces the energy of the electron–hole attraction. This effectively reduces the bandgap energy (red shift in bandgap energy). The result is opposite that of the bandgap-filling process:

$$\text{for } h\nu < \xi_G, \quad \alpha \text{ increases and } \Delta n > 0 \qquad (7.71)$$

The NLO coefficient associated with this process is dominant only at very high optical intensities. In effect, the photo-excitation process needs to modify the plasma resonance frequency of the semiconductor by the associated change in free carriers. A good rule of thumb is to expect this effect when the free-carrier concentration approaches the Mott density $[n_e = (0.2/a_e)^3$, where a_e is the electron Bohr radius in the semiconductor]. The NLO index takes the following form:

$$n_2{'} = \frac{e^2 \alpha_0 \tau_R}{2\epsilon_0 n_0 m^* \hbar \omega^3} \qquad (7.72)$$

where m^* is the electron effective mass; α_0 is the linear absorption, and τ_R is given earlier as the electron–hole recombination time.

Exciton resonance processes: Excitons were discussed in detail in Chapter 5. They are distinct at low temperatures in direct-gap bulk semiconductors, and in many quantum-confined semiconductors (quantum wells and quantum dots) at room temperature. The binding energies of excitons are low, even in quantum confined structures. Consequently, during band renormalization, there is also a screening of the exciton Coulomb interaction. This reduces the exciton binding energy and bleaches the exciton absorption. This effect only adds to the band-renormalization effect.

At low temperatures and in quantum confined semiconductors, exciton absorption is strong (see Fig. 5.14 for $CdS_x Se_{1-x}$ glass composites). Exposing these materials to light tuned to the exciton resonance leads to the creation of excitons and a bleaching of the exciton absorption. Such a bleaching process decreases the exciton absorption band with increasing light intensity. This causes the following effect in the NLO index coefficient:

$$\begin{aligned}
\text{For } h\nu < \xi_x, \quad \alpha \text{ decreases and } \Delta n < 0 \\
\text{For } h\nu > \xi_x, \quad \alpha \text{ increases and } \Delta n > 0
\end{aligned} \qquad (7.73)$$

The value of the NLO is calculated from the amount of bleaching observed and Eq. (7.69). Note that since Δn is a function of the total $\Delta \alpha$, in order to have a very large NLO coefficient, there needs to be a large absorption process in the first place, and it must be effectively bleached.

Exciton bleaching has been observed in pump-probe experiments in CdSe quantum dots, in which a tuned high-intensity pump illuminates a sample with a short pulse (10–100 fsec) and a weak, broad-band probe coincident on the sample measures the absorption spectrum as a function of pulse delay (see, for example, a classic paper by Peyghambarian et al. (1989). In the exciton bleaching experiments, when the pump is tuned to the exciton absorption wavelength, the coincident probe shows a marked decrease in absorption. The delayed probe shows a slower relaxation of the bleaching condition as excitons recombine and empty the filled states, returning the absorption to its prepump value. Exciton bleaching experiments have shown bleaching times too fast to measure and recombination times varying from 400 fsec in glass films containing RF sputtered CdTe quantum dots to 50 nsec in melt-derived CdS_xSe_{1-x} quantum dots in bulk glass.

Biexcitons are formed by the condensation of two excitons with opposite spins. Exciton nonlinearity arises from exciton absorption bleaching. Biexciton nonlinearity results from the opposite process. Under high illumination, excitons are formed until they saturate the system. This makes it possible for biexcitons to be formed. This formation process is intensity dependent and involves an optical absorption. The absorption occurs just below the exciton absorption band due to a small energy difference between the two caused by the binding energy of the biexciton. The nonlinear optical effect of the biexciton formation is opposite that of the exciton:

$$\Delta n \ (\nu < \nu_{xx}) > 0$$
$$\Delta n \ (\nu > \nu_{xx}) < 0$$
(7.74)

Plasma resonance: Optical nonlinearity has also been observed in metal colloids in glass. This process arises from a thermal shift in plasma resonance at high illumination intensities. This shift results from a smearing of the Fermi energy that changes the effective number of free carriers. The NLO index coefficient is positive.

7.4.1.5   *Other sources of nonlinearity*

Raman-induced Kerr effect
Stimulated Raman scattering is often used by spectroscopists to

enhance a weak Raman signal. The method relies on setting up a typical Raman scattering apparatus with a strong pump laser beam at ω_0. The Raman scattering process generates a Stokes scattered beam at $\omega_0 - \Omega_p$ (where Ω_p is the scattered phonon frequency). To induce stimulated Raman scattering, one uses a weak trigger beam at the frequency $\omega_0 - \Omega_p$. The presence of this beam increases the scattering of the pump beam between 10 and 100 times. The effect is coupled through the nonlinear susceptibility $\chi^{(3)}$ as follows:

$$P_{sR}^{(3)}(\omega_0 - \Omega_p) = 6\varepsilon_0\chi^{(3)}(-\omega_0 + \Omega_p; \omega_0 - \Omega_p, \omega_0, -\omega_0)E(\omega_0 - \Omega_p)|E(\omega_0)|^2$$

$$(7.75)$$

In effect, the weak stimulating beam is drawing gain from the pump beam through the Raman scattering process of the medium. This gain has been seriously considered in the design of fiber amplifiers for the region near 1.35 μm.

Electrostriction: If a spatially varying field is present, an electrostrictive optical nonlinearity will develop due to a change in density of the material from the gradient in the applied field.

Molecular Orientation: This process occurs in liquids (CS_2). It corresponds to the interaction of the electric field with the polarization resonance, shown in Fig. 3.1. Molecular reorientations will cause a significant change in index and thus a third-order optical nonlinearity.

Thermal nonlinearity: Thermal effects were discussed earlier for the IR region. They occur at all frequencies. They are particularly large near the bandgap energy, where absorption effects are significant. The resonance lines and energy gap of semiconductors can be shifted by temperature. The bandgap energy of semiconductors shifts to lower values with increasing temperature. The resulting change in refractive index is as follows:

$$\Delta n \ (h\nu < \xi_g) > 0$$
$$\Delta n \ (h\nu > \xi_g) < 0$$

$$(7.76)$$

In 2-beam experiments, the time scale for onset of thermal changes is about 10–100 nsec. In single-beam experiments, the time scale is in the subnanosecond range.

7.4.1.6 Relative values of the various processes

Different materials will exhibit different sensitivity to the various processes. As a rule of thumb, however, here are some average values to be expected for the various regimes of nonlinearity:

Optical Kerr Effect:	$n_2 \approx 20$ pico esu (best reported is 9 pico esu, silicates $= 2$ pico esu)
Two-photon absorption	$n_2 \approx 30$ pico esu (band edge in the green) ≈ 300 pico esu (band edge in the red)
Band filling in quantum dots	$n_2 \approx -13,000$ pico esu
Gold colloids	$n_2 \approx +10,000$ pico esu
CdS exciton absorption bleaching	$n_2 \approx 6 \times 10^{-3}$ esu
Band filling in GaAs (4 K)	$n_2 \approx 0.1$ esu

7.4.2 Structural defect–induced processes

Many materials can undergo changes in refractive index associated with structural changes. Often these are mediated by electronic defects in the material's structure. Next, we give an example of a process that has received much attention because of its commercial promise. At the same time, the underlying mechanisms are interesting and representative of this class of materials. The process is photosensitivity. We also briefly discuss photothermal behavior, which has received much less emphasis but also offers interesting underlying mechanisms.

7.4.2.1 Photosensitivity

Photosensitive (PS) materials are those in which exposure to optical radiation results in a permanent change in the refractive index of the base material. It is an effect that is distinguishable from photorefraction because it occurs in isotropic materials rather than in noncentrosymmetric crystals. In addition, photosensitive effects are based upon photoinduced absorption processes originating from structural defects existing in the glass, and are not dependent upon an electro-optic coefficient as in photorefractives (see Section 7.4.3.1). In PS materials, photoexcitation can cause an electron or hole to migrate and become trapped at a structural defect site in the material. Carrier trapping at the site then alters the absorption characteristic of the defect itself. In melt-derived GeO_2-SiO_2 glasses, for example, a photoexcited hole migrates to a germanium-associated oxygen-deficient center that absorbs at 242 nm. When the hole is trapped, the center becomes a Ge E' center, which now absorbs light at 200 nm. Therefore, exposure to UV light bleaches the absorption at 242 nm and induces absorption at 200 nm. These two changes in absorption (bleaching of the 242-nm absorption and creation of an absorption peak at 200 nm) induce associated refractive index changes (see Kramers–Kronig discussion in Chapter 3). If the light is incident on the sample from two directions to form a standing wave with fixed nodes, the absorptive changes will have a spatial profile matched to the intensity profile of the standing optical wave. This will produce a modulation in the refractive index of the material matching the same modulation frequency. Since these changes are permanent, photosensitive optical gratings can be formed in these isotropic glasses. In germanosilicates, photosensitivity is mediated by the concentration of oxygen-deficient defects in the structure. Hence, sensitizing treatments of the material in a hydrogen atmosphere are often used to produce a high concentration of oxygen-deficient defects with an associated absorption at 242 nm. Graphs in Fig. 7.7 show the change in absorption at 242 nm on a germanosilicate glass heat-treated in the presence of hydrogen. The graphs of Figure 7.8 show the decay of the 242-nm absorption band and the growth of the 202-nm absorption band under exposure to light at 248 nm. Figure 7.9 shows the calculated photo-induced change in refractive index resulting from the absorption band bleaching and growth observed in the ultraviolet.

It should be noted that, depending upon the oscillator strengths of the two absorption peaks, the refractive index change measured at longer wavelengths can be either positive or negative due to the competing contributions from the growing and decaying absorption bands (since absorption-band growth results in a positive change in the material refractive index while band bleaching results in a negative induced change in refractive index). Thus, the proper design of a material that exhibits only the growth or the bleach process will necessarily produce a material capable of a larger net induced change in index of refraction over materials that undergo both absorption-band growth and bleaching. This has been demonstrated in some $GeO_2 : SiO_2$ films reactively sputtered in an oxygen-deficient atmosphere. In these material, exposure of the films to UV radiation results in an overall bleaching of the UV absorption band of the material. The resulting photoinduced index change is quite large and negative and is obtained without the need for postdeposition hydrogen treatments. Figure 7.10 shows the measured UV absorption

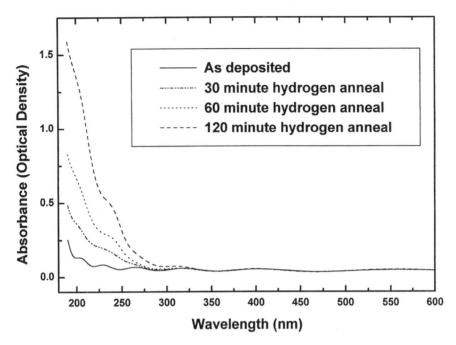

Figure 7.7: UV absorption spectra of photosensitive $GeO_2 : SiO_2$ films heat treated at 550 °C under a flowing H_2/N_2 (1:20 volume ratio) atmosphere. Heat-treatment schedules are as follows: (a) 2 hr, (b) 1 hr, (c) 30 min, (d) as-deposited sample, no heat treatment (Simmons et al. 1993).

structure before and after UV exposure and the calculated induced Δn in the material.

In general, photosensitivity can be triggered by single-photon absorption at a relevant photon energy to create the photoexcited electrons and holes or by multiphoton absorption of subbandgap light. The study of multiphoton absorption in photosensitive materials is generally conducted with argon-ion lasers, whose 488-nm beam requires either two-photon absorption to reach the relevant absorption band of the oxygen-deficient germania defect or four-photon absorption to excite electrons or hole to the bandgap (above the Urbach tail of the material) of the germanosilica glasses tested. More recent studies with excimer lasers, which utilize single and two-photon absorption processes, have produced more severe changes in the glass structure. Exposure to high energy (more than at least $0.1\,mJ/cm^2$) excimer radiation has actually resulted in densification of the regions exposed to the maximum light intensity of an

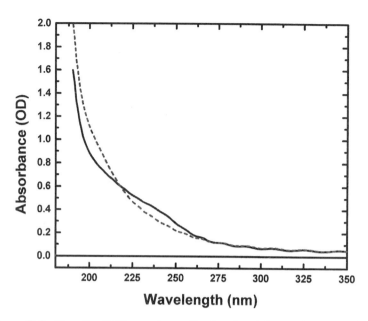

Figure 7.8: Growth of 202-nm absorption band and decay of 242-nm absorption band in photosensitive thin film following exposure to UV laser radiation. Solid line is absorption feature in as deposited material, dashed line is absorption feature in UV-irradiated material (from Simmons-Potter et al. 1998).

optical standing wave. This permanent photoinduced densification also produces a large positive index modulation in the material. Because the photoinduced index modulation in photosensitive glasses can be formed anywhere in the material where light at a relevant wavelength is focused, even deep inside the bulk, this process has very high commercial potential for application in optical grating formation for communications and spectroscopy.

Recently, photosensitive grating formation has been observed in many other glass compositions resulting from a combination of defect absorption bleaching and formation and the less selective modulated densification of the glass. Good candidate materials are typically those that have optically bleachable defects.

7.4.2.2 *Photothermal effects*

The process of precipitation of defects in transparent media under light exposure has long been studied. Clearly the photographic process is a

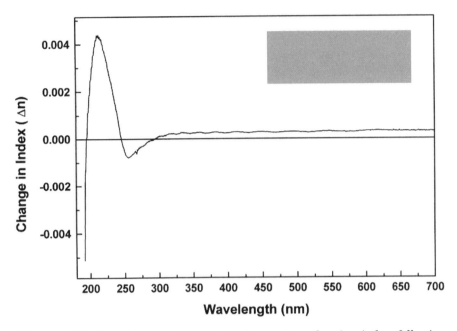

Figure 7.9: Calculated photo-induced change in refractive index following exposure of photosensitive material to excimer laser radiation at 248 nm. Maximum induced $\Delta n \sim 5 \times 10^{-4}$ (after Potter et al. 1997).

well-used example of this. Older and recent studies of glasses containing fluoride and bromide ions have led to the use of this type of behavior for recording photoexposure through a refractive index distribution. Briefly, the process starts with essentially alkali aluminosilicate glasses that contain sodium fluoride and silver bromide in solution. The glasses also contain CeO as a sensitizer. Upon exposure to UV light for the first time, the Ce^{+2} ion gives up an electron to form Ce^{+3}. The electron is trapped by the Ag^{+} ion to form a metallic silver precipitate. The subsequent heat treatment nucleates NaF crystals around the silver precipitates, which are present only in the exposed region of the glass. The precipitated NaF crystal is cubic and nucleates a bromide pyramid on one of the faces. With heat treatment, the NaBr crystal grows. From this point, different routes have been followed. Corning investigators Stookey, Beall, and Pierson (1978) and R. Araujo (private discussion) have exposed the glass a second time to precipitate the remaining silver. During the second exposure, the silver precipitates on the points of the NaBr pyramids. A second heat treatment can get the silver particles to grow to different anisotropies and

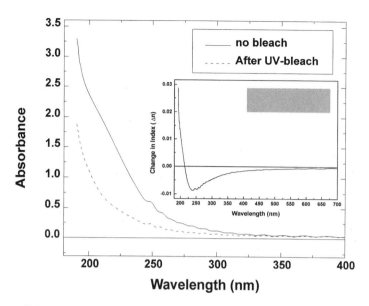

Figure 7.10: UV absorption bands before (solid line) and after (dashed line) optical exposure at 248 nm. Inset shows the calculated induced change in index of refraction for this sample. Maximum induced Δn in these samples is typically 5×10^{-3} (Simmons-Potter, Potter Jr., and Sinclair 1998).

aspect ratios. These have strong directional structural birefringence that produces a wide variety of colors. Unfortunately, the colors are dependent on the second heat treatment and not on the color of the exposing light. So the process is not useful for color photography. Glebov and coworkers (1993) conduct the second exposure at UV wavelengths, but they keep the color to a pale yellow by avoiding the second heat treatment. The precipitates essentially have a larger refractive index than the base glass, so holographic imaging and storage are possible. High diffraction efficiencies have been reported (Efimov et al. 1999) with illumination from a He–Cd laser. The gratings are stable to 400 °C. The reported index changes produced by this process have been around 2×10^{-3}.

7.4.3   DC Field–induced processes

Two examples of processes that operate by the internal formation of a DC electric field are photorefractive behavior in materials and second harmonic generation in optical fibers.

7.4.3.1 Photorefraction

Photorefractive materials (generally crystals) are particularly interesting nonlinear optical materials due to their ability to display strong nonlinear effects at milliwatt incident power levels. As in photosensitive processes, photorefractive effects arise from the photoexcitation of carriers from defect donor states. Unlike the case in PS models, however, carriers in photorefractive materials are not thought to trap at adjacent defect sites but to migrate under the influence of the incident optical excitation to "dark" or low-light regions of the crystal. There they are trapped by multivalent ions or by thermally ionized donors. The charge separation that results gives rise to an optically induced space charge field that then modulates the material's refractive index through the linear electro-optic effect. If the incident optical irradiation has a periodic intensity distribution, as from the interference of two quasi-plane waves, then the induced space charge field will also be periodic, thus producing a phase matched holographic image of the incident optical beams. The development of the space charge field is depicted in Fig. 7.11.

The development of the space charge field is only the first part of the process in permanent photorefractive effects. The spatially induced DC field also causes a diffusion of ions or protons and oxygen vacancies through the structure. This process is much slower than the charge carrier diffusion. When the field is removed, the small charge carriers (electrons and holes) back-diffuse to even out any charge imbalance. Thus, there is no electrical field driving force to redistribute the diffused ions and vacancies. The resulting index modulation is now long-lived and is caused essentially by a mixture of electronic charge defects and the displaced ionic cores, with the latter causing the greatest index change through local density variations.

Photorefractive materials include the ferroelectric $\chi^{(2)}$ materials discussed earlier, generally doped with a multivalent transition metal, such as Fe, Cr, Cu, or Ti. Iron is the most used dopant. Its reduced valence (Fe^{+2}) is the active ion. It absorbs in the near IR and near 490 nm. Under photoexcitation, it is ionized and gives up an electron to form an Fe^{+3} ion and a conduction electron. The Fe^{+3} ion absorbs in the UV. The electron follows the local field and forms a spatially modulated space charge. This is sufficient to form a photorefractive index grating. However, if alkali ions are present, they will diffuse to form a counter space charge. The addition of the ionic diffusion process will increase the holographic

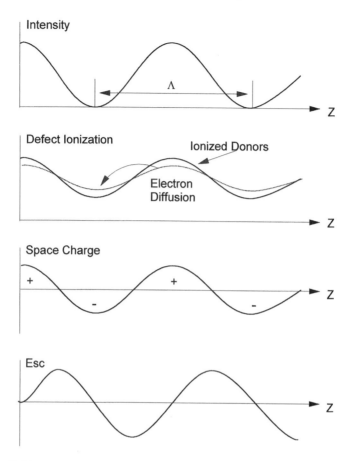

Figure 7.11: Characterization of the buildup of the space charge field (E_{sc}) in photorefractive materials. The induced refractive index will be 180° out of phase with E_{sc}.

fixation time by a factor of over 1,000. The photorefractive effect is easily erased by heating or by oxidizing the material, effectively removing the Fe^{+2} donors.

Materials that exhibit a large photorefractive effect include the ferroelectrics ($BaTiO_3$, $LiNbO_3$, $KNbO_3$), some semiconductors (InP:Fe, GaAs:Cr, GaP, CdTe—with the highest figure of merit), some polymers (TCNQ and COANP), and bismuth–column IV oxides ($Bi_{12}SiO_{20}$, $Bi_{12}GeO_{20}$, and $Bi_{12}TiO_{20}$).

Applications include holographic storage, coherent-light amplification, optical phase conjugation, associative memory, and adaptive optics (since their refractive index can be modified by optical exposure). A good review of photorefractivity can be found in T. J. Hall et al. (1985).

7.4.3.2   Second harmonic generation in optical fibers

Second harmonic generation (SHG) is generally thought to be a $\chi^{(2)}$ effect, which is prohibited in centrosymmetric materials. As such, it was not thought to be possible in isotropic optical fibers. Thus, it came as something of a surprise when optical fibers pumped by 1.06-μm light from a Nd:YAG laser were seen to produce 530-nm green radiation with an impressive conversion efficiency of about 5% (Osterberg and Margulis 1986). The effect arises from the existence of defects in the fiber that are produced and oriented in such a way as to yield a noncentrosymmetric core. This enables the process to occur but does not account for the high efficiency of the generated light.

Highly efficient SHG in fibers must rely on phase matching of the fundamental and second harmonic waves. In general, the polarization wave radiating at the harmonic frequency typically gets out of phase with the fundamental after a short distance, because the refractive index is not the same at the fundamental and harmonic frequencies. The corresponding dissipation of the SHG effect with propagation distance can be avoided by matching the phase velocities of the fundamental and harmonic waves. This process is called *phase matching*. In fibers, efficient SHG is obtained by alternating the sign of the $\chi^{(2)}$ nonlinearity with exactly the right periodicity to compensate for the mismatch in wave vectors between the fundamental and harmonic waves. This is called *quasi–phase matching* (QPM). There are numerous models explaining the mechanisms necessary to produce the QPM structure. One of the current favorites proposes that intense fields at the fundamental and harmonic frequencies mix to orient structural defects that produce the direct dipole-allowed response. In this model, two fundamental waves mix with the harmonic wave by a third-order nonlinear process to form a DC polarization component. The orientation of the DC polarization alternates with the correct periodicity to achieve phase matching.

7.4.4   Materials

There are so many processes of interest in the $\chi^{(3)}$ behavior of materials that it is difficult to give a set of rules for materials testing, selection, or optimization. Thus, when an NLO index measurement is made, there are often several processes underlying the value obtained. For example, the third-order polarizability contributes in all wavelength ranges. However, its value is smaller than the resonance processes, so that it becomes negligible near resonances. If the test frequency, $h\nu$, is below $^1/_2\xi_G$, the greatest likelihood is that the measurement addresses the third-order polarizability and thermal effects. At very high intensities, electrostriction may play a role. Above $^1/_2\xi_G$, the measurement is susceptible to two-photon absorption, and any analysis must calculate the two-photon absorption coefficient. Near the bandgap, electronic resonances are important. At low powers with short pulses, one can examine bandfilling. At high power, one examines bandgap renormalization. Over all frequency ranges, one must measure the absorption in the sample before and after the test to determine if any defects have been ionized or have trapped free-carriers, for this will change the behavior of the sample. For example, band-filling studies in glasses containing semiconductor quantum dots observed order of magnitude differences in carrier recombination times during photodarkening of the samples. The latter appears to be due to iron traps at the semiconductor–glass interface. Finally, thermal effects are always present and must be identified.

The applications dictate the most suitable or desirable set of properties needed. For example, optical switching requires an intensity-dependent refractive index. Thus if a high-intensity signal is present, or if two signals add, the local index will be different from the case with a weaker or a single signal. If the index changes, one can build a variety of Fabry–Perot cavities in which a small change in index will drastically affect transmission and reflectivity.

Clearly, the magnitude of the NLO coefficient required is determined by the size of the Fabry–Perot cavity. If size is no object (e.g., optical fibers), then nonresonant hyperpolarization is the most desirable. Even though the process produces a low NLO index coefficient, the associated loss is low, so the signal can be used in many series switches before amplification. However, if the application requires thin films, then resonant NLO processes are more desirable, since the NLO index coefficient is much larger (but so is the associated loss). A number of authors have developed figure-of-merit ($\Im$) formulae to define desirable characteristics. For example, one may consider only the refractive index change $\Im = \Delta n/n_0$. However, it must be pointed out that loss is important and one would prefer a material with reduced loss: $\Im = n_2'/(n_0\alpha)$. This question is again dependent on application, so we will discuss the various materials in terms of their behavior in a specific NLO process.

In the transparency region, the important NLO process is the third-order polarizability (optical Kerr effect). Here the best materials have the highest linear refractive index and dispersion (lowest Abbe number). Figure 7.12 shows the results of a BGO equation calculation for a variety of refractive index and Abbe number values. From this, it is obvious that massive elements are desirable. Dispersion is increased by the proximity of the use frequency to the electronic excitation resonances (bandgap). So, if you take Bi_2O_3 and PbO and add components that help form a glass, the best solution is to decrease the UV edge. Long ago we took a glass developed by Hall et al. (1989) with the highest NLO optical Kerr effect measured at that time. The composition included Bi_2O_3, PbO, and Ga_2O_3. We expected that replacing gallium with more massive indium would raise the index further, and it did. But the increase in NLO coefficent was

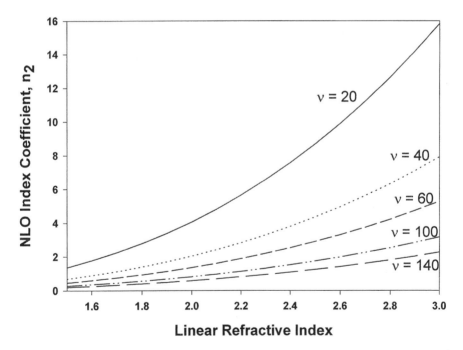

Linear Refractive Index

Figure 7.12: Variation in NLO index coefficient (in pico-esu) as a function of linear index and Abbe number, according to Eq. (7.61) (Boling, Glass, and Owyoung 1978).

greater than could be calculated from the change in linear index. The reason was that the In_2O_3-substituted glass had an appreciably lower UV edge and thus a higher dispersion. The model suggests that heavier is better, as is a greater number of valence electrons. Changing anions is more effective than changing cations. Covalent bonding is better than ionic, since the latter tends to form filled electronic shells with lower polarizabilities. The best solution from our experiments is to lower the bandgap energy. However, one must keep it high enough so that the application frequency remains below the two-photon absorption region. For this, high-quality semiconductors are the best, since they have a sharp band edge and one can get closest to the two-photon region with the least absorption. Finally, Eq. (7.68) shows that the lowest bandgap energy that will satisfy the foregoing condition will also yield the highest NLO value.

Another method for increasing dispersion is to add color centers. This can bring the use frequency close to an absorption band and yield a large dispersion. However, from looking at Fig. 7.12, it is clear that changes are possible, but they are reasonably small.

Larger NLO indices are possible with two-photon absorption, and suitable applications are in the area of optical limiting. The largest NLO values are obtainable near electronic resonances in semiconductors. Here, population redistribution processes offer the best potential. These include bandfilling and exciton bleaching. In theory, the latter promises the highest NLO values. Bulk semiconductors require low temperatures to bleach excitons, but tests with quantum dots have proven that this effect can take place at room temperature.

7.5   Test methods

There are a number of test methods used regularly to measure non-linear optical properties. Some common techniques are degenerate four-wave mixing, z-scan measurements and pump–probe spectroscopy. The first two give the change in NLO index, the last measures population redistribution. These will be discussed briefly next. Other techniques, including more detail on them, are covered in many books; some of these are listed at the end of the chapter (Butcher and Cotter, 1990; Chase and Van Stryland, 1995; Peyghambarian, Koch and Mysyrowicz, 1993, Prasad and Willams, 1991; Sheih-Bahae *et al.*, 1990b; Yariv, 1976; and especially Sutherland, 1996.)

7.5.1 Degenerate four wave mixing (DFWM)

Figure 7.13 shows the four-beam arrangement for backward geometry four-wave mixing. The forward, backward, and probe beams mix in the sample and produce a phase-conjugate fourth wave whose intensity is dependent on the coupling coefficient in the material. This coupling coefficient is a function of the third-order susceptibility. Essentially, the forward and probe beams form a transient index grating in the material from which the backward beam diffracts into the phase conjugate wave. The appropriate polarization component is written as

$$P_c^{(3)} = 6\varepsilon_0 \hat{e}_c \chi^{(3)}(-\omega; \omega, \omega, -\omega) : E_{0f}\hat{e}_f E_{0b}\hat{e}_b E_{0p}^*\hat{e}_p e^{i[(k_f+k_b-k_p)\cdot r]} \qquad (7.77)$$

The samples require little preparation, except for a good face, so this makes a good measurement for comparing different materials. Alignment is obviously critical and must be made carefully. However, the sign of the index non-linearity is not measurable.

7.5.2 *z*-scan measurement

Figures 7.14 and 7.15 show a typical set up for a *z*-scan measurement with the data on CS_2 which is used as a standard. The test is conducted with

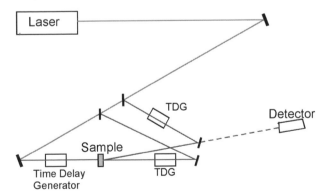

Figure 7.13: Simplified schematic for a backward degenerate four-wave mixing experiment (Sutherland 1996).

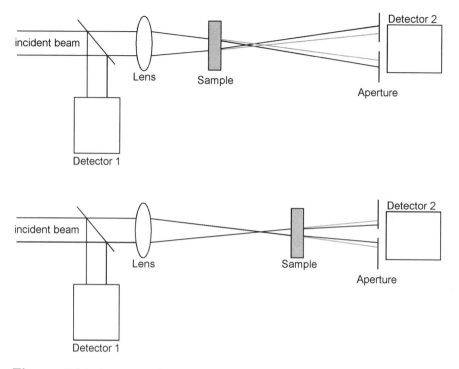

Figure 7.14: Apparatus for a z-scan experiment showing the cases where the sample is before and after the focal point of the focusing lens. The beam path with a purely linear sample is shown with light lines; the actual beam is shown with dark lines for a sample with a positive NLO index.

closed or open aperture. The sample is usually a thin slab with well-polished faces. In the closed-aperture test, the sample is simply scanned across the focal point of a clean optical beam past a short-focal-length lens. As the sample passes through the focal point of the beam, the power density increases to a value determined by the experimental design. If the sample has a positive index nonlinearity, the sample will focus the beam too early when it is placed between the lens and focal point. This produces a decrease in intensity transmitted through a narrow aperture. When the sample is scanned past the focal point, it will act to narrow the outgoing beam and raise the transmitted intensity, as shown in the drawing. If the sample has a negative index nonlinearity, the opposite will happen as the sample first pushes the focal point further toward the aperture when it is located between the lens and focal point and then diverges the beam more

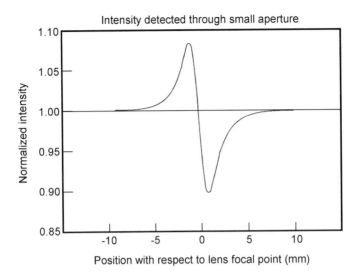

Figure 7.15: z-scan trace from a measurement on CS_2 (after Sheik-Bahae, Said, and Van Stryland (1989)). The data show the reverse behavior to that described in Fig. 7.14. The sample is exhibiting thermal self-defocusing due to expansion of the liquid at the focal point.

widely when the sample is placed past the focal point. The peak-to-valley variation in transmitted intensity is related directly to the NLO index coefficient (Sheik. Bahae *et al.*, 1990b).

In the open aperture test, the sample is scanned through the focal point of the beam: however the aperture in front of the detector is fully open. This measures the total transmitted intensity. As the sample passes through focus, the optical power density increases: if there is any nonlinear absorption, it will show up as a decrease in transmitted intensity.

7.5.3 Pump-probe spectroscopy

Figures 7.16 and 7.17 illustrate the principle of pump–probe spectroscopy. Essentially, one has a high-intensity, monochromatic pump beam that arrives at the sample first. The pump distorts the equilibrium carrier population. A delayed broad-band probe beam propagates through the sample and is detected by a grating spectrometer. The spectrometer yields a measurement of the transmissivity of the sample as a function of

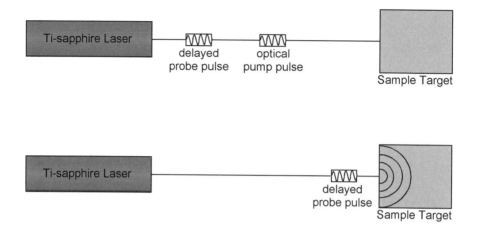

Probe pulse measures the effect of the pump pulse on the target

Figure 7.16: Schematic of the operation of time-resolved pump–probe experiments. Essentially, two pulses are incident on the sample, with an adjustable delay between their arrival. The strong pump arrives first and modifies the state of the sample under photoexcitation. The weak probe arrives later and measures any changes in the absorption of the sample as a function of delay from the arrival of the pump pulse. If the probe is broad-band, then a spectrum can be collected in one measurement.

wavelength. If the pump beam is tuned to a carrier excitation resonance, then band filling and/or absorption bleaching will occur, and the probe beam will detect either a change in absorption or a shift in a band edge. If one uses femtosecond pulses, time resolution is usually fast enough to show carrier recombination and bleaching recovery processes. However, the test is not fast enough to measure carrier thermalization or absorption bleaching. Often, if the absorption band is inhomogeneously broadened, then the bleaching effect appears as hole burning, because only a portion of the band is bleached. The resulting hole generally has the homogeneous linewidth of the process. So this method is excellent for measuring the homogeneous linewidth, even in a sample with a broad inhomogeneous population of states.

7.5.4 Third-harmonic generation

We close this chapter with a brief mention of third-harmonic generation measurements. Here one measures the $\chi^{(3)}(-3\omega; \omega, \omega, \omega)$ component of the

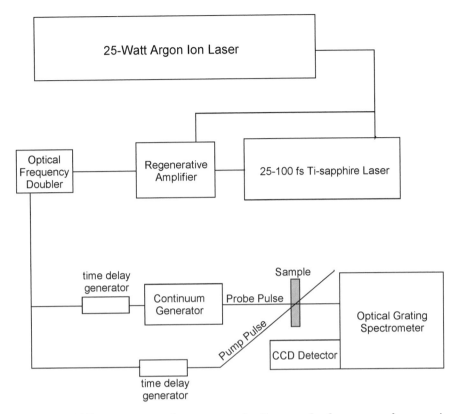

Figure 7.17: Experimental apparatus for time-resolved pump–probe experiments.

third order susceptibility. Consequently, this test only measures the electronic hyperpolarizability and is independent of thermal effects. Effectively, the measurement consists of recording the change in intensity of the 3ω beam per change in intensity of the incident 1ω beam. However, because of the need for phase matching of the two waves at widely separated frequencies (n(3ω) and n(ω)), the test is best conducted on very thin samples (films), otherwise the interference effects from the mismatch of the respective indices will cause significant measurement error. We refer the reader to Sutherland (1996) for a detailed description of the test method.

References

Alley, T. G., S. R. J. Brueck, and R. A. Myers. 1998. Space charge dynamics in thermally poled fused silica. *J. Non-Cryst. Solids* 242:165–176.

Bloembergen, N. 1992. *Nonlinear Optics*. Addison-Wesley, Redwood City CA.

Bloembergen, N., and P. S. Pershan. 1962. Light waves at the boundary of nonlinear media. *Phys. Rev.* 128:606.

Boling, N. L., A. J. Glass, and A. Owyoung. 1978. Empirical relationship for predicting nonlinear refractive index changes in optical solids. *IEEE J. Quant. Electron.* QE14:601.

Butcher, P. N., and D. Cotter. 1990. *The Elements of Non-Linear Optics*. Cambridge; Cambridge University Press.

Chase L. L., and E. W. Van Stryland. 1995. Non-linear refractive index: Inorganic materials. p 269, In *Handbook of Laser Science and Technology*. Suppl. 2: *Optical Materials*, edited by M. J. Weber. Boca Raton, FL: CRC Press.

Chiang, Y.-M., D. Birnie III, and W. D. Kingery. 1997. *Physical Ceramics*. New York: Wiley.

Davies, J. H. 1998. *The Physics of Low Dimensional Semiconductors*. Cambridge: Cambridge University Press.

Efimov, O. M., L. B. Glebov, L. N. Glebova, K. C. Richardson, and V. Smirnov. 1999. High-efficiency Bragg gratings in photothermorefractive glass. *Applied Optics* 38:619–627.

Franken, P. A., A. E. Hill, C. W. Peters, and G. Weinreich. 1961. Generation of optical harmonics. *Phys. Rev. Lett.* 7:118.

Geusic, J. E., H. J. Levinstein, S. Singh, R. G. Smith, and L. G. Van Uitert. 1968. Continuous 0.53 μm solid state source using $Ba_2NaNb_5O_{15}$. *IEEE J. Quant. Electron.* OE4:352.

Gibbs, H. M. 1985. *Optical Bistability: Controlling Light with Light*. Orlando, FL: Academic Press.

Glebov, L. B., N. V. Nikonorov, G. T. Petrovskii, and M. V. Kharchenko. 1993. Formation of optical elements on glasses by ion exchange methods and photo-thermo-induced crystallization. *SPIE Proc.* 1751:169–183. The original work was in: V. A. Borgman, L. B. Glebov, N. V. Nikonorov, G. TT. Petrovskii, V. V. Savin, and A. D. Tsvetkov. 1989. Photothermal refractive effect in silicate glasses. *Sov. Phys. Dokl.* 34:1011, and *Dokl. Akad. Nauk. SSSR* 309:336–339.

Hall, D. W., M. A. Newhouse, N. F. Borrelli, and D. Weidman. 1989. Non-linear susceptibility of high-index glasses. *Appl. Phys. Lett.* 54:1293–1295.

Hall, T. J., R. Jaura, L. M. Connors, and P. D. Foote. 1985. The photorefractive effect—A review. *Prog. Quant. Electr.* 10:77–146.

Malyj, M., and J. E. Griffiths. 1983. Stokes/Anti-Stokis Raman Vibrational Temperatures. *Appl. Spectrosc.* 37:315.

Medrano, C., and P. Gunter. 1995. Photorefractive materials. In *Handbook of Laser Science and Technology.* Suppl. 2 *Optical Materials,* p. 431 edited by M. J. Weber. Boca Raton, FL:CRC Press.

Mizrahi, V., and J. E. Sipe. 1988. Phenomenological treatment of surface second harmonic generation. *J. Opt. Soc.* B5:660.

Osterberg, U., and W. Margulis. 1986. Efficient Second Harmonic Generation in an Optical Fiber. XIV International Quantum Electronics Conference, San Fracisco, paper WBB1.

Peyghambarian, N., S. W. Koch, and A. Mysyrowicz. 1993. *Introduction to Semiconductor Optics.* Englewood Cliffs, NJ: Prentice Hall.

Peyghambarian, N., B. Fluegel, D. Hulin, A. Minges, M. Joffre, A. Antonetti, S. W. Koch, and M. Lindberg. 1989. Femtosecond optical nonlinearities of CdSe quantum dots. *IEEE J. Quant. Electron.* 25:2516–2522.

Potter, Jr., B. G., K. Simmons-Potter, W. L. Warren, J. A. Ruffner, and D. C. Meister. 1997. Structural defect control and photosensitivity in reactively sputtered germanosilicate glass films. *SPIE Proc.* 2998:146.

Prasad, P. N., and D. J. Williams. 1991. *Introduction to Nonlinear Optical Effects in Molecules and Polymers.* New York: Wiley.

Sheik-Bahae, M., D. J. Hagan, and E. W. Van Stryland. 1990. Dispersion and bandgap scaling of the electronic Kerr effect in solids associated with two-photon absorption. *Phys. Rev. Lett.* 65:96–99.

Sheik-Bahae, M., A. A. Said, and E. W. Van Stryland. 1989. High sensitivity, single-beam n_2 measurement. *Opt. Lett.* 14:955.

Sheik-Bahae, M., D. J. Hagan, A. A. Said, J. Young, T. H. Wei, and E. W. Van Stryland. 1990a. Kramers–Kronig relation between n_2 and two photon absorption. *SPIE Proc.* 1307:395.

Sheik-Bahae, M., A. A. Said, T. H. Wei, D. J. Hagan, and E. W. Van Stryland. 1990b. Sensitive measurements of optical non-linearities using a single-beam. *IEEE J. Quant. Electron.* 26:760–769.

Sheik-Bahae, M., D. C. Hutchings, D. J. Hagan and E. W. Van Stryland. 1991. Dispersion of bound electronic nonlinear refraction in solids. *IEEE J. Quant. Electron.* 27:1296–1309.

Shen, Y. R. 1984. *The Principles of Non-Linear Optics.* New York: Wiley.

Simmons, J. H., O. R. Ochoa, and B. G. Potter. 1995. Non-linear optical processes in quantum-confined cluster-insulator composites. In *Handbook of Laser Science and Technology*, Suppl. 2: *Optical Materials*, p 250, edited by M. J. Weber. Boca Raton, FL; CRC Press.

Simmons, K. D., G. I. Stegeman, B. G. Potter, Jr., and J. H. Simmons. 1993. Photosensitivity of solgel-derived germanosilicate planar waveguides. *Opt. Lett.* 18:25.

Simmons-Potter, K., B. G. Potter, Jr., and M. B. Sinclair. 1998. Photosensitive thin films: Manipulation of defects through synthesis control. *Jap. J. Appl. Phys.* 37–1:8.

Simmons-Potter, K., B. G. Potter, Jr., D. C. Meister, and M. B. Sinclair. 1998. Photosensitive thin film materials and devices. *J. Non-Cryst. Solids* 239:96.

Stookey, S. D., G. H. Beall, and J. E. Pierson. 1978. Full color photosensitive glass. *J.Appl. Phys.* 49:5114–5123.

Sutherland, R. L. 1996. *Handbook of Non-Linear Optics.* New York: Dekker.

Van Stryland, E. W., and L. L. Chase. 1995. Two-photon absorption: Inorganic materials. In *Handbook of Laser Science and Technology*, Suppl. 2: *Optical Materials*, p. 299 edited by M. J. Weber. Boca Raton, FL: CRC Press.

Yariv, A. 1976. *Introduction to Optical Electronics.* New York: Holt, Rinehart and Winston.

Index